T0387572

Damjili Cave

Investigating the Late Pleistocene to Holocene human history
in the Southern Caucasus

Edited by

Yoshihiro Nishiaki, Azad Zeynalov, and Yagub Mammadov

OXBOW | books
Oxford & Philadelphia

Published in the United Kingdom in 2025 by
OXBOW BOOKS
81 St Clements, Oxford OX4 1AW

and in the United States by
OXBOW BOOKS
1950 Lawrence Road, Havertown, PA 19083

Hardback Edition: ISBN 979-8-88857-180-4
Digital Edition: ISBN 979-8-88857-181-1 (epub)

A CIP record for this book is available from the British Library

Library of Congress Control Number: 2024946507

Printed in Malta by Melita Press

For a complete list of Oxbow titles, please contact:

UNITED KINGDOM
Oxbow Books
Telephone (0)1226 734350
Email: oxbow@oxbowbooks.com
www.oxbowbooks.com

UNITED STATES OF AMERICA
Oxbow Books
Telephone (610) 853-9131, Fax (610) 853-9146
Email: queries@casemateacademic.com
www.casemateacademic.com/oxbow

Oxbow Books is part of the Casemate Group

Front cover: Photo by Yoshihiro Nishiaki

Preface

Archaeological research carried out since the late 19th century has sufficiently demonstrated that the beginning of the food production economy and the establishment of a farming society had a significant impact on the shaping of subsequent human history. Accordingly, these processes, coined Neolithization in the literature, have attracted a great deal of interest from archaeologists and anthropologists worldwide. The South Caucasus, i.e. the region discussed in this book, has remained one of the least studied regions, in a modern sense, for a long time. However, the research situation has improved remarkably since the 2000s, for a number of reasons, above all the increasing efforts of local and international archaeologists in collaboration.

The Azerbaijan-Japan Archaeological Mission, organized under the authorization of the Institute of Archaeology and Anthropology of the Azerbaijan National Academy of Sciences, Azerbaijan, and the University Museum, the University of Tokyo, Japan, has played a major role in elucidating the origins and developments of the earliest farming communities in the South Caucasus. Since its first season in 2008, the mission has conducted a series of field campaigns in West Azerbaijan. The remarkable achievements made thus far include the establishment of a secure chronological framework for understanding the Neolithization processes in the early sixth millennium BC through the excavations of two important Neolithic sites: Göytepe and Hacı Elamxanlı Tepe in the Tovuz region, representing the early and late phases of the Neolithic of the South Caucasus, respectively. Particularly important was the documentation of one of the earliest occurrences of the Neolithic economy at Hacı Elamxanlı Tepe, dating from around 6000 cal BC.

In the current volume, we present a set of archaeological evidence from another site, Damjili Cave, also found in West Azerbaijan, which contained cultural layers from the Mesolithic period. We believe that, through combining the records of the late (Göytepe), early Neolithic (Hacı Elamxanlı Tepe), and Mesolithic (Damjili Cave) periods, our understanding of the Neolithization processes of the South Caucasus will be greatly promoted. In fact, data from a combination of three chronologically-different sites provide the first opportunity to observe Neolithization processes with secure stratigraphic evidence in a small region of West Azerbaijan.

It is also worth mentioning that Damjili Cave has attracted Paleolithic researchers since the 1950s. We acknowledge the existence of Middle Paleolithic artifacts in the collection. Although our renewed excavations have failed to identify the primary layers of the Middle Paleolithic to date, the recovered materials indicate that this cave was occupied by hominins of the Middle Paleolithic. The aforementioned finding will further increase the potential of this cave to inform prehistoric research on the South Caucasus.

Fieldwork always requires plenty of logistic support from a wide range of collaborators. We deeply acknowledge that our fieldwork was realized through a mutual understanding of collaboration between the Institute of Archaeology and Anthropology of the Azerbaijan National Academy of Sciences, Republic of Azerbaijan, represented by Dr. Maissa Ragimova, then by Dr. Abbas Qadirnoglu Seyidov, currently by Dr. Farhad Guliyev, and the University Museum, the University of Tokyo, Japan, directed by Prof. Gen Suwa and currently by Prof. Yoshihiro Nishiaki. Local logistics were prepared by the staff of the Avey State Historical-Cultural Reserve of the Ministry of Culture and Tourism, represented by Dr. Seadat Aliyeva. Our sincerest gratitude goes to these institutions, which made our fieldwork possible.

In addition, the important contributions from the participants of the investigations in both the field and the laboratory are also greatly appreciated. These participants included, to mention only a few: Dr. Musa Mursaqulov, Dr. Mansur Mansurov, Dr. Azad Zeynalov, Dr. Farhad Guliev, Ms. Ulviya Safarova, Mr. Mhmmad Musaquliev, Mr. İmaş Hajiyev, Mr. Pərviz Qasımov, Mr. Ejder Babazade, Mr. Orkhan Zamanov, and Mr. Shahin Salimbayov from the Azerbaijan side, in addition to Dr. Seiji Kadowaki, Dr. Toru Tamura, Dr. Hiroto Nakata, Mr. Kazuya Shimogama, Ms. Chie Akashi, Dr. Saiji Arai, Dr. Yuichi S. Hayakawa, Ms. Fumika Ikeyama, Mr. Masato Hirose, Mr. Hironobu Kirihara, Dr. Yoshiki Miyata, Dr. Akiko Horiuchi, Mr.

Kaoru Bokuhan, Prof. Yuichiro Wakano, Dr. Takehiro Miki, Mr. Kantaro Tanabe, Ms. Shizuka Miyai, and Mr. Masuto Ebina from the Japan side.

Likewise, the research team received practical support whenever necessary from the local workers in association with the Avey State Historical-Cultural Reserve of the Ministry of Culture and Tourism.

Financial support for travel and research expenses was provided by the Institute of Archaeology and Anthropology of the Azerbaijan National Academy of Science, the University Museum, the University of Tokyo, and a grant from the Ministry of Education, Culture, Sports, Science and Technology (16H06408, 16K21721, 23H00690, and 24H00001). We are grateful to all of these institutions and individuals.

Yoshihiro Nishiaki
The University Museum, The University of Tokyo, Japan

Azad Zeynalov
The Institute of Archaeology and Anthropology, The Azerbaijan National Academy of Sciences, Azerbaijan

Yagub Mammadov
The Institute of Archaeology and Anthropology, The Azerbaijan National Academy of Sciences, Azerbaijan

July 2024

Group photo of the excavation team at Damjili Cave, 2018 season.

CONTENTS

Appendix: Research at Damjili Cave in 1953–1957

List of Contributors

Chie Akashi
Research Institute of Cultural Properties, Teikyo University, Isawa, Japan

Saiji Arai
Department of Archaeology, Faculty of Letters, The University of Tokyo, Tokyo, Japan

Masuto Ebina
Forestry Research Institute, Hokkaido Research Organization, Sapporo Japan; Graduate School of Environmental Science, Hokkaido University, Sapporo Japan

Yuichi S. Hayakawa
Faculty of Environmental Earth Science, Hokkaido University, Sapporo, Japan

Fumika Ikeyama
Department of Archaeology, Faculty of Letters, The University of Tokyo, Tokyo, Japan

Yagub Mammadov*
Institute of Archaeology and Anthropology, The Azerbaijan National Academy of Sciences, Baku, Azerbaijan

Takehiro Miki
Department of Archaeology and Ethnology, Faculty of Letters, Keio University, Tokyo, Japan

Mansur Mansurov
Institute of Archaeology and Anthropology, The Azerbaijan National Academy of Sciences, Baku, Azerbaijan

Yoshihiro Nishiaki*
The University Museum, The University of Tokyo, Tokyo, Japan

Ulviya Safarova
Institute of Archaeology and Anthropology, The Azerbaijan National Academy of Sciences, Baku, Azerbaijan

Shahin Salimbayov
Institute of Archaeology and Anthropology, The Azerbaijan National Academy of Sciences, Baku, Azerbaijan

Kazuya Shimogama
Institute for Geo-Cosmology, Chiba Institute of Technology, Chiba, Japan

Toru Tamura
National Institute of Advanced Industrial Science and Technology, Institute of Geology and Geoinformation, Tsukuba, Japan

Orkhan Zamanov
Institute of Archaeology and Anthropology, The Azerbaijan National Academy of Sciences, Baku, Azerbaijan

Azad Zeynalov*
Institute of Archaeology and Anthropology, The Azerbaijan National Academy of Sciences, Baku, Azerbaijan

* Co-editors

List of Figures

List of Tables

Chapter 1

Introduction

Yoshihiro Nishiaki, Azad Zeynalov, and Yagub Mammadov

The establishment of the Neolithic farming socio-economy is one of the greatest events in human history. The Neolithization in the South Caucasus has also attracted a community of archaeologists and anthropologists for years. Given that the first farming socio-economy was founded in the twelfth millennium cal BP or earlier in the Fertile Crescent of Southwest Asia, our research on the origins of the Neolithic farming in the South Caucasus inevitably involves an analysis of its relationship(s) with the earlier farming of Southwest Asia. As in other regions neighboring the Fertile Crescent, such as the Southern Zagros and West Anatolia, the Neolithization processes in the South Caucasus may well have included population immigration and knowledge transmission as two extreme cases, or a mixture of the two processes. Indeed, both extreme models have been provided in research history (Kushnareva 1997: 156; Sagona 2018). It is time to evaluate the rather intuition-based old ideas in light of the solid datasets obtained through modern scientific strategies.

Research on the Neolithization process in the South Caucasus has entered a new stage since the early 2000s. A rapid increase in local and international teams conducting well-organized multidisciplinary field campaigns has greatly enriched our knowledge on this subject (e.g., Lyonnet et al. 2012; Chataigner et al. 2015; Helwing et al. 2017). Our research at Göytepe (Nishiaki and Guliyev 2020) and Haji Elamhanlu Tepe (Nishiaki et al. 2021) has been fully in line with recent research developments. One of the most significant results was that the earliest farming socio-economy was not brought into the South Caucasus as a package. In other words, the oldest known Neolithic culture of the region, the Shomutepe Culture, developed gradually since the introduction of the Southwest Asia-originated farming culture (Nishiaki et al. 2022). The earliest Neolithic cultural assemblages, as documented at Haji Elamhanlu Tepe, consisted of both new elements most likely imported from the Fertile Crescent of Southwest Asia and unique elements potentially derived from indigenous societies from the beginning, while crops and animal herds had exogenous origins (Kadowaki et al. 2017; Nishiaki 2021).

To further develop the Neolithization research in the South Caucasus, we needed to investigate Mesolithic settlements that existed before the introduction of the earliest farming socio-economy. Otherwise, the cultural relationship between the incoming farming societies/cultures and the local indigenous hunter-gatherer societies/cultures during the Mesolithic-Neolithic interface period would not be elucidated with tangible archaeological evidence. However, there have been no Mesolithic sites precisely dated to the period immediately before the Neolithic, that is, the late ninth millennium cal BP (Varoutsikos 2015: 109). Many of the dozen sites assigned to the Mesolithic on both sides of the Lesser Caucasus (Fig. 1.1) have no certain radiocarbon dates (Meshveliani 2013), and others are dated to much earlier periods, in the eleventh–tenth millennia cal BP (e.g., Arimura et al. 2022; Varoutsikos 2015). Consequently, the late ninth millennium cal BP has been a "missing link" in our understanding of the transition or replacement processes between the Mesolithic and Neolithic societies in the South Caucasus (Chataigner et al. 2015: 18).

Consequently, we conducted a series of site reconnaissance surveys and re-evaluated available lithic artifact collections from the Middle Kura Valley. Our first site reconnaissance surveys focused on Neolithic sites near Göytepe. The regions investigated include the valleys of the Zayam Cay, Esrik Cay, and Tovuz Cay, limited to Tovuz County (Shimogama and Alibayov 2020). Although they resulted in the discovery of Hacı Elamxanlı Tepe, which pushed the Neolithic origins in the region back to the beginning of the sixth millennium BC, no Mesolithic settlements preceding the first Neolithic society were found. Taking into consideration the geological background of the research region, we decided to turn our attention to the limestone mountains of the Gazakh Region, west of Tovuz, where flint rocks, which are indispensable raw materials for lithic manufacturing, were available.

The possibility of Mesolithic occupations in the

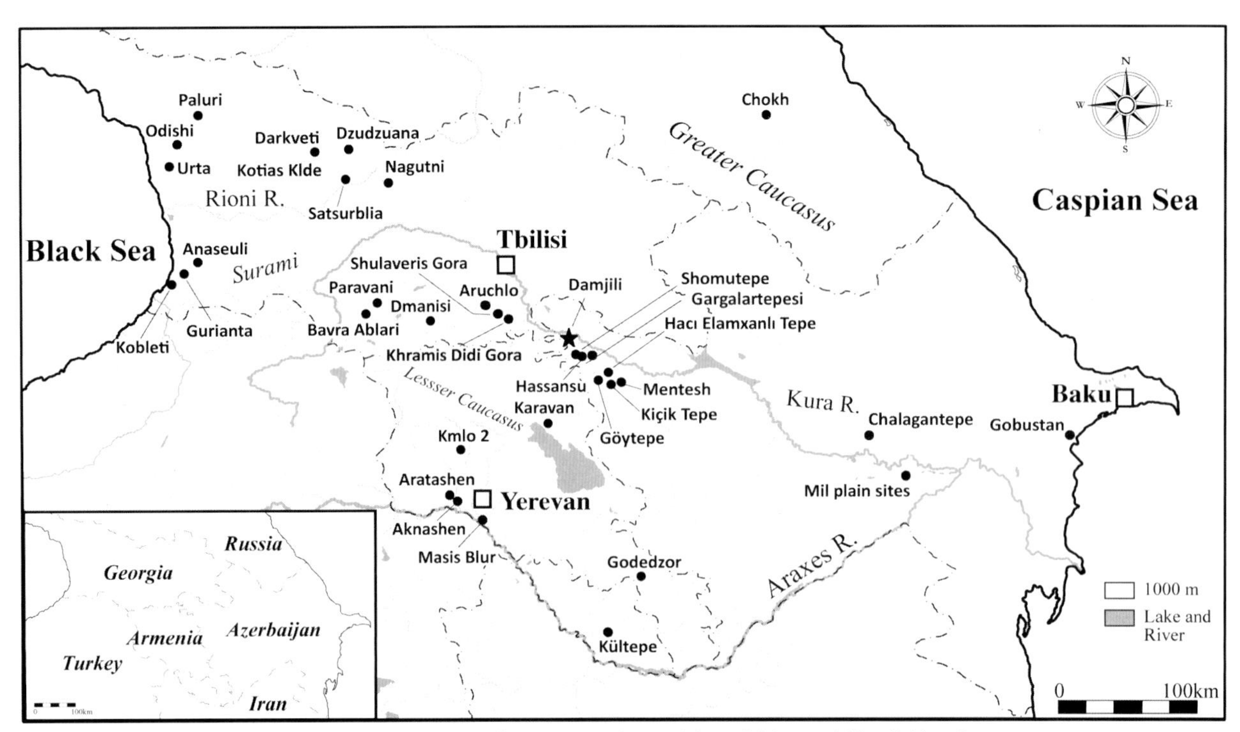

Fig. 1.1 Map showing the location of Damjili Cave and related Mesolithic and Neolithic sites.

Gazakh Region was already suggested by excavations in the 1950s; the principal excavator stated that the Mesolithic remains were recovered from disturbed contexts at Damjili Cave (Huseynov 2010: 212). Our repeated visits to this cave did not reveal any evidence of Mesolithic occupations. This is because of the extensive pavement with concrete and asphalt used in this cave site as an important recreational location. Nevertheless, our preliminary study of the lithic materials at the National Academy of Sciences of Azerbaijan in 2015 strongly suggested the occurrence of Mesolithic layers in this cave. Accordingly, we organized a new session of our research to investigate Damjili Cave and the Avey Mountains to fill an important gap in our knowledge of the Neolithization in the South Caucasus.

Damjili Cave is situated in the eastern foothills of Avey Mountain, approximately 1 km west of the modern town of Dash Salahly, and approximately 20 km northwest of Gazakh. The mountain consists mainly of Upper Cretaceous limestone/chalk rocks, with the highest peak being 1400 m above sea level (m asl) (Fig. 1.2). The cave is open at the head of the Damjili Valley at 650 m asl, which features a tributary of the Salahly River running eastward to the Kura River (Fig. 1.3). The cave is located at one of the largest groundwater sources available at an extensive notch developed horizontally for over 100 m. Before archaeological excavation was conducted there, we undertook a topographic survey of this site (Chapter 3 of this volume). The survey clearly demonstrates that the site in fact consists of two distinguishable caves (rock shelters), developed along the left side of the Damjili Valley (Figs. 1.4; 1.5). Both caves, designated as Damjili Caves 1 and 2 for convenience, were subjected to excavations in 2016.

The cave commonly known as Damjili Cave (Figs. 1.6; 1.7) is Damjili Cave 1, situated to the west. It is one of the most popular recreational parks for locals, offering a comfortable roofed area measuring approximately 17 m wide, 7.3 m long, and 4.2 m high, along with an extensive flat terrace. In addition, spring water still comes out of the underground aquifers of the cave wall, serving as the water source for a small valley rich in vegetation.

Damjili Caves 1 and 2 were investigated by S. N. Zamyatnin and M. M. Huseynov in 1953, 1956, and 1957. Although the published accounts are limited (Appendix of this volume), Huseynov (2010: 212) writes: "In 1956–1957, there were excavated and examined sediments in Grotto Damjili where there were discovered partially destructed layers of the Middle and Late Paleolithic as well as of Mesolithic and Neolithic." (Huseynov 2010: 212). In short, the remaining archaeological sediments containing materials from different chronological periods were reportedly disturbed. However, the alleged existence of "Mesolithic" materials draws our attention, thus resulting in our resumed investigations. According to unpublished reports of the 1956 and 1957 seasons in the archives of the Institute of Archaeology, Ethnography, and Anthropology, the excavations in the 1950s took place in two major parts. The larger excavation area was set up in front of the major

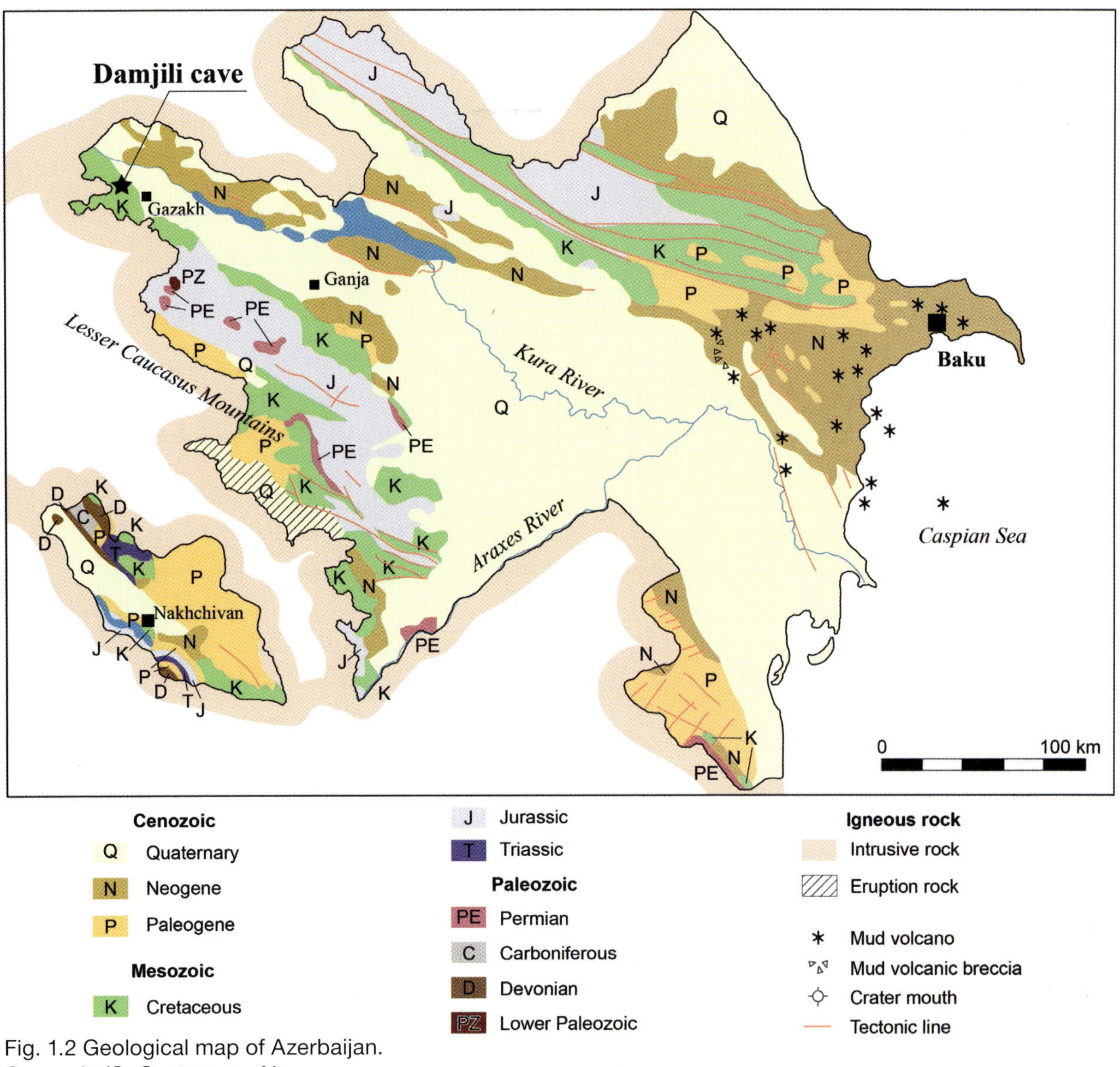

Fig. 1.2 Geological map of Azerbaijan. Cenozoic (Q: Quaternary, N: Neogene, P: Paleogene); Mesozoic (K: Cretaceous, J: Jurassic, T: Triassic); Paleozoic (P: Permian, C: Carboniferous, D: Devonian, PZ: Lower Paleozoic); Igneous rock (Intrusive rock, Eruption rock); Star: Mud volcano; Triangle: Mud volcanic breccia; Cross: Crater mouth; Dashed line: Tectonic line. Modified from a map made by the State Land and Cartography Committee of the Republic of Azerbaijan in 2000.

Fig. 1.3 Contour map of Avey Mountain and its vicinity. Modified from a map made in the Soviet era, 1990.

Fig. 1.4 Distant view of the Damjili Cave area from the east.

Fig. 1.5 General view of Damjili Cave.

water spring at the valley head, while the smaller area was based along the wall situated approximately 10 m to the northeast. A spring is located near the latter area as well. It is notable that the excavation areas, both more than 4 m by 5 m, were quite extensive in terms of the size of the cave, and the remain-

Fig. 1.6 Close view of Damjili Cave, looking east.

Fig. 1.7 Close view of Damjili Cave, looking north.

ing parts of the cave terrace were almost completely covered with concrete owing to recent construction works for the recreation park. Accordingly, our re-excavations were carried out away from these previous excavation areas and the paved areas. The results are presented in the following chapters.

Damjili Cave has enjoyed publicity for archaeologists and the public for decades since the 1950s. However, its scientific significance has not yet been fully established. Our research between 2016 and 2022 has demonstrated the strong potential of this cave for studying the transition or replacement between the Mesolithic and Neolithic in the South Caucasus. Moreover, it provided a unique opportunity to study occupational history starting from as early as the Middle Paleolithic, enlarging our research perspective.

The present volume aims to document our findings from the excavations of Damjili Cave between 2016 and 2022. While our focus has been on the evidence of Holocene human occupations, which significantly enriched our understanding of the Neolithization in the South Caucasus, we also present data on Pleistocene occupations assignable to the Middle Paleolithic. The description of the latter will be of significance when excavating areas other than the present trenches at this large site complex.

References

Arimura, M., K. Martirosyan-Olshansky, A. Petrosyan, and B. Gasparyan (2022) Exploring the changes in lithic industries during the Neolithisation in Armenia (7th–6th millennium BCE): A comparison of chipped stone tools from Lernagog-1 and Masis Blur. In: *Tracking the Neolithic of the Near East*, edited by Y. Nishiaki, O. Maeda, and M. Arimura, pp. 489–502. Leiden: Sidestone Press.

Chataigner, C., R. Badalyan, and M. Arimura (2015) The Neolithic of the Caucasus. *Oxford Handbooks Online*. Oxford: Oxford University Press. URL: <doi:10.1093/oxfordhb/9780199935413.013. 13>

Helwing, B., T. Aliev, B. Lyonnet, F. Guliyev, S. Hansen, and G. Mirtskhulava (2017) *The Kura Projects: New Research on the Later Prehistory of the Southern Caucasus*. Reimer: Dietrich.

Huseynov, M. (2010) *The Lower Paleolithic of Azerbaijan*. Baku: National Academy of Sciences of Azerbaijan (in Russian with English summary).

Kadowaki, S., K. Ohnishi, S. Arai, F. Guliyev, and Y. Nishiaki (2017) Mitochondrial DNA analysis of Neolithic goats in the southern Caucasus: Implications for the domestication of goats in west Asia. *International Journal of Osteoarchaeology* 27: 245–260.

Kushnareva, K. K. (1997) *The Southern Caucasus in Prehistory: Stages of Cultural and Socioeconomic Development from the Eighth to the Second Millennium BC*. Philadelphia: University of Pennsylvania Museum of Archaeology.

Lyonnet, B., F. Guliyev, B. Helwing, T. Aliyev, S. Hansen, G. Mirtskhulava, L. Astruc, K. Bastert-Lamprichs, W. Bebermeier, F. Becker, N. Benecke, L. Bouquet, G. Bruley-Chabot, A. Courcier, M. B. D'anna, A. Decaix, I. Fassbinder, M. Fontugne, F. Geitel, A. Goren, C. Hamon, J. Koch, G. L. Dosseur, A. Lincot, R. Link, R. Neef, D. Neumann, V. Ollivier, P. Raymond, A. Ricci, A. Samzun, S. Schorr, F. Schlütz, I. Shillito, M. Ullrich, and J. Wahl (2012) Ancient Kura 2010–2011: The first two seasons of joint field work in the Southern Caucasus. *Archäologische Mitteilungen aus Iran und Turan* 44(1): 1–189.

Meshveliani, T. (2013) On Neolithic origins in western Georgia. *Archaeology, Ethnology and Anthropology of Eurasia* 41(2): 61–72.

Nishiaki, Y. (2021) Mobility and sedentism in the Mesolithic-Neolithic contact period of the Southern Caucasus. *Senri Ethnological Studies* 106: 111–125.

Nishiaki, Y. and F. Guliyev (2020) *Göytepe: Neolithic Excavations in the Middle Kura Valley, Azerbaijan*. Oxford: Archaeopress.

Nishiaki, Y., F. Guliyev, and S. Kadowaki (2021) *Hacı Elamxanlı Tepe: The Archaeological Investigations of an Early Neolithic Settlement in West Azerbaijan*. Berlin: ex oriente.

Nishiaki, Y., A. Zeynalov, M. Munsrov, and F. Guliyev (2022) Radiocarbon Chronology of the Mesolithic-Neolithic Sequence at Damjili Cave, Azerbaijan, Southern Caucasus. *Radiocarbon* 64(2): 309–322.

Sagona, A. (2018) *The Archaeology of the Caucasus: From Earliest Settlements to the Iron Age*. Cambridge: Cambridge University Press.

Shimogama, K. and V. Alakbarov (2020) Archaeological reconnaissance survey around *Göytepe, Tovuz-Qovlar region*. In: *Göytepe: Neolithic Excavations in the Middle Kura Valley, Azerbaijan*, edited by Y. Nishiaki and F. Guliyev, pp. 137–166. Oxford: Archaeopress.

Varoutsikos, B. (2015) *The Mesolithic-Neolithic Transition in the South Caucasus: Cultural Transmission and Technology Transfer*. Ph.D. dissertation. Harvard: Harvard University.

Part I:

Fieldwork at Damjili Cave in 2016–2022

Chapter 2

Site reconnaissance survey in Avey Mountain and its vicinity

Yoshihiro Nishiaki and Kazuya Shimogama

2.1 Introduction

When we started excavations at Göytepe in 2008, there were no known Mesolithic sites in the Ganja-Gazakh Plain. Consequently, the origin of the Shomutepe culture, that is the earliest Neolithic cultural entity in the region remained unclear (Nishiaki and Guliyev 2020). To explore earlier cultural occurrences, which will contribute to an understanding of the emergence of the Shomutepe culture, we carried out the first set of archaeological surveys between 2011 and 2014, focusing primarily on the Tovuz region in western Azerbaijan (Shimogama and Alakbarov 2020). One of the most significant findings of this survey was the discovery of Hacı Elamxanlı Tepe. The excavations between 2012 and 2015 demonstrated this site as one of the earliest Neolithic settlements thus far known in the South Caucasus (Nishiaki et al. 2021).

Nevertheless, the surveys did not identify any prehistoric sites dating earlier than the Neolithic period in the Tovuz region. We expanded our scope to target sites other than mounds, such as caves within the hilly and mountainous areas west of Tovuz, between 2013 and 2014. However, no important sites suitable for our research were recognized.

Accordingly, in 2015 we focused on the Gazakh region, approximately 40 km west from Göytepe, Tovuz. Previous reports by local archaeologists, mentioning some prehistoric cave sites including Paleolithic and Mesolithic sites caught our attention (Appendix of this volume). Our major target was the Avey Mountain, where Damjili and Dash Salahli Caves were known to have contained pre-Neolithic occupations (Huseynov 2010). As the Dash Salahli Cave was inaccessible because of its proximity to the Armenian border, we visited the Damjili Cave to study its potential for further investigations (Fig. 2.1). Simultaneously, we studied, the lithic artifacts collected from excavations at the cave in the 1950s, at the National Academy of Sciences, Baku. These studies convinced us that the cave deserves intensive investigation to understand the Mesolithic period of this region.

Accordingly, we conducted a site reconnaissance survey between 2014 and 2016 focusing on the Avey Mountain region and other selected spots in the Gazakh Region. An outline of this survey research is presented below.

2.2 Avey Mountain, Gazakh

The area we focused on is the eastern foothills of the Avey Mountain, located west of the modern town of Dash Salahli Gazakh region in Western Azerbaijan (Fig. 2.2). It covers an area of c. 21 km^2 (7 km north-south by 3 km east-west), dissected by many valleys, most of which lack perennial water flow. The mountain consists of Upper Cretaceous limestone/chalk rocks, some parts of which are under exploitation as quarries for limestone of high quality. Several lines of cliffs have developed on the anticline slopes of the eastern Avey, where springs emerge at some places. A series of rockshelters is visible from afar along these cliff bottoms, which were our main targets.

2.2.1 Survey methods

As survey methodology, we adopted an intensive pedestrian survey. To find cave/rockshelters with anthropogenic deposits for the above area, we walked along each valley's cliffs and mountain ridges. In some cases, we encountered tracking problems as thick vegetation made some locations invisible and inaccessible. Although this hindered us from covering the entire cliff surface, most of the valleys in the eastern foothill were sufficiently visited.

Once cave/rockshelters or artifact scatters were identified in the field, geographic information of each visited point (latitude, longitude, and altitude) was recorded using a handheld GPS navigator (GPSmap 60CSx, Garmin Ltd.). In the case of rockshelters we measured basic dimensions and terrace size according to the measurement protocols of Suzuki and Kobori (1970). The distribution ranges (length and width) were roughly measured for artifact scatters where possible. We utilized a laser rangefinder for measurement (TruPulse 200™, Laser Technology Inc.).

Fig. 2.1 Distant view of the Damjili Cave area.

Fig. 2.2 Map showing the locations visited during our survey of Avey Mountain.

2.2.2 Survey results

We recorded a total of 28 archaeological find spots in Avey Mountain as follows (Fig. 2.2; Table 2.1).

QA001, Damjili Cave 1 (Figs. 2.1; 2.3: 1)
This site was excavated by S. N. Zamyatnin and M. Huseynov in the 1950s and reported as a multilayered site ranging from the Paleolithic and Mesolithic to the Neolithic. It is located at the deepest end of a valley where spring water comes from an underground aquifer, making the small valley rich in vegetation. For this reason, the site has become a popular recreation park for the local communities. The dimensions we measured do not contradict with the reported ones (Huseynov 2010: 53), that is, 17 m wide, 7.3 m long, and 4.2 m high (Table 2.1). As the floor of the rockshelter was almost covered with concrete due to recent construction works, we failed to observe any anthropogenic deposits in the survey.

QA002, Damjili Cave 2 (Fig. 2.3: 2)
This rockshelter, located c. 100 m east of the Damjili Cave 1, is one of the largest recorded in our survey. It measures c. 45 m wide, 8 m long, and 3.2 m high, with a sizeable terrace of 8.4 m (Table 2.1). This site was also investigated in the 1950s, and the previous test soundings revealed mixed multilayered cultural deposits (for medieval pottery from the soundings at Cave 2, see Chapter 16 of this volume). Although no prehistoric artifact was recovered on the surface, it likely has a considerable depth of anthropogenic deposits.

QA003 (Fig. 2.3: 3)
This is an artifact scatter without any visible structures like a kurgan, located on the saddle-like terrace of a hill ridge. Given the wide range of collected artifact categories (pottery fragments, ground stone, glass bracelet, and obsidian flakes) and estimated ages (Late Bronze Age and Antique period), this terrace may have been used recurrently as a pastoralist camp or station site. Although, the absence of mounded kurgans do not rule out the possibility that an ancient graveyard being situated here.

QA004 (Fig. 2.3: 4)
A rock-cut monastery is located next to the better-known Ay Mabadi (QA005), close to the top of the mountain. Although there is a possibility that the cut was made using extended natural caves, the recognizable evidence shows that this monument is irrelevant to our research.

QA005, Ay Mabadi (Fig. 2.3: 5)
Ay Mabadi (QA005), also locally called Avey Mebedi, is a monumental hilltop complex located on the summit of Avey mountain at an altitude of 922 m. The architectural complex principally consists of two buildings built with well-cut limestone blocks. Although the larger building has been preserved to show a rectangular plan with two domed halls, the smaller one, which lies to the south and has a single domed structure (chapel?), is damaged. They are believed to have been parts of a fortified Christian sanctuary or monastic complex built on the foundation of a former pagan temple, probably dating from the Late Antique period (mid-first millennium AD). There are inscriptions on the building walls, but they are too damaged to be recognizable. Notably, there are rock-cut caves (QA004) on the limestone cliffs near the mountain summit. Owing to their proximity to the Azerbaijan-Armenian border, intensive archaeological survey, collecting surface artifacts, and documentation was not allowed during our visit. The exact dating of the architectural complex is unknown, with no datable surface finds. However, the monumental architecture and rock-cut caves may have been linked to increased human occupation in the Avey and surrounding areas in the later part of the first millennium AD, including the Damjili Cave (Chapters 4 and 16 of this volume).

QA006 (Fig. 2.3: 6)
A rockshelter identified in the southeastern part of Avey Mountain is referred to as QA006. It is located at the deepest end of a broad valley where aquifer water gushes out, reminiscent of the topographical setting of the Damjili Cave. Its dimensions (over 30 m wide and 4 m long) are large and we suggest that it may have long-term occupational deposits, although no artifacts were recovered. The present surface is covered with thick recent limestone clastics from the roof (Fig. 2.4).

QA007 (Fig. 2.3: 7)
This find spot represents an artifact scatter area on a low hill near the plain. The pottery fragments sampled here are assignable to the Late Bronze-Early Iron Ages, associated with a few pieces of obsidian artifacts. There was no evidence to suggest the use of this area for a burial field.

QA008 (Fig. 2.3: 8)
At a kurgan field located on the bottom of a pediment beside a dried riverbed, we identified a dozen kurgans with similar construction, but with larger dimensions of c. 4–5 m diameter and 1–2 m height. Some parts of the kurgans were destroyed by mod-

ern construction works or deliberate looting. It is difficult to date these kurgans without controlled excavations because the collected artifacts are insufficient or absent. A handful of red-brown ware fragments were collected from the surface, which dated to the late medieval period.

QA009 (Fig. 2.3: 9)
Small rockshelter inaccessible due to thick vegetation.

QA010 (Fig. 2.3: 10)
Another small rockshelter with a deep cave, at the base of the wall. The present surface is covered with modern fills and thick vegetation. No artifacts were identified.

QA011 (Fig. 2.3: 11)
Relatively large rockshelter, sufficient for habitation. However, there are hardly any sediments; therefore, bedrocks are exposed in parts on the surface.

QA012 (Fig. 2.3: 12)
Relatively large rockshelter, with sufficient area for habitation. The surface is heavily covered with vegetation and limestone rocks fallen from the wall.

QA013 (Fig. 2.3: 13)
Inaccessible due to thick vegetation.

QA014 (Fig. 2.3: 14)
A shelter made by fallen rocks. No artifacts were identified on the surface.

QA015 (Fig. 2.3: 15)
Sizable rockshelter with a terrace showing plenty rocks fallen from the wall.

QA016 (Fig. 2.3: 16)
Small rockshelter showing fresh breaks on the wall. The ground surface is heavily covered with vegetation. No artifacts were recovered.

QA017 (Fig. 2.3: 17)
An open-air site with a sparse distribution of pottery fragments and flaked stone artifacts. The small number of collected artifacts do not allow us to estimate their chronological position.

QA018 (Fig. 2.3: 18)
Rockshelter made by a large limestone block. The ground surface is heavily covered with vegetation.

QA019 (Fig. 2.3: 19)
Small rockshelter. The ground surface is heavily covered with vegetation and limestone rocks fallen from the roof.

QA020 (Fig. 2.3: 20)
A gentle hill at a mountain ridge, where a sparse distribution of possible medieval ceramics were collected. No identifiable structures were recorded on the surface.

QA021 (Fig. 2.3: 21)
This is a large rockshelter with the maximum width of more than 20 m. It has no overhanging ceiling but a broad precipitous cliff surface facing a flat natural terrace northward. A single piece of Late Bronze Age potsherd and a chipped obsidian flake was found in the vicinity of the location. However, the association of these artifacts with this rockshelter has not been confirmed.

QA022 (Fig. 2.3: 22)
Kurgan field located on a gentle slope looking down the plain. Flaked stone artifacts were recovered on the surface.

QA023 (Fig. 2.3: 23)
Another kurgan field on a gentle slope looking down the plain. Approximately ten kurgans were identified on the surface.

QA024 (Fig. 2.3: 24)
Shelter made by a large rolling limestone from the basal formation with heavy vegetation on the ground surface.

QA025 (Fig. 2.3: 25)
A slope looking down the plain. A small number of potsherds were recovered on the surface.

QA026 (Fig. 2.3: 26, not shown on Fig. 2.2 map)
A find spot near a modern limestone quarry is situated on the northern foothill of Avey Mountain. A small number of obsidian flaked artifacts were recovered. Their chronological position has not been identified.

QA027, Qara Bulaq (Fig. 2.3: 27)
This spot, already known in the 1950s (Chapter A3 of this volume), is near a popular recreational area with fresh spring water, approximately one kilometer southeast of Damjili Cave. While the original topography has been considerably modified near the spring water area, the slope toward the modern recreation zone appears to have retained its original condition. We briefly surveyed the slope, which yielded a few obsidian flakes. We also conducted

Fig. 2.3 General views of surveyed sites at Avey Mountain. The photos numbered 1–28 correspond with QA001–QA028 (see Table 2.1).

soundings. One of the three sounding pits was scattered on the slope (Fig. 2.5), while the other two were near the huge rock fall to the west. The results show that the slope was probably included in a habitation

Fig. 2.3 Continued.

zone of a certain period in the past, likely because of a freshwater spring nearby. The most intensive use was undoubtedly made in the medieval period, as indicated by the abundant occurrence of pottery from that period, and perhaps from the Bronze Age, although less frequent, as the occurrence of the small

Fig. 2.3 Continued.

number of obsidian artifacts indicates. However, no traces of Mesolithic or Paleolithic occupations have been discovered.

QA028, Kör Bulaq (Fig. 2.3: 28)
Another site recorded in our survey from the 2016 season worth mentioning is a spring known as Kör

Fig. 2.3 Continued.

Bulaq, located about 2 km west of Damjili Cave. This spring, invisible nowadays because of the construction of an underground channel to take water downwards, was noted by M. Huseynov in the 1950s as a possible prehistoric site (Chapter A3 of this volume). It is open at the bottom of a huge rock fall on the slope, approximately 4 m high and 8 m wide. A flat terrace of 4 m in length has developed from the rock. Although M. Huseynov mentioned the occurrence of some possible Paleolithic artifacts, our survey failed to identify any comparable finds. Instead, medieval sherds were collected.

2.2.3 Discussion

It is said that almost 30 rockshelters and caves have been identified in the Avey Mountains, which preserve prehistoric remains, including those from the Paleolithic period. Our survey identified a total of 15 rockshelters and caves. However, apart from the Damjili Cave 1, we could not confirm occupation traces of the Paleolithic period. This difference should be interpreted, taking into consideration that our survey covered only the eastern foothills of the mountain. Furthermore, as the region has been declared a national conservation region of natural environments, Mount Avey is densely covered with vegetation, making a survey of artifact collection considerably difficult. Therefore, the possibility of the presence of Paleolithic caves and rockshelters is open to more intensive investigations in the future. Some of the rockshelters/caves, such as QA006 and QA021, for example, seem to have dense deposits.

Our survey located a series of kurgan fields and artifact scatter spots, which clearly demonstrated the exploitation of this mountain in prehistoric times. The kurgans of QA008, QA022, and QA023 are particularly interesting in their good preservation. The latter two kurgan fields lie on the high plateau of the southern piedmont overlooking the Gazakh plain, consisting of at least several to ten kurgans in each. The kurgan mounds, constructed by piled rubble (cairn), typically show an elliptical or round shape, with moderate sizes of 1–2 m length or 2.5 m diameter and 0.5–0.6 m height.

The artifact scatters indicate prehistoric and later human exploitations of this mountain environment. The distribution of artifactual remains was sparse, indicating that the find spots represent temporary camps or transitory stations for exploitation of mountain resources. Animal herding is an interpretation. The obsidian artifacts may indicate Bronze Age occupations. However, the pottery sherd scatters seem to belong to the medieval period, which may have been related to the construction

Table 2.1 Survey data of the visited sites and locations at Avey Mountain.

Site number	Site name	Type	Period	Finds	N			E			Altitude (m)	Width (m)	Length (m)	Height (m)	Terrace Width (m)	Terrace Length (m)	Remarks
					°	′	″	°	′	″							
QA001	Damjili 1	rockshelter	Paleolithic, Mesolithic to medieval	none	41	8	51.6	45	13	95.1	581	17	7.3	4.2	-	-	Damjili Cave
QA002	Damjili 2	rockshelter	Medieval	potsherd, ground stone?	41	8	54	45	13	99.4	585	44.8	8	3.2	45+*	8.4	
QA003	-	artifact scatter	Late Bronze-Antique	potsherd, obsidian, ground stone, glass bracelet	41	8	76.6	45	14	21.4	589	35	65	-	-	-	
QA004	-	rock-cut monastery	Antique-medieval?	-	41	8	83.8	45	13	21.9	902	-	-	-	-	-	
QA005	Ay Mabadi	architectural monument	Antique-medieval	-	41	8	85.9	45	13	21.3	922	-	-	-	-	-	located next to the modern military garrison
QA006	-	rockshelter	-	none	41	9	0.07	45	14	34.4	662	31.6	4.2–8	1.2–2	35+*	3?	
QA007	-	artifact scatter	Late Bronze-Early Iron	potsherd, obsidian	41	8	91.6	45	14	76.8	510	34	87	-	-	-	
QA008	-	kurgans	Late medieval	potsherd, obsidian	41	9	14.4	45	14	79.9	502	30	40	-	-	-	kurgan diameter c.< 5 m
QA009	-	rockshelter	-	none	41	9	17.3	45	14	47.6	652	5–6*	N/A	N/A	-	-	inaccessible due to thick vegetation
QA010	-	rockshelter	-	none	41	9	0.84	45	14	40.3	679	3*	N/A	N/A	-	-	
QA011	-	rockshelter	-	none	41	9	28.4	45	14	42.4	645	8.8	2.9	2.6	9+*	0.3	c. 5 cm of sediment?
QA012	-	rockshelter	-	none	41	9	44.6	45	14	38.3	648	8	6	1.8	N/A	N/A	
QA013	-	rockshelter	-	none	41	9	44.6	45	14	38.2	651	N/A	N/A	N/A	N/A	N/A	inaccessible due to thick vegetation
QA014	-	rockshelter	-	none	41	9	45.1	45	14	37.7	653	3*	5+*	1.2*	N/A	N/A	fallen ceiling rocks
QA015	-	rockshelter	-	none	41	9	53.3	45	14	65.2	583	8.1	1–2.1	1.6–2.5	10+*	5*	
QA016	-	rockshelter	-	none	41	9	58.6	45	14	75.7	543	4	0.7	2	N/A	N/A	
QA017	-	artifact scatter	Unknown	obsidian	41	9	63.3	45	14	84.9	539	-	-	-	-	-	campsite?
QA018	-	rockshelter	-	none	41	9	83.5	45	14	70.6	555	5.3	0.5	2	N/A	N/A	
QA019	-	rockshelter	-	none	41	9	79.7	45	14	40.9	634	4.9	2.9	2.3	-	-	
QA020	-	artifact scatter	Medieval	potsherd	41	8	45	45	13	93	621	-	-	-	-	-	on a ridge
QA021	-	rockshelter	Late Bronze?	potsherd, obsidian	41	8	28.6	45	13	83.2	659	26.5	-	4–5*	27+*	1–2*	
QA022	-	kurgans	Unknown	flint	41	8	0.81	45	13	78.9	691	10*	20*	-	-	-	4–5 kurgans
QA023	-	kurgans	Unknown	?	41	8	17.6	45	13	71.1	718	20*	30*	-	-	-	approximately 10 kurgans
QA024	-	rockshelter	-	none	41	8	47	45	13	58.2	735	8.5	4	1.5	-	-	
QA025	-	artifact scatter	Unknown	potsherd	41	8	55.9	45	13	32.8	685	-	-	-	-	-	
QA026	-	artifact scatter	Unknown	obsidian	41	11	76.7	45	13	58	443	-	-	-	-	-	
QA027	Qara Bulaq	spring	Unknown	obsidian, potsherds	41	8	0.51	45	14	55.1	479						see Chapter A3
QA028	Kör Bulaq	spring	Unknown	potsherds	41	8	14	45	14	23.4	557						see Chapter A3

Fig. 2.4 Closer view of the rockshelter of QA006.

Fig. 2.5 Sounding areas of the rockshelter of QA027.

of the early Christian monastery site of Ay Mabadi (QA005) on the summit of the Avey Mountain. The extensive rock-cut residences below the monastery building (QA004) likely promoted a more intensive use of Avey Mountain and its surrounding areas in the Late Antique period (early first millennium AD).

2.3 Investigations at Yataq Yeri

In addition to the research at Avey Mountain, in 2015 and 2016, we conducted preliminary surveys at West Gazakh and soundings at Yataq Yeri (Fig. 2.6; Table 2.2). It is an open-air site located between Shikhli I and II villages (for the location, see Mansurov 2020: 206), which M. Huseynov first discovered in 1963. According to Mansurov (2020), who visited this site repeatedly in the last decades, the surface materials contain lithic artifacts dating from the Neolithic, Mesolithic, and Upper Paleolithic periods. Accordingly, this site deserves investigation as our mission aims to study Mesolithic issues.

The site of Yataq Yeri lies on the northern edge of one of the old terraces of the Kura River, which is cut to the east by a small wadi called Ajidara that flows into the Kura plain (Fig. 2.7). This small wadi has formed some lower terraces, where archaeological materials can be collected. Accordingly, we identified at least three terraces containing archaeological materials: the lowest terrace is about 12 m higher than the wadi bed, the middle terrace 19 m, and the highest, 23.5 m. According to the typological dating, the lowest terrace can be dated to the Holocene, and the highest to the Middle Paleolithic period. The middle terrace should represent a period in between, but no artifacts showing a definite chronological period were collected.

We opened sounding pits on these three groups of terraces to examine the sedimentological history (Fig. 2.8). The pit on the highest terrace yielded evidence of an alteration of several sand and mud

Fig. 2.6 General view of the site of Yataq Yeri.

Table 2.2 Data of open-air site Yataq Yeri.

Season	GPS Point	Site name	Type	Finds	N °	N ′	N ″	E °	E ′	E ″	Altitude (m)	Remarks
2015	447	Yataq Yeri	open air site	lithic	41	17	16.8	45	8	27.5	291	
2015	448	Yataq Yeri	open air site	lithic	41	17	22	45	8	26.3	309	Sampling square A; sounding
2015	449	Yataq Yeri	open air site		41	17	21	45	8	26.1	310	Sampling square B
2015	450	Yataq Yeri	open air site	lithic	41	17	17.5	45	8	27.3	301	Sampling square C; sounding
2015	451	Yataq Yeri	open air site	lithic	41	17	17.5	45	8	27.5	293	
2015	453	Yataq Yeri	open air site	lithic	41	17	20.5	45	8	30	290	Geological section
2015	454	Yataq Yeri	open air site	lithic	41	17	25.1	45	8	35.2	287	Geological section

Fig. 2.7 Photo image showing the location of sounding pits at Yataq Yeri. The sounding trenches are indicated by white rectangles, where A and B lie on the high terrace, while C lies on the middle terrace, and D on the low terrace.

Fig. 2.8 Sounding excavation at Yataq Yeri.

layers, indicating the past fluvial activities, and a layer of river gravel on the middle terrace. The lowest terrace was found to be covered with thick eolian loess deposits; however, neither of our test pits yielded artifacts. As for the Mesolithic research, our intensive surface survey did not yield any materials indicating the Mesolithic Age.

2.4 Conclusions

During the 2015–2016 surveys, we discovered a small number of prehistoric sites at Avey Mountain in the form of kurgans and artifact scatters. The ceramics may indicate the Bronze Age and the later periods. However, the collected obsidian and flint artifacts may include some earlier materials. Although, our survey did not locate any new rockshelters and caves. It should be noted that most of those sites are covered with heavy vegetation and/or thick later limestone rubble. Accordingly, it is challenging to identify small lithic artifacts through preliminary surface surveys. We want to reserve a possibility that future investigations may lead to the discovery of prehistoric rockshelters and caves. In this regard, the Damjili Cave stands out as an exceptional site containing rich prehistoric deposits, whose potential for prehistoric research certainly deserves further pursuit.

Along with the surveys at Avey Mountain, complementary surveys covering other parts of the Gazakh plain to search for Neolithic and earlier sites are expected to be fruitful. The region consists of Cretaceous limestone formations with flint raw materials suitable for lithic manufacturing. There will be no surprise in discovering promising prehistoric sites in this favorable region.

References

Huseynov, M. (2010) *The Lower Paleolithic of Azerbaijan. Baku: National Academy of Sciences of Azerbaijan* (in Russian with English summary).

Mansurov, M. M. (2020) *Lower Paleolithic of Western Azerbaijan.* Baku: Apostrof-A (in Russian with English summary).

Nishiaki, Y. and F. Guliyev (2020) *Göytepe: The Neolithic Excavations in the Middle Kura Valley, Azerbaijan.* Oxford: Archaeopress.

Nishiaki, Y., F. Guliyev, and S. Kadowaki (2021) *Hacı Elamxanlı Tepe: The Archaeological Investigations of an Early Neolithic Settlement in West Azerbaijan.* Berlin: ex oriente.

Nishiaki, Y., A. Zeynalov, M. Mansrov, and F. Guliyev (2022) Radiocarbon chronology of the Mesolithic-Neolithic sequence at Damjili Cave, Azerbaijan, Southern Caucasus. *Radiocarbon* 64(2): 309–322.

Shimogama, K. and V. Alakbarov (2020) Archaeological reconnaissance survey around Göytepe, Tovuz-Qovlar region. In: *Göytepe: The Neolithic Excavations in the Middle Kura Valley, Azerbaijan*, edited by Y. Nishiaki and F. Guliyev, pp. 137–166. Oxford: Archaeopress.

Suzuki, H. and I. Kobori (1970) *Report of the Reconnaissance Survey on Palaeolithic Sites in Lebanon and Syria.* Bulletin of the University Museum, The University of Tokyo 1. Tokyo: The University of Tokyo.

Chapter 3

Geomorphological settings around Damjili Cave

Yuichi S. Hayakawa and Masuto Ebina

3.1 Geographic environment around Damjili Cave

The Damjili Cave is located at the base of steep cliffs on the left bank of the east-facing Damjili Valley in the northern foothills of the Lesser Caucasus Mountains (Fig. 3.1). The site is designed as a part of the State Historical and Cultural Reserve Avey and has an elevation of around 582 m above sea level (Fig. 3.2). The substrate lithology of the site entrance consists mainly of carbonaceous sedimentary, particularly limestone, from the Upper Cretaceous period (United Nations 2000). The area has a warm and humid climate, with an average annual precipitation of approximately 300 mm and monthly-mean daily temperatures ranging from −2.6–6.5°C in winter (January) and from 19.5–31.7°C in summer (July) (data from Ganja; World Meteorological Organization 2023). According to Köppen's classification, the modern climatic type is supposed to be BSk (arid steppe and cold) to Cfa (warm temperature, fully humid, and hot summer) (Kottek et al. 2006). The predominant vegetation cover in the region is characterized by steppe, and the primary land use in the foothill area is agriculture and meadow with the use of groundwater and irrigation water. There are also some mining sites for minerals in the mountain sides.

The occupations at Damjili Cave are located at the bottom of steep cliffs, where notches are developed for several meters exhibiting a cave-like feature particularly at the head of the sculpted valley. These notches are supposed to be formed by combined processes of fluvial erosion, groundwater sapping, and weathering. These notches are favorable for people to stay because of the availability of water constantly supplied as springs, as well as shade areas from sunlight.

Here we describe the available data sources of

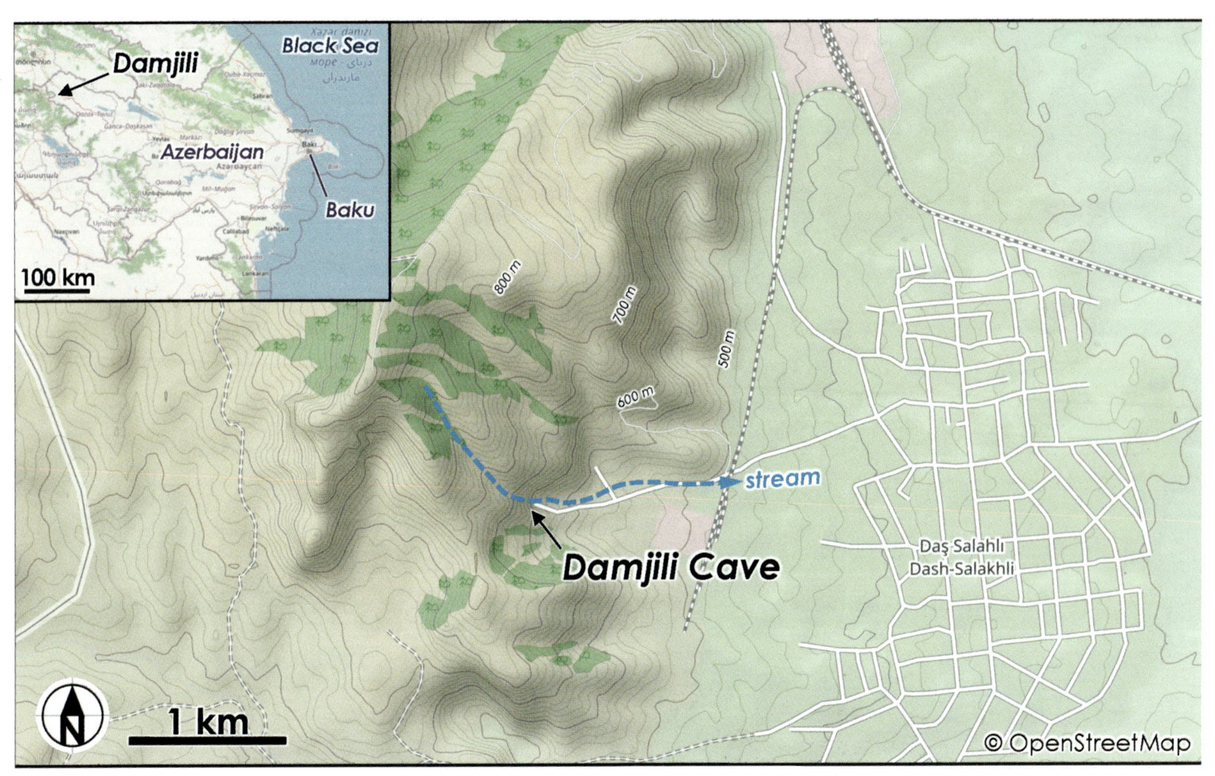

Fig. 3.1 Overview map of the study area. Background map data is from OpenStreetMap.

Fig. 3.2 Pictures of the study area (taken in August 2016). a) Panorama view of the Damjili Valley; b) Head of the amphitheater valley viewed from the left side; c) The valley head viewed from the right side. White arrow indicates the position of waterfall lip, where pictures of d and e were taken; d) Downstream view from the waterfall lip; e) Upstream view from the waterfall lip.

topographic information in the site below. Because of the limitation of the availability of existing maps, we present the acquisition of detailed topographic data performed in the site by ourselves in 2016. Based on the topographic information presented, descriptions on geomorphological conditions of the Damjili Valley follow.

3.2 Topographic data

3.2.1 Existing maps

Although some small-scale (1:250,000) topographic maps were published in this region (e.g., U.S. Army Map Service 1954; U.S. National Imagery and Mapping Agency 1978), finer, larger-scale maps, as well as detailed topographic digital data, have been unavailable for the study site. It is therefore necessary to carry out an on-site topographic measurement of the site, including the cave (notches as rock-shelters) and surrounding cliffs and slopes.

3.2.2 Detailed local mapping of topography

In these decades, topographic measurement methods using Global Navigation Satellite System (GNSS) and laser rangefinders have been developed for an efficient topographic measurement method and applied at several archaeological sites including Tell Seker al-Aheimar (Syria) and Göytepe (Azerbaijan) (Hayakawa et al. 2007; Hayakawa and Tsumura 2009; Hayakawa 2021). This was to obtain topographic information over an area of several hundred meters in length with a resolution of several meters to several tens of centimeters. More recently, technological innovations have made it possible to obtain even higher-resolution topographic information more quickly at a low cost. Among else, laser scanning and photogrammetry approaches are recognized as the modern high-definition topographic mapping methods. Terrestrial Laser Scanning (TLS) is to acquire the distance and angle from a measurement point to a target object using a laser beam, similar to total stations and laser rangefinders, but it is capable of faster measurement (tens of thousands of points per second) with an accuracy of a few millimeters to a few centimeters, and the number of measurement points obtained can range from several million to several tens of thousands. TLS was originally developed in engineering and construction fields, but has also been applied to natural landforms (Hayakawa and Oguchi 2016). The measurement by Structure-from-Motion Multi-View Stereo (SfM-MVS) photogrammetry, hereafter simply referred to as SfM, automatically reconstructs the camera position from a large number of stereo-pair photographs and acquires high-density point cloud data based on the principle of photogrammetric algorithms. Since aerial photographs taken by small uncrewed aerial system (UAS) can be used to obtain a wide range of topographic information, UAS-based SfM is becoming increasingly popular for landscape measurements (Hayakawa et al. 2016). SfM can also be applied for topographic measurements even for the areas where the use of UAS is limited, using ground-based platforms of a camera. The three-dimensional (3D) point cloud data obtained by TLS or SfM can be further processed to create datasets in various kinds of representations, such as two-dimensional plans, cross sections, Digital Elevation Models (DEMs), and 3D mesh models. It is important to use each of these measurement methods according to the purposes and the environmental or practical conditions in which they are used.

Because the use of aerial measurements by UAS was limited around the site that is close to the nation borders, we employed ground-based measurement methods for the Damjili Site: TLS as the main method for the cave and excavation sites, and SfM with a handheld pole-mounted digital camera as a supplementary method for a wider area measurement across the valley (Fig. 3.3).

Fig. 3.3 Equipment used for topographic measurements. a) Terrestrial laser scanner, Trimble TX5; b) Digital camera mounted on a pole (extended up to 4-m long) for ground-based SfMMVS photogrammetry.

The TLS device used was a Trimble TX5, which is capable of ranging of up to tens of meters (maximum 120 m) from the machine location, and capturing full color information by an RGB (red-green-blue) sensor equipped. Multiple scans with different sight of views were necessary to acquire the entire area of interest. The point clouds of individual scans were merged by post-processing using the iterative closest point algorithm to a single point cloud covering the entire area. Since the three-dimensional point cloud data of the scans includes reflections from all materials surrounding, the ground surface needs to be extracted by filtering. The filtered point cloud data were then converted into a DEM, and shaded relief and contour lines were calculated using a GIS software (QGIS).

To conduct the SfM measurement, we used a RICOH GR II digital camera with a wide-angle conversion lens and a monopod (maximum length of 4 m) (Fig. 3.3: b). Many photos were taken by interval shooting of photos at 2 seconds while moving around the valley, including surrounding ridges, to cover the entire valley. To ensure accuracy, we placed several ground control points (GCPs) in the range of photo shooting. The geographic coordinates of the GCPs were obtained with the differential positioning (post-processed kinematic correction) by GNSS receivers (Trimble GeoXH). Furthermore, the GNSS log data of the local base station at the cave entrance zone were corrected using a SOPAC station data at Zelenchukskaya (Russia) (43°47′18.21131″N, 41°33′54.22925″E, 420 km away), providing its coordinates of 41.141758°N, 45.232435°E, and 581.9 m a.s.l. with horizontal and vertical accuracies of 0.39 m and 0.61 m, respectively. The SfM photogrammetric processing was then carried out using Agisoft PhotoScan software to produce three-dimensional point clouds, DEMs, and orthorectified mosaic image.

As a result, we obtained 446,588,156 points by TLS scans from 102 scan positions in total (Fig. 3.4). The TLS-based point cloud data were used to generate DEMs and shaded relief maps, so that we can recognize the detailed topography around the excavation sites and springs beneath the cliffs (Figs. 3.5; 3.6). From the SfM photogrammetry of the pole-mounted camera images, point cloud, 3D model, DEM (48 cm resolution), and orthorectified mosaic image (3 cm resolution) of the entire area were generated (Figs. 3.7–3.9). For this, successfully matched 572 images, out of more than a thousand images in total, were used.

3.3 Geomorphological conditions of the Damjili Valley

3.3.1 Formation of the Damjili Valley

The Damjili site is located at the base of cliffs of an amphitheater-headed valley formed by fluvial erosion. The upstream end of the amphitheater valley forms a prominent knickpoint (waterfall) of the stream (Fig. 3.9). Although permanent flow is not observed under current climatic and hydrological conditions, surface water flow appears as a waterfall during rainfall seasons. Such water flows may have been steadily present in the wet conditions in the past, causing the progressive erosion at the waterfall. In general, the recession of waterfalls is one of the most significant processes of bedrock erosion by rivers, where waterfalls may retreat at a rate of a few centimeters to several meters per year in wet regions (Wohl 1998; Hayakawa and Matsukura 2003a, 2009). However, under the present dry climate of the study site (annual precipitation of about 300 mm at Ganja), continuous surface flow is not expected to cause drastic waterfall recession at least under the present climate. Intermittent stream flow along the channel may cause some sediment transport and slight abrasion on the bedrock surface, which is evidenced as the smooth rock surface along the channel. In fact, we observed bare surface of bedrock along the channel just upstream of the waterfall, where sparse vegetation covers the channel bed with little soil layer (Fig. 3.2: d, e). Although there is no evidence for the estimation of the erosion rate, if the annual erosion rate is on the order of millimeters, the form of the waterfall, including the steep cliff and underneath notches, could have been different for several meters in the timescale of thousands of years. Moreover, the current cliffs show degraded slopes on the both sides of the waterfall. This may be due to the higher weathering rates on the side slopes where freeze-thaw actions may occur in the autumn and winter seasons. In that case, the area of the stream flow may be rather protected by the presence of unfreeze water.

3.3.2. Topographic characteristics around the cliff base

The main spring near the present recreation area is located at the lower part of the cliff face, which corresponds to the plunge pool of the waterfall. It is common to observe notches at the lower part of waterfall faces due to differences in bedrock hardness or susceptibility to weathering, as seen in other famous waterfalls like Niagara Falls in North America and Nikko Kegon Falls in central Japan (Hayakawa and Matsukura 2003b, 2009). At the Damjili Cave,

Fig. 3.4 Three-dimensional point cloud data around the Damjili Cave obtained by TLS. The data can be explored on the Sketchfab website: https://sketchfab.com/3d-models/damjili-0436f5e744e3470ab1381670afea0b1f

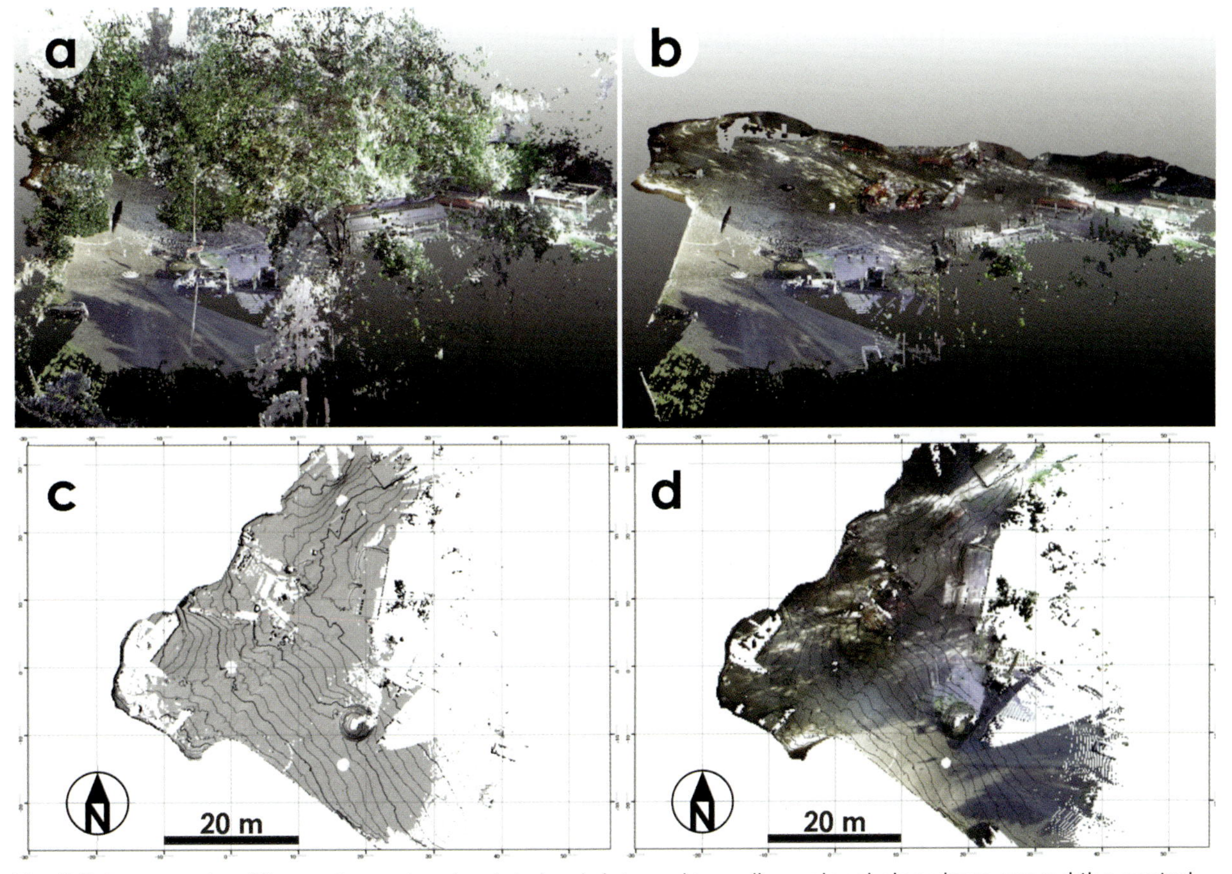

Fig. 3.5 An example of three-dimensional point cloud data and two-dimensional plan views around the central portion of the plunge pool. a) Original point cloud; b) Point cloud after filtering; c) Shaded relief image and topographic contour lines by 10 cm interval; d) Colored surface map with 10-cm topographic contour lines.

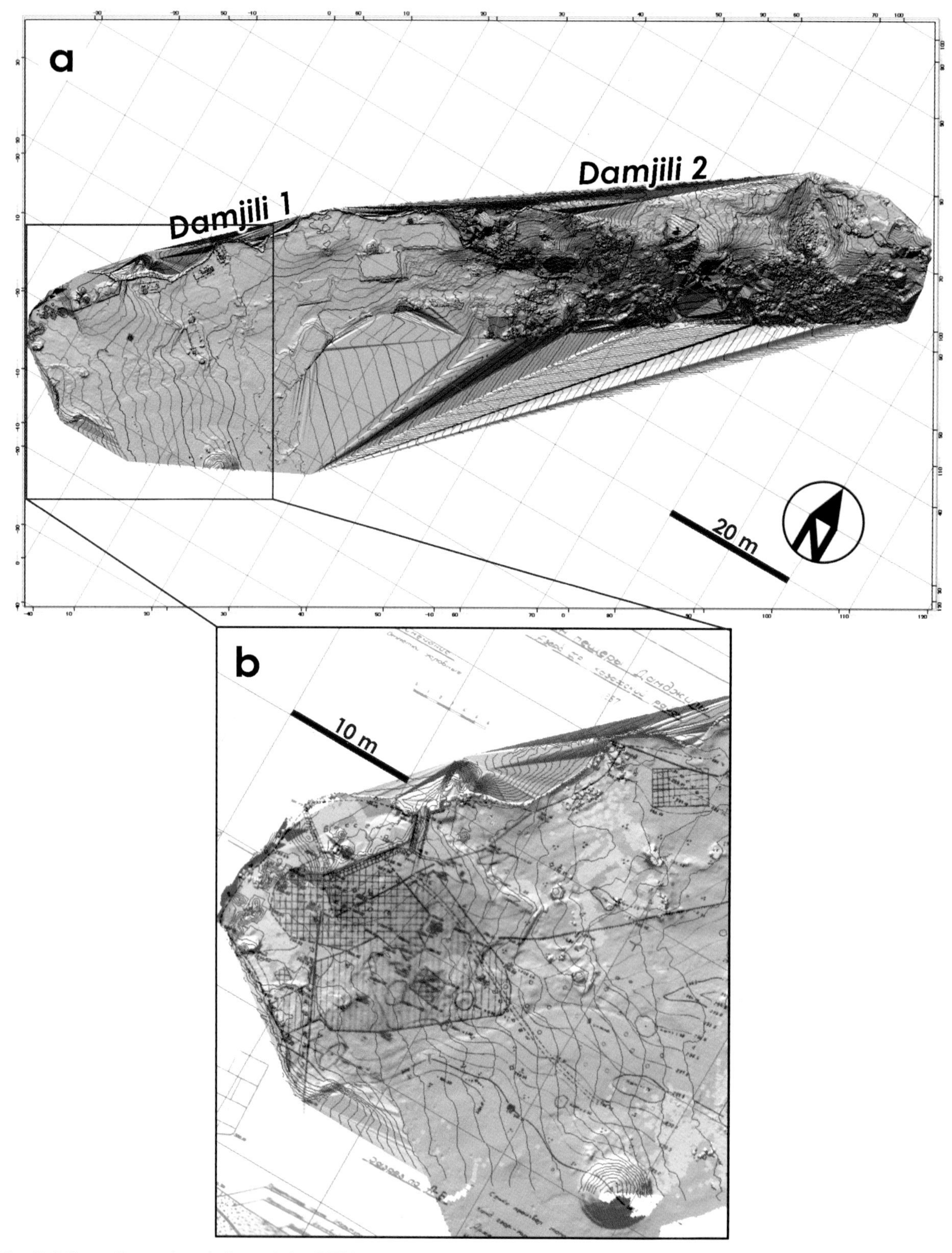

Fig. 3.6 Two-dimensional view of the DEM-derived shaded relief image and topographic contour lines by 10 cm interval. a) Entire view; b) Magnified view around Damjili 1 with an overlay of an old map of the past excavation in the site.

the notches are composed of conglomerates, which are more susceptible to weathering and erosion than the overlying limestone layers, and thus appear to have formed the notches. However, differences in rock types were not clearly observed at the notches near the excavation area. It was difficult to confirm this because the floor below the notch is covered with thick sediments, under which a weak layer may exist.

As noted, the form of the cliff face and notches

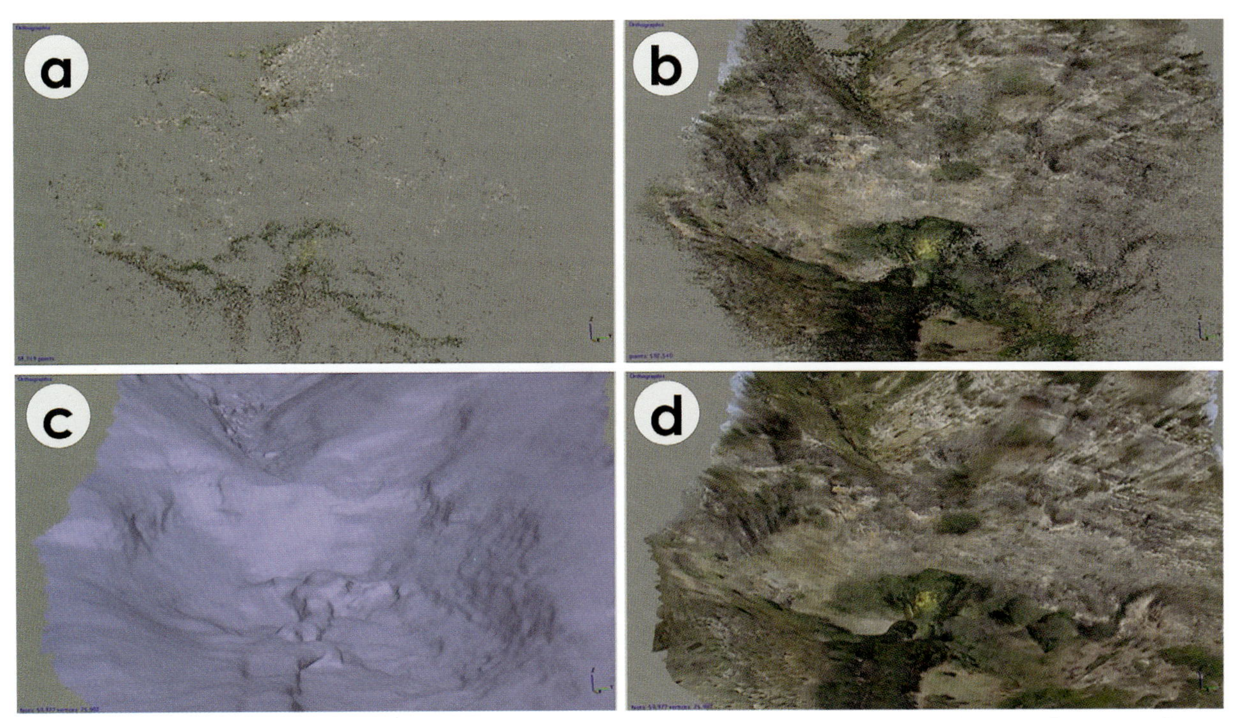

Fig. 3.7 Processing of three-dimensional data by SfM-MVS photogrammetry. The view is set at the center front of the waterfall. a) Sparse cloud, i.e., tie points of photo matching by SfM; b) Dense cloud, i.e., point clouds derived by MVS photogrammetry after SfM; c) 3D TIN model; d) Textured 3D TIN model, showing RGB color.

Fig. 3.8 Three-dimensional model of the entire area of measurement in the Damjili Valley. The model can be explored on the Sketchfab website: https://sketchfab.com/3d-models/damjili-site-overview-94330078e9a8457a82b6b0c4af8323b6.

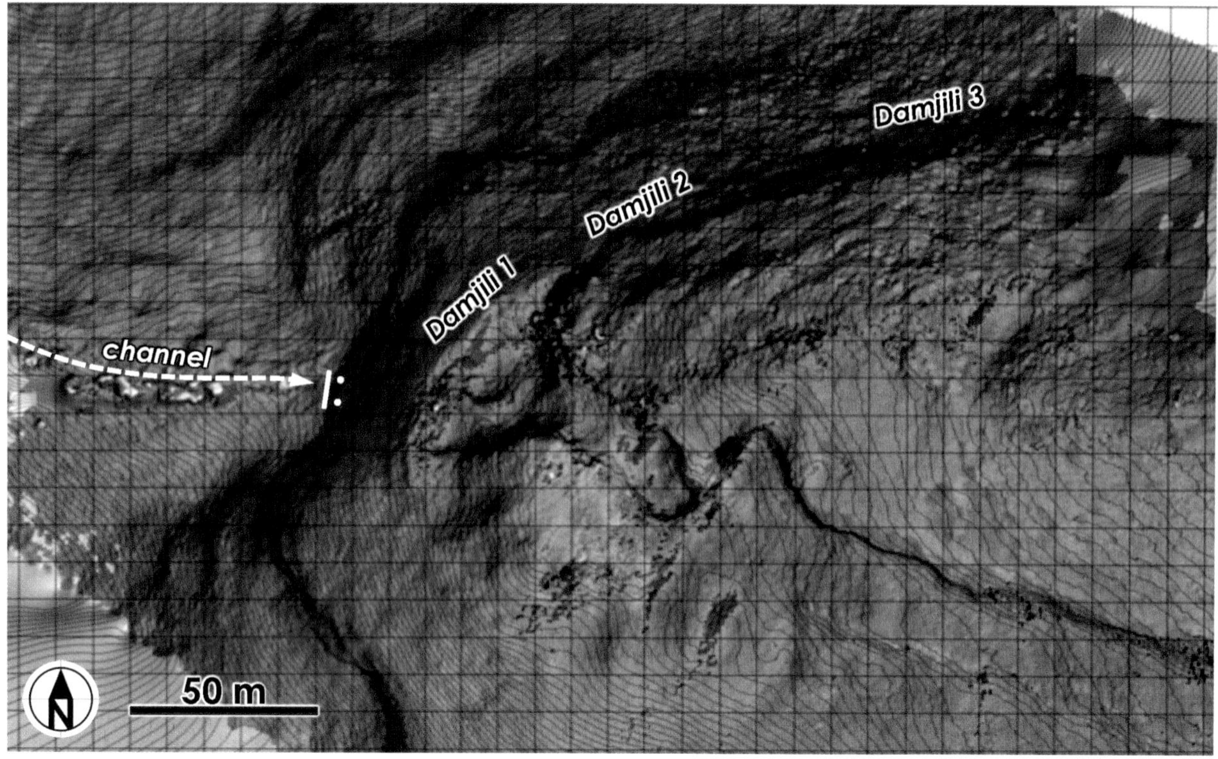

Fig. 3.9 Colored shaded-relief image of the central Damjili Valley. Black lines are topographic contour lines by 1 m interval.

may have been different for meters of scale in the thousands of years past, the environmental conditions might have been so different for the lifetime of the archaeological sites. Not only the climatic and hydrological conditions, differences in the topographic conditions, including the sunlight shadings, could have affected the formation of the settlements of the sites. Further detailed investigations would be recommended for exploring such long-term geoenvironmental changes and human activities in the site.

References

Hayakawa, Y. S. (2021) Geomorphological settings of Göytepe. In: *Göytepe: Neolithic Excavations in the Middle Kura Valley, Azerbaijan*, edited by Y. Nishiaki and F. Guliyev, pp. 11–14. Oxford: Archaeopress.

Hayakawa, Y. and Y. Matsukura (2003a) Recession rates of waterfalls in Boso Peninsula, Japan, and a predictive equation. *Earth Surface Processes and Landforms* 28(6): 675–684.

Hayakawa, Y. and Y. Matsukura (2003b) Recession rates of Kegon Falls in Nikko, Tochigi Prefecture, Japan. *Journal of Geography* (Tokyo) 112(4): 521–530 (in Japanese with English abstract).

Hayakawa, Y. and Y. Matsukura (2009) Factors influencing the recession rate of Niagara Falls since the 19th century. *Geomorphology* 110: 212–216.

Hayakawa, Y. S., H. Obanawa, H. Saito, and S. Uchiyama (2016) Geomorphological applications of structure-from-motion multi-view stereo photogrammetry: A review. *Transactions, Japanese Geomorphological Union* 37(3): 321–343 (in Japanese with English abstract).

Hayakawa, Y. S. and T. Oguchi (2016) Applications of terrestrial laser scanning in geomorphology. *Journal of Geography (Chigaku Zasshi)* 125(3): 299–324 (in Japanese with English abstract).

Hayakawa, Y. S., T. Oguchi, J. Komatsubara, K. Ito, K. Hori, and Y. Nishiaki (2007) Rapid on-site topographic mapping with a handheld laser range finder for a geoarchaeological survey in Syria. *Geographical Research* 45(1): 95–104.

Hayakawa, Y. S. and H. Tsumura (2009) Utilization of laser range finder and differential GPS for high-resolution topographic measurement at Hacitürul Tepe, Turkey. *Geoarchaeology* 24(2): 176–190.

Kottek, M., J. Grieser, C. Beck, B. Rudolf, and F. Rubel (2006) World map of the Köppen-Geiger climate classification updated. *Meteorologische Zeitschrift* 15: 259–263.

United Nations (2000) Geology and mineral resources of Azerbaijan. *Atlas of Mineral Resources of the ESCAP Region, Vol. 15*. New York: United Nations Publications.

U.S. Army Map Service (1954) Tbilisi. *Eastern Europe 1:250,000 Series N501, NK38-8*.

U.S. National Imagery and Mapping Agency (1978) *Joint Operations Graphic 1:250,000 Series 1501 Air, NK38-8*.

Wohl, E. E. (1998) Bedrock channel morphology in relation to erosional processes. In: *Rivers over Rock*, edited by K. J. Tinkler and E. E. Wohl, pp. 133–151. Washington, DC: American Geophysical Union.

World Meteorological Organization (2023) *World Weather Information Service: Ganja, Azerbaijan*. Retrieved from https://worldweather.wmo.int/en/city.html?cityId=19 (Last accessed 10 February 2023).

Chapter 4

Renewed excavations at Damjili Cave (2016–2022)

Yoshihiro Nishiaki, Azad Zeynalov, Yagub Mammadov, Mansur Mansurov, Ulviya Safarova, Kazuya Shiomogama, Shahin Salimbayov, Orkhan Zamanov, Takehiro Miki, Saiji Arai, and Fumika Ikeyama

4.1 Introduction

Damjili Cave is situated at the west end of a large cave-rockshelter complex developed along the Damjili Valley, the eastern foothills of Avey Mountain in West Azerbaijan (Chapter 3 of this volume). When this cave was excavated for the first time in 1953, it was described as consisting of two caves (Zamyatnin 1958). However, unfortunately, no detailed map showing their locations was published. Yet, as far as the available records are concerned (Appendix of this volume), the first excavators, or Sergei Zamyatnin and Mammadali Huseynov, apparently defined the waterfall basin at the west end of the cave complex as Damjili Cave 1. This interpretation is supported by a couple of the extant descriptions including the reported size of Damjili Cave 1, which measurred approximately 17 to 27 m by 6 m (Chapter A2 of this volume). On the other hand, the location of Zamyatnin's Damjili Cave 2 is unclear. According to Zamyatnin (1958), future exploration is promising because of the abundance of prehistoric artifacts such as obsidian. However, no drawings are provided. Furthermore, Huseynov's later research in 1956 and 1957 did not include any description of it, which requires further discussion.

Zamyatnin (1958) states that Damjili Cave 2 has the following features: it (1) is not a cave but a rockshelter, (2) is located "next door" to Damjili 1, (3) is about 70 m long, (4) has yielded many prehistoric remains such as obsidian artifacts from a test-pit, and (5) may contain deposits at least 12–15 m thick. Considering these descriptions, despite the absence of a plan map, we wonder if Zamyatnin's Damjili 2 may correspond to the eastern part of our Damjili Cave 1, which was the main excavation area for our team between 2016 and 2022. Features (1), (2), and (4) nicely match with our observations (Chapters 3 and 4 of this volume). On the other hand, features (3) and (5) require careful evaluation.

Assuming that Zamyatnin's Damjili Cave 2 is a rockshelter extending to the east from Cave 1, we made a cross-section of the main excavation area of our investigations. The results revealed that the present cave ground surface is approximately 12–15 m high from the valley bottom, suggesting that feature (5) could refer to the height of the cave terrace from the bed of the Damjili Valley (Figs. 4.2; 4.3). If so, the remaining issue is interpretation of feature (3), it being 70 m wide from the waterfall basin (Damjili Cave 1). It is not as straightforward to interpret. Indeed, a large rockshelter extends to the east from the waterfall basin at least 50 m within Damjili Cave 1. To the eastward, Damjili Cave 2 is 45 m wide (Chapters 4 and 5 of the present volume). Therefore, Zamyatnin's Damjili Cave 2 does not match either of our Damjili 1 and 2. It is also important to note that the absence of any prehistoric objects from the sounding trenches at our Damjili Cave 2, which is a strong support that Zamyatnin's Damjili Cave 2 is not our Damjili Cave 2.

Considering these, our current interpretation is that Zamyatnin's Damjili Cave 2 mainly corresponds to the eastern part of our Damjili Cave 1, where our main excavation area was carried out. When measuring from the eastern edge of the waterfall basin, a point 70 m away was located in the area of Trench 6 of our Damjili Cave 2, which was delineated further east by the huge boulder derived from a wall collapse (Fig. 4.4: 3). In short, we tentatively interpret that the suggestion made by Zamyatnin (1958), as a promising area for future excavations, his Damjili 2, has been tested by our excavations of the eastern part of Damjili Cave 1 (see below).

In the present study, accordingly, we refer to these two "caves" collectively as Damjili Cave, or Damjili Cave 1 when appropriate. This redefinition is made following our understanding that the two "caves" represent the same formation process. Additionally, this cave system extends further to the east, allowing us to define different "caves" more easily. As described in Chapter 3 of this volume (fig. 3.9), extensive rockshelters that can be called Damjili Caves 2 and 3 are situated to the east. Judging from the available records (Chapter A2 of this volume: fig. A2: 5), Damjili Cave 2, so defined in the present study, was in fact investigated in the 1950s, although the details have not been published. There is no record on previous archaeological investigations at Damjili Cave 3.

Our excavations took place at the redefined Damjili Caves 1 and 2. Damjili 3, unlikely to accommodate prehistoric cultural deposits due to its location at a steep limestone cliff, was not investigated. We opened a total of ten sounding trenches in 2016 to evaluate the stratigraphy of Damjili Caves 1 and 2, which guided substantial excavations from 2017 to 2022 (Fig. 4.1). This chapter presents an outline of the fieldwork. First, it reports the results of the 2016 soundings at Damjili Caves 1 and 2, and second, the excavations conducted in Damjili Cave 1 between 2017 and 2022.

4.2 Soundings in the 2016 season

4.2.1 Soundings at Damjili Cave 2

Our investigations started firstly at Damjili Cave 2, which is situated approximately 100 m east of the waterfall basin of Damjili Cave. It is a rockshelter of approximately 45 m wide, 8 m long, and 3.2 m high, with a sizeable terrace 8.4 m long. Six sounding trenches were opened to determine whether this part of the cave complex contained cultural deposits dating back to the Neolithic or earlier periods. The trenches were numbered according to the order of excavation, roughly starting from the eastern end (Fig. 4.4).

Trench 1

Trench 1 is located in the easternmost area of Damjili Cave 2 (Fig. 4.4: 3). There is a plenty of fallen limestone rubble in this part of the cave and a few large blocks remain standing. Excavation was conducted in a 1 × 1 m trench that revealed stratified lithostratigraphic layers (Fig. 4.5). The bedrock was not reached, although the trench was excavated approximately 1.5 m below the surface. A hearth associated with black ash deposits and charcoal (Context P1.4) was found approximately 35 cm below the surface (Fig. 4.5: 1). Gray ash deposits presumably related to firing activities were also identified beneath this hearth; for instance, Context P1.11 (Fig. 4.5: 2)–indicating that similar hearths seem to have been repeatedly installed around this area.

All deposits below the topsoil were dry-sieved, and several archaeological materials were retrieved. A large part of the material consisted of ceramic fragments of various ware groups, occurring from the topmost contexts to the P1.14 yellow clayey layer, with the highest frequency in P1.11 (Fig. 4.5: 2–4). Most of the datable ceramics indicate a date from the medieval (Islamic) period. No prehistoric findings were observed during the excavation, with the exception of a possible flint core from P1.11.

Trench 2

Trench 2, which is also a small 1 × 1 m sounding trench, is located 4 m west of Trench 1 (Fig. 4.4: 3). The sounding reached approximately 1.5 m below the thin layer of modern animal dung (Fig. 4.6: 1, 2). The stratigraphy is clearly divided into two parts (Fig. 4.6: 3): the upper part consists of loose grayish-brown soil approximately 70 cm thick, whereas the lower part represents light-yellowish-gray, clayey, sticky soil. The upper section provides evidence of human habitation, including ash and ceramic remains. A fireplace containing burned animal bones was also recovered (Context P2.5; 5 of Fig. 4.6: 3). The associated ceramics indicate occupations of the medieval period. In contrast, the underlying light yellowish-gray clay layer was found to be archaeologically sterile.

Trench 3

We opened another 1 × 1 m trench, Trench 3, 4 m west of Trench 2 (Fig. 4.4: 3). The sounding showed that, except for a topsoil layer approximately 10 cm thick that contained modern and medieval artifacts, this area of Damjili Cave 2 did not contain any evidence of human occupation (Fig. 4.7: 1). The main matrix of the trench was brown soil containing yellowish-gray clay blocks (Fig. 4.7: 2), which was comparable to the lower part of Trench 2. This trench was archaeologically sterile.

Trench 4

The stratigraphy of Trench 4, 4 m west of Trench 3 (Figs. 4.4: 3; 4.8: 1), shows the same pattern as Trenches 2 and 3. Loose grayish-brown soil layers were recovered from the upper part, which contained abundant fresh limestone slabs and a small number of modern and medieval artifacts. Below this part was a sterile yellowish-gray clay layer (Fig. 4.8: 2).

Trench 5

Damjili Cave 2 looks down the valley slope, sharply descending east of the Gazakh plain. We opened Trench 5 at its northeastern terrace, approximately below the drip line, located 4 m south of Trench 1 (Figs. 4.4: 3; 4.9: 1). This trench is 1 m wide and 3 m long. The general stratigraphic pattern was the same as that mentioned above: loose grayish-brown soil

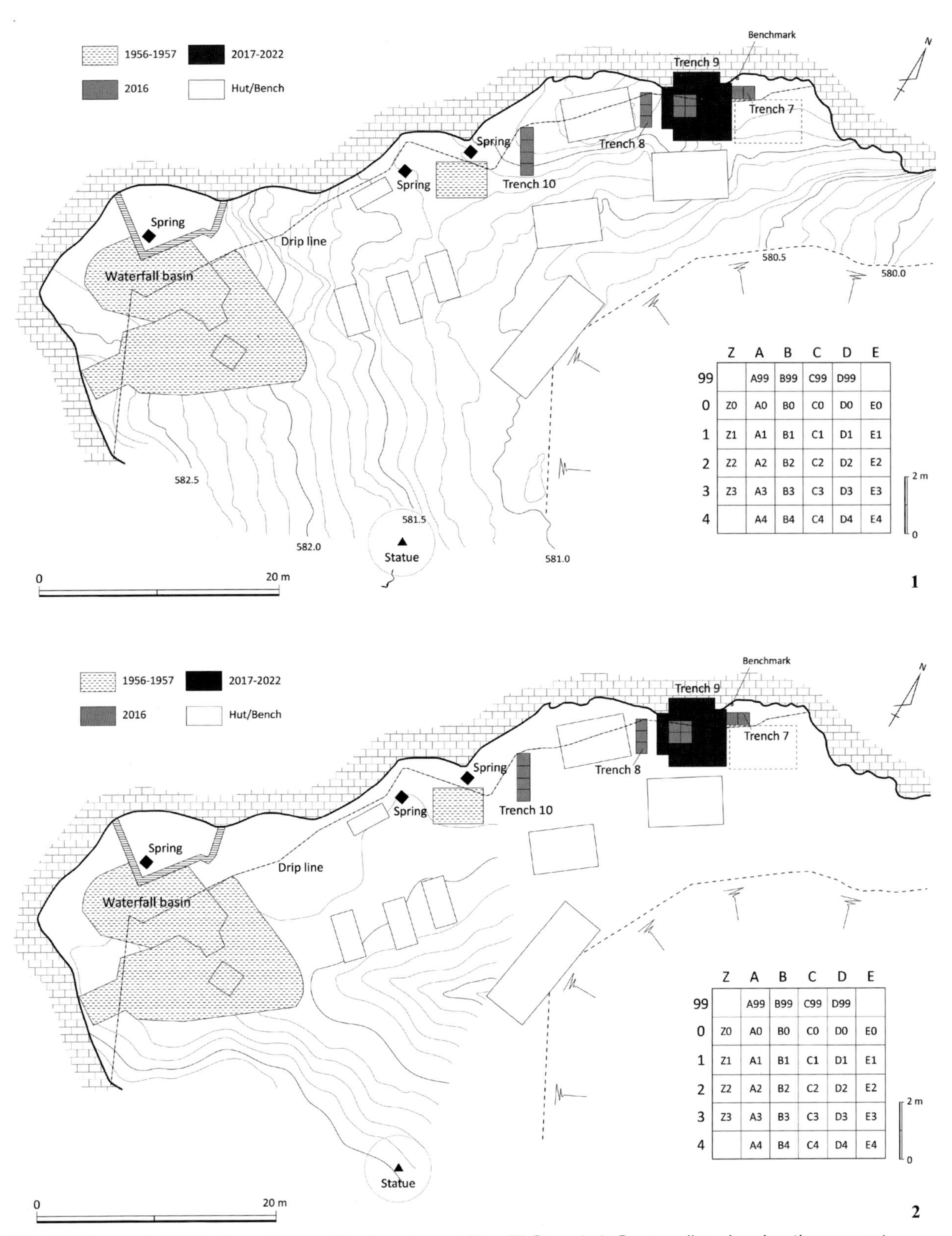

Fig. 4.1 Plan and distribution of excavation trenches at Damjili Cave 1. 1: Contour line showing the present topography; 2: Contour line showing the topography at the time of the 1953–1957 excavations (see Chapter A2 of this volume).

containing medieval remains overlying a sterile yellowish-gray clay layer (Fig. 4.9: 3). A unique discovery of Trench 5 is a flask-shaped pit dug from the upper stratigraphic unit. The pit measures at least 1.5 m in diameter and 1 m deep, whereas its mouth is smaller and less than 1 m in diameter. There is a concentration of limestone slabs at the mouth that may have served to seal the pit (Fig. 4.9: 2, 3). The filling of this pit did not yield any recognizable remains that would indicate a particular function for

Fig. 4.2 General view of the main excavation area from the bottom of Damjili Valley, 2023.

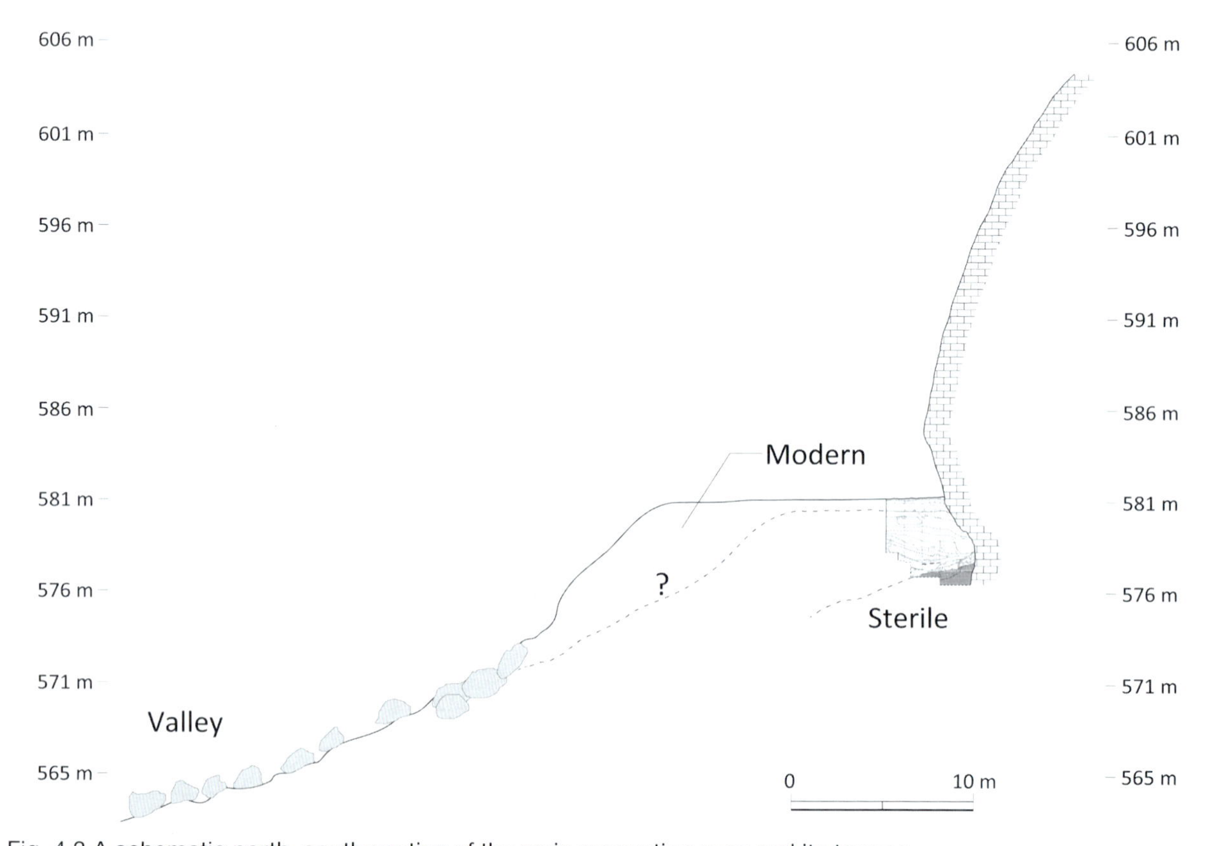

Fig. 4.3 A schematic north–south section of the main excavation area and its terrace.

Fig. 4.4 Plan and distribution of excavation trenches at Damjili Cave 2. 1: Distant view (yellow arrow); 2: Closer view; 3: Distribution of Trenches 1–6.

the pit such as human remains or storage objects. However, the stratigraphic position of the pit indicates a medieval date.

Trench 6

Another sounding trench with an area of 1 × 2 m was opened at the terrace edge near the western end of Damjili Cave 2 (Fig. 4.4: 3). The excavation reached approximately 1.4 m deep from the present surface, whose stratigraphy again showed the same pattern as at other trenches of this cave: a loose grayish-brown soil layer containing medieval arti-

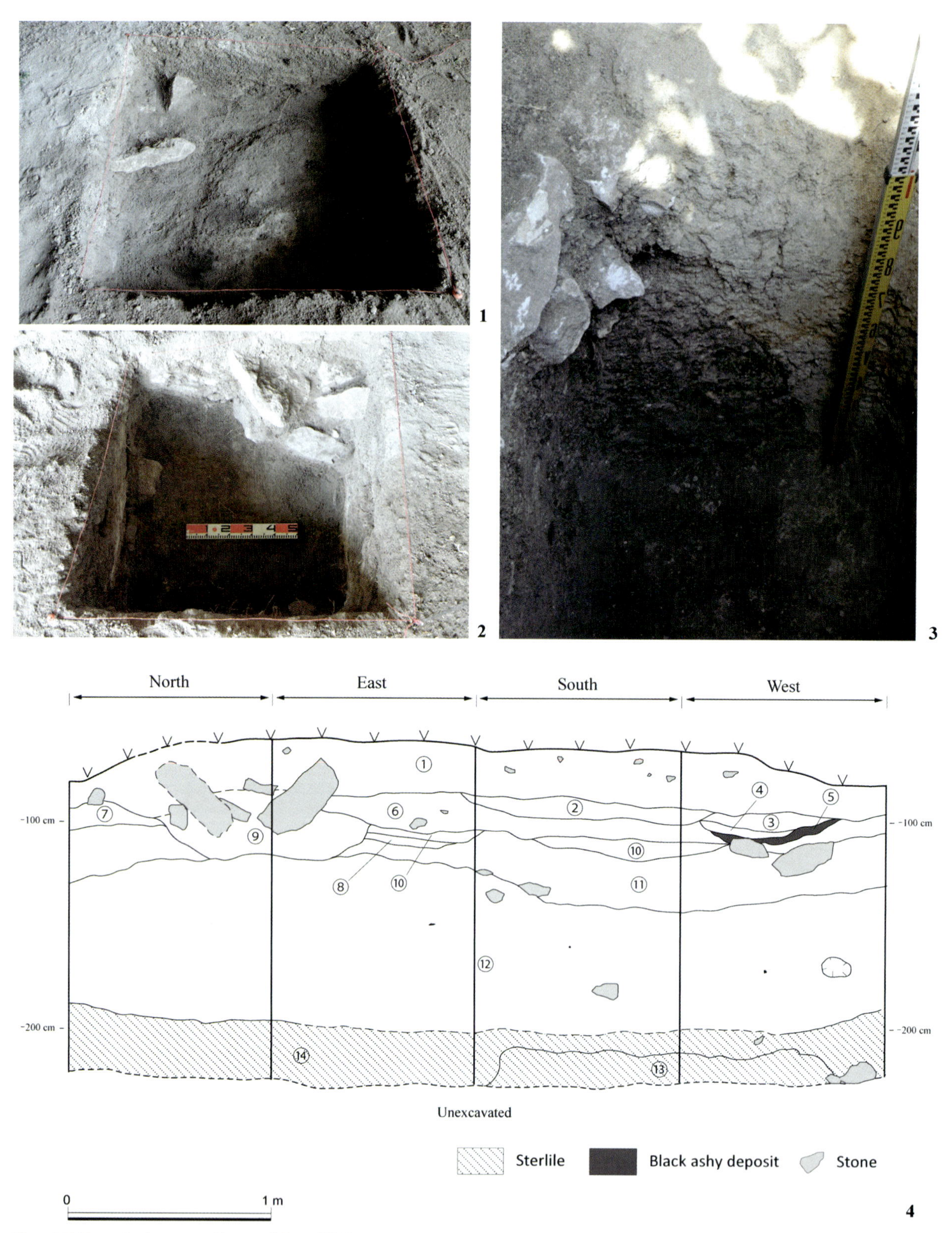

Fig. 4.5 Trench 1 excavations of Damjili Cave 2. 1: Hearth P1.4 looking north; 2: Hearth P1.11 looking east; 3: Section of Trench 1 looking north; 4: Trench 1 stratigraphy (1. Topsoil, gray-brown loam with angular limestone fragments; 2. Topsoil, gray-brown loam; 3. Fine light-gray/brown soil; 4. Hearth with black ash deposit; 5. Whitish gray ash deposit; 6. Light-brown soil including sparse dark-soil block; 7. Sandy dark-brown soil with rubble; 8. Compacted light-gray sediment (ash?) containing charcoal pieces; 9. Fine dark-brown soil containing large cobbles and glazed sherds; 10. Large limestone boulders and loose brown soil; 11, 12. Yellow clayey sediment containing limestone rubble, charcoal, many Islamic period sherds, and a single flint core (?); 13, 14. Light-yellow sediment, including brown clayey portions containing no artifacts).

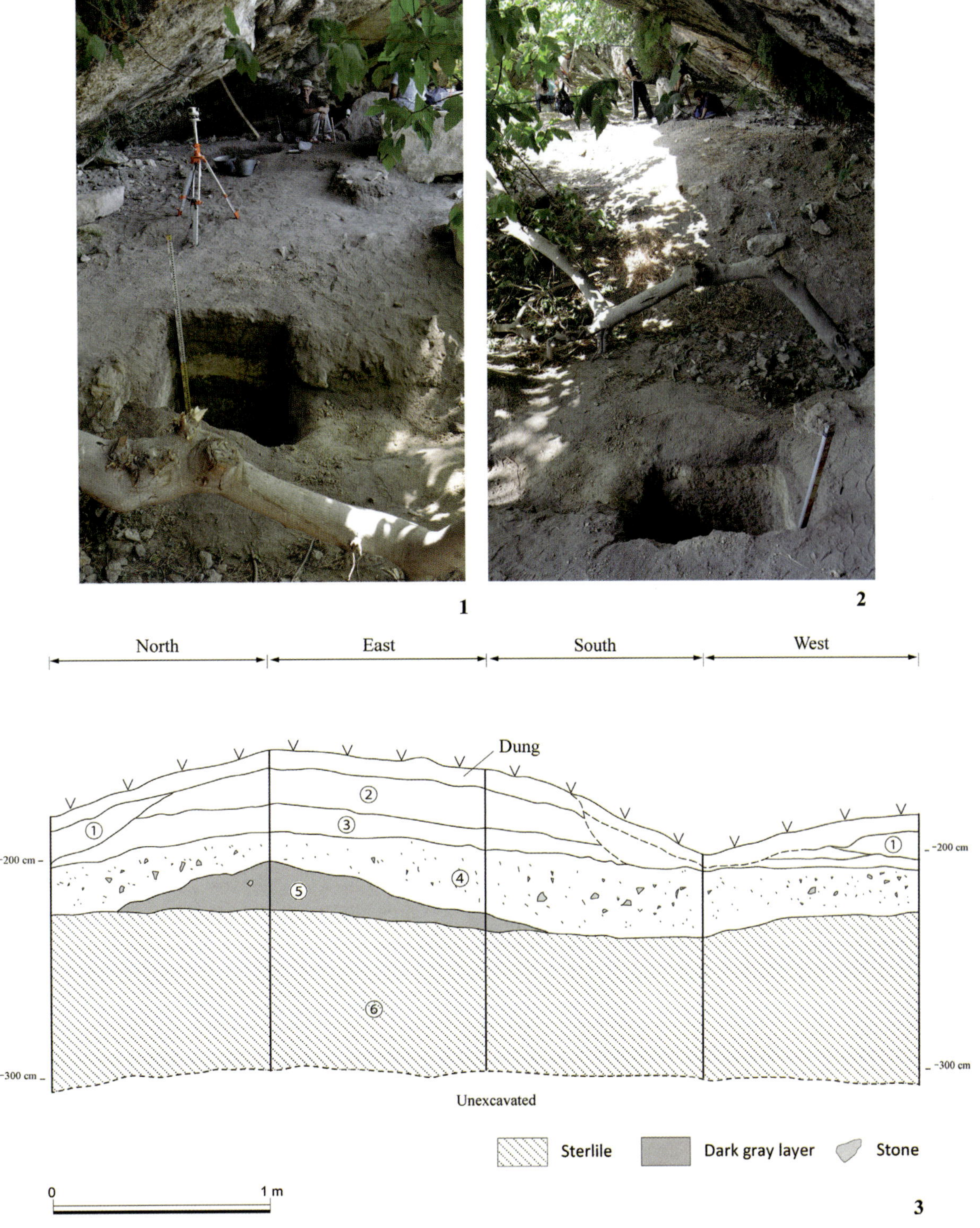

Fig. 4.6 Trench 2 excavations of Damjili Cave 2. 1: General view of Trench 2 looking east; 2: General view of Trench 2 looking west; 3: Trench 2 stratigraphy (0. Dung; 1. Topsoil, loose sediments, yellowish-gray soil, numerous yellow spots; 2. Light-yellow soil, loose, a few yellow spots; 3. Gray soil, loose, a few yellow spots, darker; 4. Yellowish-gray soil, compact, φ2–5 cm limestone rubble, sterile; 5. Dark-gray soil with occasional limestone rubble, a few bones (including burnt pieces); 6. Light-yellowish-gray soil, sticky, sterile, φ2–5 cm limestone rubble).

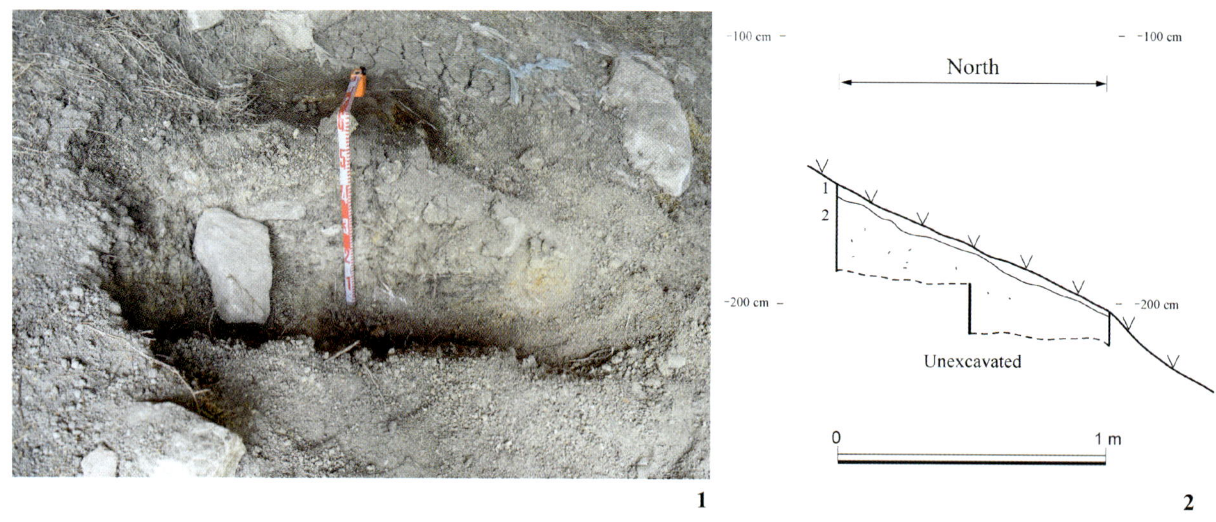

Fig. 4.7 Trench 3 excavations of Damjili Cave 2. 1: General view of Trench 3 looking north; 2: Trench 3 stratigraphy (1. Loose black soil with yellowish clay blocks containing modern and medieval sherds; 2. Brown soil with yellow clay blocks).

facts above a sterile layer with yellowish-brown clay blocks (Fig. 4.10: 1–3). The pottery sherds uncovered from this trench show features very similar to those of other trenches, as exemplified by the red slipware fragments. However, no prehistoric stone tools were identified.

4.2.2 Soundings at Damjili Cave 1

Much of the ground surface of Damjili Cave 1 has been covered with asphalt and concrete to provide a recreational space for visitors. However, there were uncovered areas in the eastern part away from the waterfall basin. Therefore, we opened four sounding trenches in those parts, avoiding the asphalt areas (Fig. 4.1). The soundings revealed a very different stratigraphic pattern from that identified in Damjili Cave 2.

Trench 7

Trench 7 represents a 1 × 2 m trench opened at the northeastern edge of Damjili Cave 1 (Fig. 4.1) in a narrow space between the cave wall and a rest area covered with asphalt (Fig. 4.11: 1). The trench was excavated down to a level of 3.6 m from the surface. Remarkably, the recovered deposits show excellent preservation without any major signs of natural or artificial disturbance, representing a superimposed accumulation of cultural levels covering the prehistoric to medieval periods.

The cultural deposits are divisible into several units. First, the top part, approximately 1 m thick, represents a loose brownish-gray soil layer containing medieval remains. A large fireplace at the western end of the trench (Fig. 4.11: 3) is a notable feature of this unit. The underlying unit does not contain any features of habitation. However, the recovered artifacts indicate a date from the Bronze Age. Neolithic deposits were discovered at levels lower than 1.5 m from the ground surface and consisted of ashy dark-brownish soil sediments. No hearths or construction features were identified, likely owing to the limited excavation area. The materials identified here include a small number of pottery sherds comparable to those of the Neolithic settlement of Göytepe and other Shomutepe sites on the Ganja-Gazakh plain (Nishiaki and Guliyev 2020). It is noteworthy that pottery was found only in the upper layers of the Neolithic unit and that the underlying deposits, nearly 1 m thick, were aceramic. The earlier artifact assemblage consisted of flaked stone artifacts such as small-sized trapezes, burins on pressure-produced blades, thick pebble scrapers, grinding and ground stones, and a well-polished pestle (Fig. 4.11: 2). The general features of the lower part of the Neolithic layers resemble those of Hacı Elamxanlı Tepe, one of the earliest Neolithic records known in this region (Nishiaki et al. 2019, 2021).

Below the Neolithic level was a yellowish-gray layer of clay, approximately 30 cm thick, containing plenty of round river cobbles, approximately 10–15 cm in diameter (Fig. 4.11: 3). A comparable layer with grayish-yellow clay color was situated below. Part of the large rock, gently sloping east, reaches approximately 3.2 m below the surface. Although it was not determined whether this rock represented the bedrock, the excavation was suspended because of the presence of groundwater at this level.

Trench 8

This trench was opened as a 1 × 3 m trench situated roughly perpendicular to the cave wall near another rest area paved with cement (Fig. 4.1). The

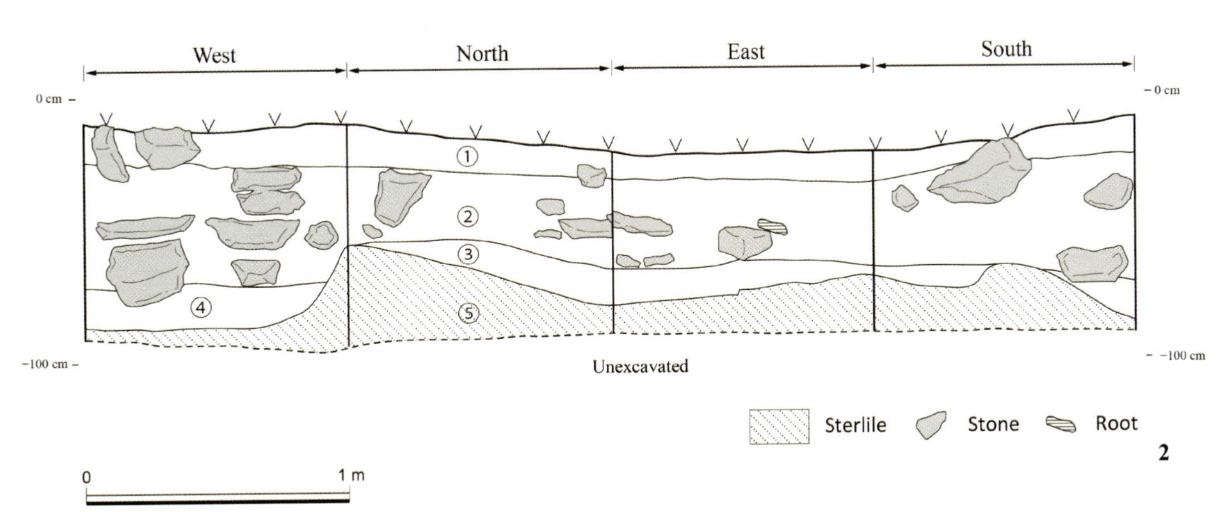

Fig. 4.8 Trench 4 excavations of Damjili Cave 2. 1: East section of Trench 4; 2: Trench 4 stratigraphy (1. Topsoil, loose sandy gray soil containing a small amount of limestone rubble; 2. Loose sandy gray soil containing a larger amount of limestone rubble; 3. Loose organic gray soil with numerous plant roots; 4. Loose organic gray soil containing reddish-brown soil spots (burnt soil?); 5. Yellowish-gray clay, compact, round river pebbles).

excavations were suspended approximately 1 m below the surface because of the discovery of at least two well-preserved human burials (Fig. 4.12: 1, 2). Each burial contained an individual enclosed in flat rectangular limestone slabs covered with large slabs. As these burials, situated almost parallel to the direction of the cave wall, were cut by the excavation limits at both ends, they were left unexcavated. Therefore, the details have not been elucidated. However, the bone samples from these tombs were subjected to radiocarbon dating (Chapter 5 of this volume).

The stratigraphy of the excavated part of the trench can be divided into two major units (Fig. 4.12: 3). The upper unit is a compact gray soil layer approximately 70 cm thick containing many limestone fragments, and the lower unit is a rather loose reddish-brown sandy layer with large and small angular limestone blocks. The burials were dug into the lower unit.

Trench 9

This trench was opened to investigate the cultural stratigraphy noted in Trench 7 over a larger area of 2 × 2 m (Figs. 4.1; 4.13: 1). Excavations were conducted approximately 4 m from the surface, but the bedrock was not reached (Fig. 4.13: 2). The stratigraphy exhibits a remarkable cultural sequence. Below the later deposits of approximately 1.5 m in thickness, a succession of Neolithic and earlier cultural layers was discovered (Nishiaki et al. 2019). As in Trench 7, the Neolithic layer contains pottery,

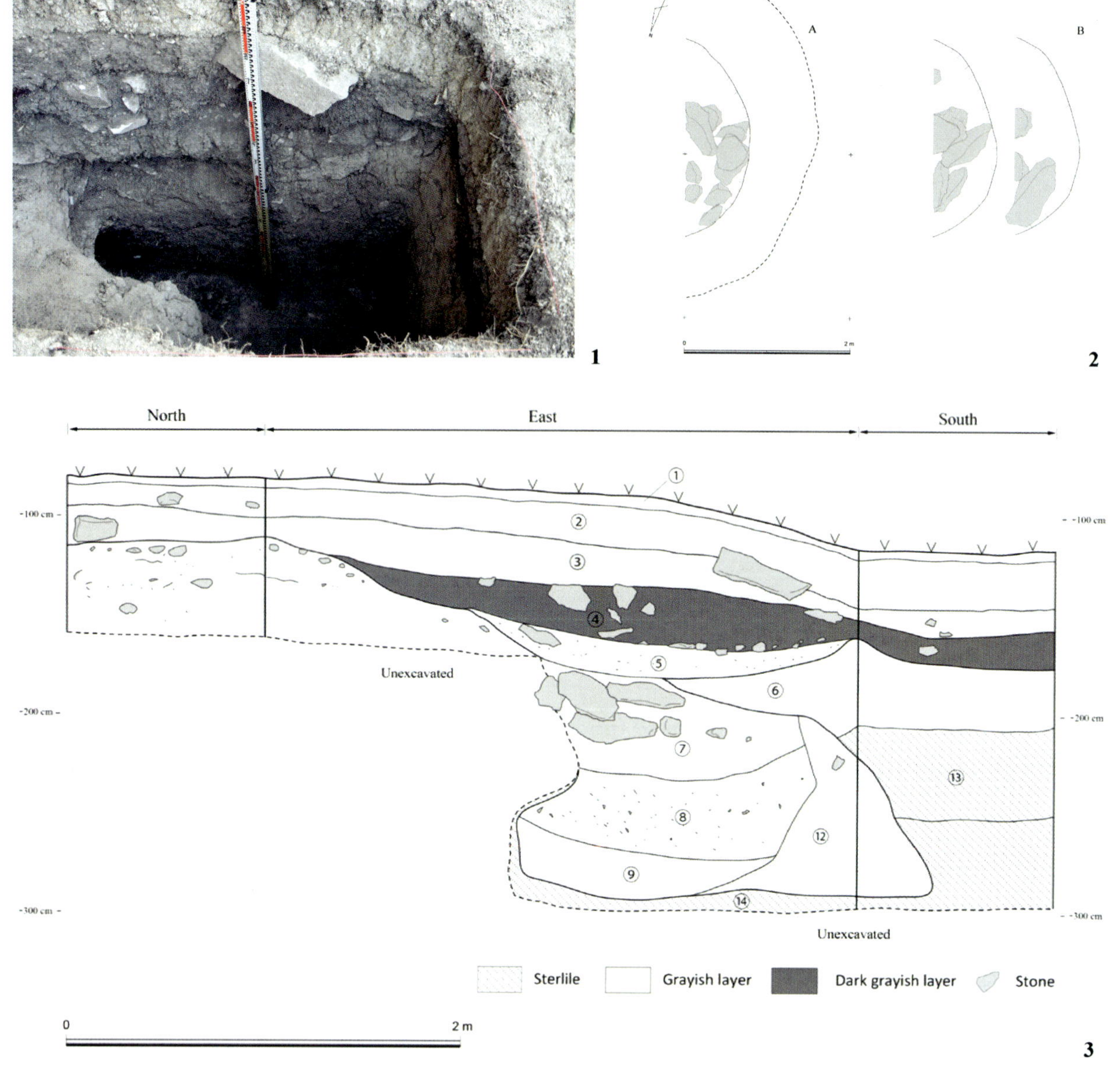

Fig. 4.9 Trench 5 excavations of Damjili Cave 2. 1: East section of Trench 5; 2: Plan of the medieval pit at different depths; 3: Trench 5 stratigraphy (1. Topsoil, loose sandy gray soil; 2. Loose sandy yellowish-gray soil containing lime spots; 3. Loose sandy gray soil containing relatively large pieces of limestone rubble; 4. Pit fill, loose dark-gray soil with relatively large pieces of limestone rubble; 5. Pit fill, loose gray soil with smaller pieces of limestone rubble; 6. Yellow clayish soil, loose; 7–12. Pit fill, sticky dark-brown soil containing yellowish spots; 13. Yellowish-gray clay, hard, reddish-brown spots; 14. Yellowish-gray clay, hard, with small pieces of limestone rubble).

and its lower part is almost aceramic. In addition, the underlying unit yielded a distinct lithic assemblage consisting of microlithic elements, such as microblades and geometrics, suggesting a Mesolithic date. Moreover, this intriguing layer was identified above a practically sterile layer of weathered limestone cobbles containing a small number of Middle Paleolithic artifacts, characterized by the use of Levallois technology on top. Thus, this trench was excavated more substantially (see below).

Trench 10

A 1 × 4 m trench was opened 8 m southwest of Trench 8, perpendicular to the cave wall (Figs. 4.1; 4.14: 1). Its stratigraphy is almost identical to that of Trench 8; the upper unit shows a compact gray soil layer and many limestone fragments, whereas the lower unit has a loose reddish-brown sandy layer and angular limestone blocks (Fig. 4.14: 2, 3). Similarly, the uppermost part of the lower unit contained human burials constructed from limestone slabs. While leaving the burials unexcavated, we further dug the southern part of the trench to lower levels. The excavations finally reached the bedrock approximately 3 m below the surface (Fig. 4.15). It slopes steeply toward the cave wall. Generally, the

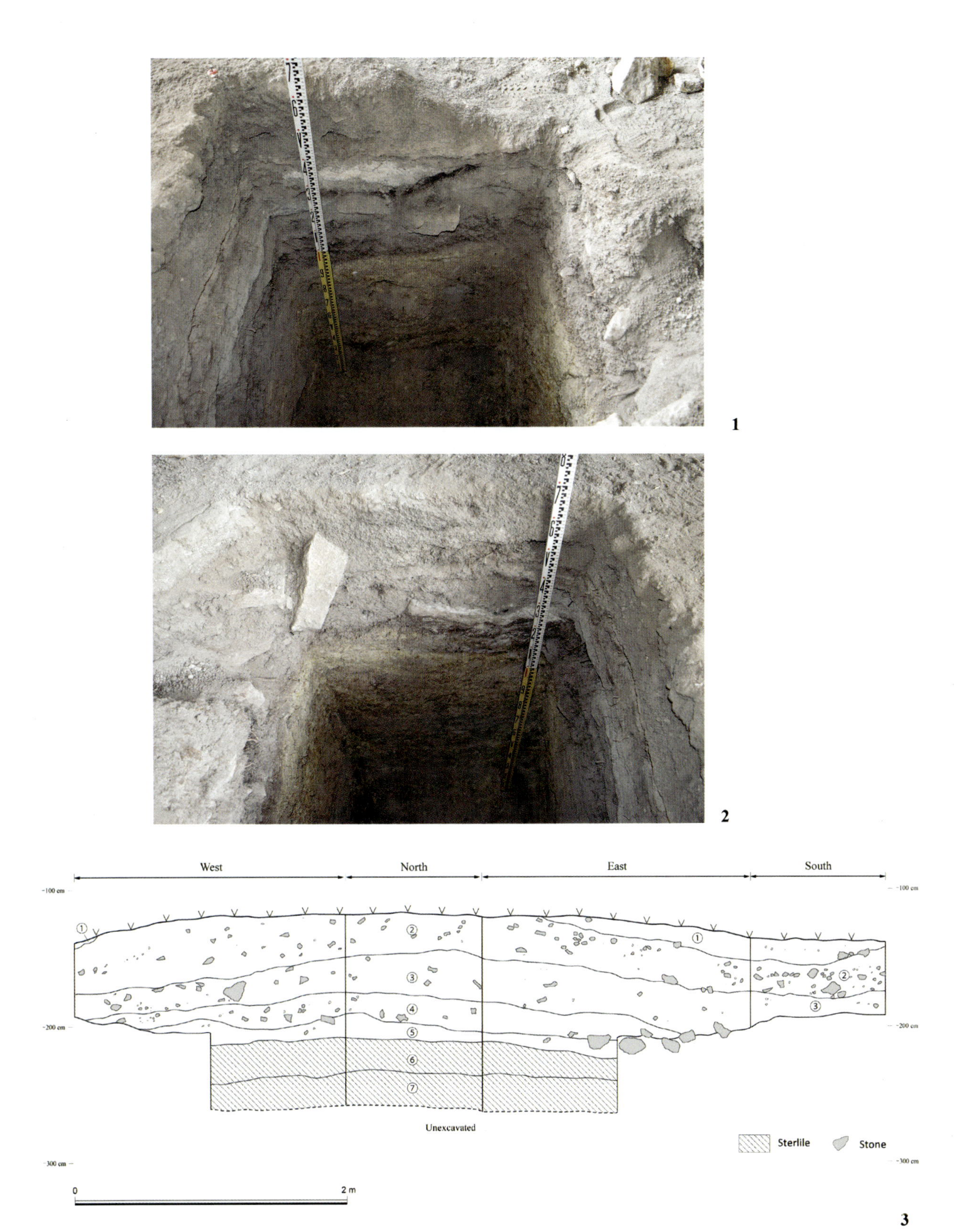

Fig. 4.10 Trench 6 excavations of Damjili Cave 2. 1: North section of Trench 6; 2: West section of Trench 6; 3: Trench 6 stratigraphy (1. Gray-brown soil with a large amount of limestone rubble, heavily disturbed by modern arboreal vegetation, containing glass fragments; 2. Yellowish-gray-brown soil with inclusions of 1 to 3 cm in diameter of angular limestone, with occasional larger pieces exceeding 30 cm in size; 3. Grayish-brown soil, a small amount of limestone rubble; 4. Sandy grayish-brown soil; 5. Clayish gray-brown soil, a small amount of limestone rubble; 6. Clayish gray-yellow soil with white lime spots; 7. Clayish light-gray-yellow soil with white lime spots).

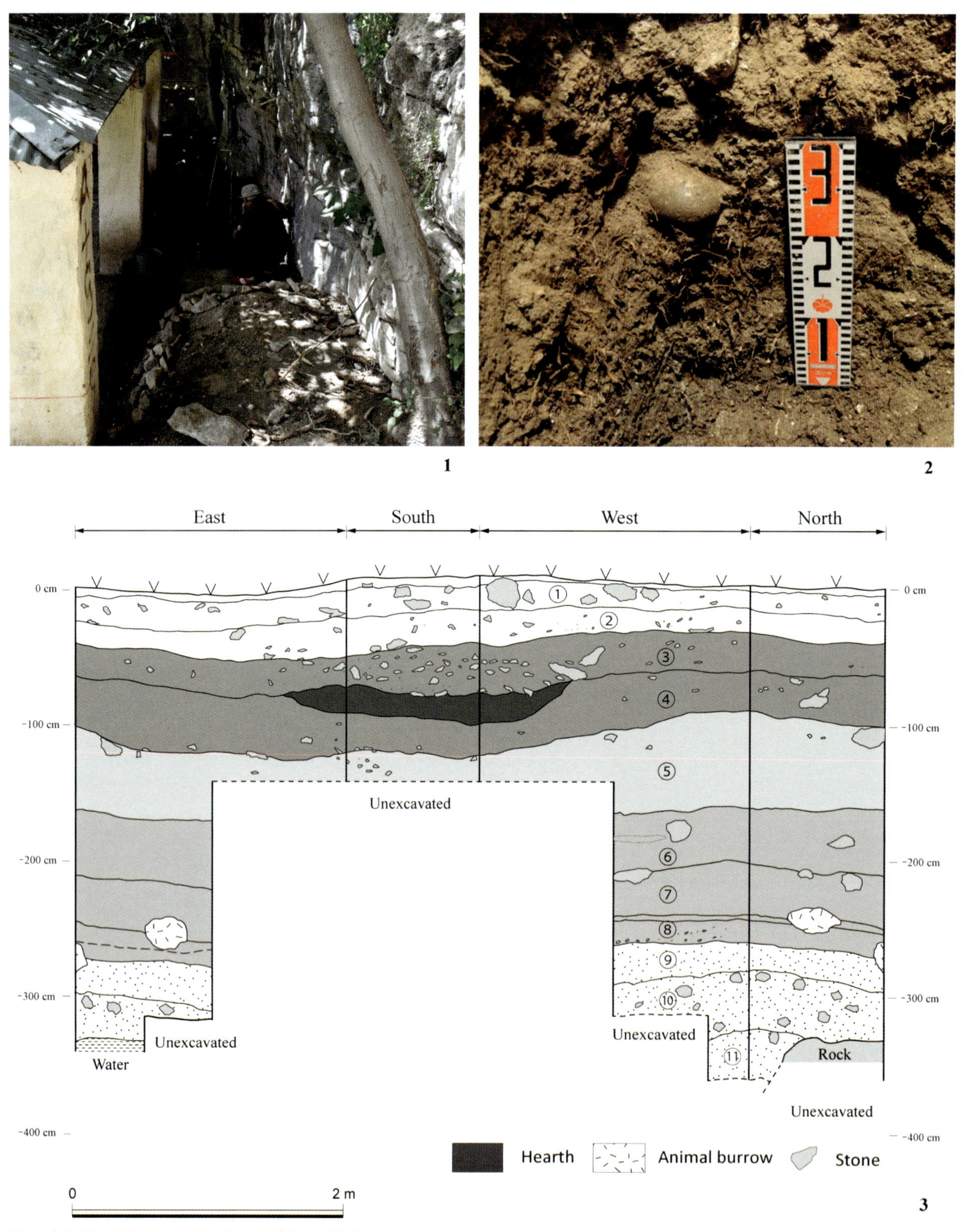

Fig. 4.11 Trench 7 excavations of Damjili Cave 1. 1: General view of Trench 7 looking west; 2: Ground stone visible on an excavation wall of the Neolithic deposits; 3: Trench 7 stratigraphy (1. Brownish gray, loose; 2. Brownish gray, loose, more limestone rubble (small); 3. Brownish gray, compact, plenty of limestone rubble; 4. Grayish brown, compact; 5. Dark grayish brown, soft, rare limestone rubble; 6. Brownish gray, soft; 7. Light-brownish gray, soft; 8. Brown layer, compact, a few horizontal floors, limestone, soft, dark brown, sticky; 9. Yellowish-gray clay, round river pebbles; 10. Grayish-yellow clay, round river pebbles; 11. Very soft, dark-gray ash layer).

Fig. 4.12 Trench 8 excavations of Damjili Cave 1. 1: General view of Trench 8 looking north; 2: Plan of Trench 8; 3: Trench 8 stratigraphy (1. Dark-grayish-brown soil with plenty of limestone rubble, modern objects; 2. Reddish-brown soil with plenty of limestone rubble, pottery; 3. Reddish-brown soil, loose, few limestone rubble).

deposits overlying the bedrock are almost sterile. The sandy matrix and limestone gravel suggest their involvement in the water activity. At least two fissures, from which water still emerges today, are situated on the cave wall. Therefore, it is likely that this part of the cave was not a favorable habitat.

4.3 Excavations of the main trench (Trench 9) in 2017–2022

One of our conclusions from the 2016 soundings is that the sheltered area of Damjili Cave 2 was occupied mainly during the medieval period. Moreover, the cultural deposits are generally thin, approximately 1.5 m at maximum (Trenches 1 and 5), situated on a sterile layer of yellowish-gray clay with abundant limestone rubble. The most likely interpretation is that the sheltered area of Damjili Cave 2 was formed by a geological formation during the late Holocene. However, it is uncertain whether this interpretation is also applicable to the terrace area overlooking the river valley, where our excavations were not carried out. The nature of terrace deposits heavily covered with vegetation is worthy of careful future research.

On the contrary, our 2016 soundings produced promising results at Damjili Cave 1. They demonstrate the presence of stratified prehistoric cultural

1

2

Fig. 4.13 Trench 9 soundings of Damjili Cave 1. 1: Beginning of Trench 9 soundings looking north; 2: Bottom of Trench 9 sounding in 2016 looking west.

records that survive in the eastern region. As shown in Trench 10, the deposits in the western area of the cave closer to the waterfall basin were apparently disturbed by water activities, confirming the findings from the 1950s excavations (Appendix of this volume). However, well-preserved prehistoric sediments were recovered in Trenches 7 and 9 at the east end. Accordingly, we decided to investigate this area more extensively during the subsequent seasons. With the permission of the local authorities, one of the recreational huts was removed so that a sufficiently large area was made available for our research.

Fig. 4.14 Trench 10 excavations of Damjili Cave 1. 1: General view of Trench 10; 2: Deep sounding of Trench 10 looking west; 3: East section of Trench 10.

4.3.1 Excavation methods

A 1 × 1 m grid system was installed in this area of Trench 9 or the main trench (Fig. 4.1). Each square was defined by a combination of alphabetic designations along the east-west axis and Roman numbers along the north-south axis. The excavations were conducted according to "contexts," each of which represents the smallest excavation unit recognizable in the field (Nishiaki et al. 2001: 49). All sediments of the prehistoric layers (Units 3–6) were sieved using an approximately 3 mm mesh. Water sieving was employed when necessary. Finally, six basic cultural units were defined for Trench 9 (Figs. 4.16; 4.17).

For safety reasons, we confined deeper excavations to smaller areas. Therefore, while the excavation area measured 6 × 6 m in Units 1 and 2, it was reduced to 3 × 4 m in Units 3–5 (Fig. 4.16). The lowest deposits (Unit 6) were investigated over a smaller area of 1 × 3 m. The deepest point reached in the 2022 season was approximately 4.4 m deep from the present surface. However, the bedrock was not exposed.

4.3.2 Excavation results

Unit 1

The latest archaeological unit, approximately 50 cm thick, is situated below the topsoil and modern debris from architectural activities. One of the unique findings in this unit is a set of massive stratified walls made of carved limestone blocks, some 60 cm wide, and running at least 2 m parallel to the cave wall. The walls likely represented part of a rectangular building for at least two chronological phases (Figs. 4.18; 4.19). The associated ceramics suggest that these walls date back to the medieval period (Chapter 16 of this volume).

The notable features of the upper phase are a few fire pits full of ash, including a large pit in the upper phase of Squares D0–E0 (Fig. 4.18: 3), which extended to Trench 7 (Fig. 4.11); a few burials were also recovered in this unit (Fig. 4.18: 1, 2). Considering that burial pits were also recovered in 2016 at the soundings of Trenches 8 and 10, Damjili Cave 1 might have been used extensively as a cemetery during the medieval period. It should also be noted that our investigations of Damjili Cave 2 revealed no such burial pits but instead found traces of living activities during this period such as fireplaces

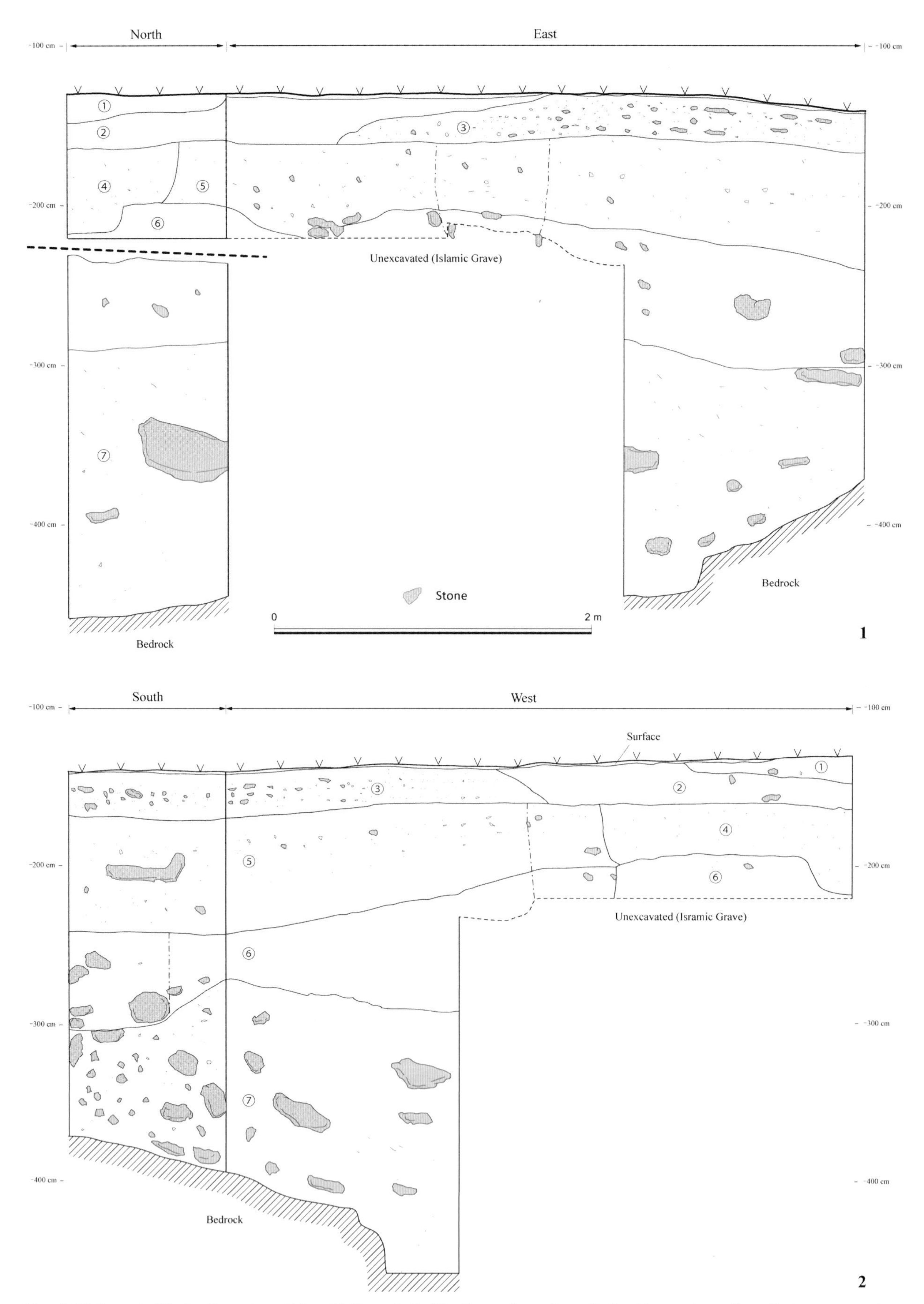

Fig. 4.15 Trench 10 stratigraphy at Damjili Cave 1. 1: North-east section; 2: South-west section (1. Loose sandy gray soil containing modern artifacts; 2. Relatively compact dark-brown soil, sandy, almost no limestone fragments; 3. Layer of limestone fragments most likely fallen from the cave wall; 4. Loose dark-brown soil containing a small amount of limestone rubble; 5. Similar to Layer 4 but containing more limestone rubble; 6. Loose brown soil layer in which medieval graves were identified; 7. Gravel layer with sandy sediments, wholly sterile).

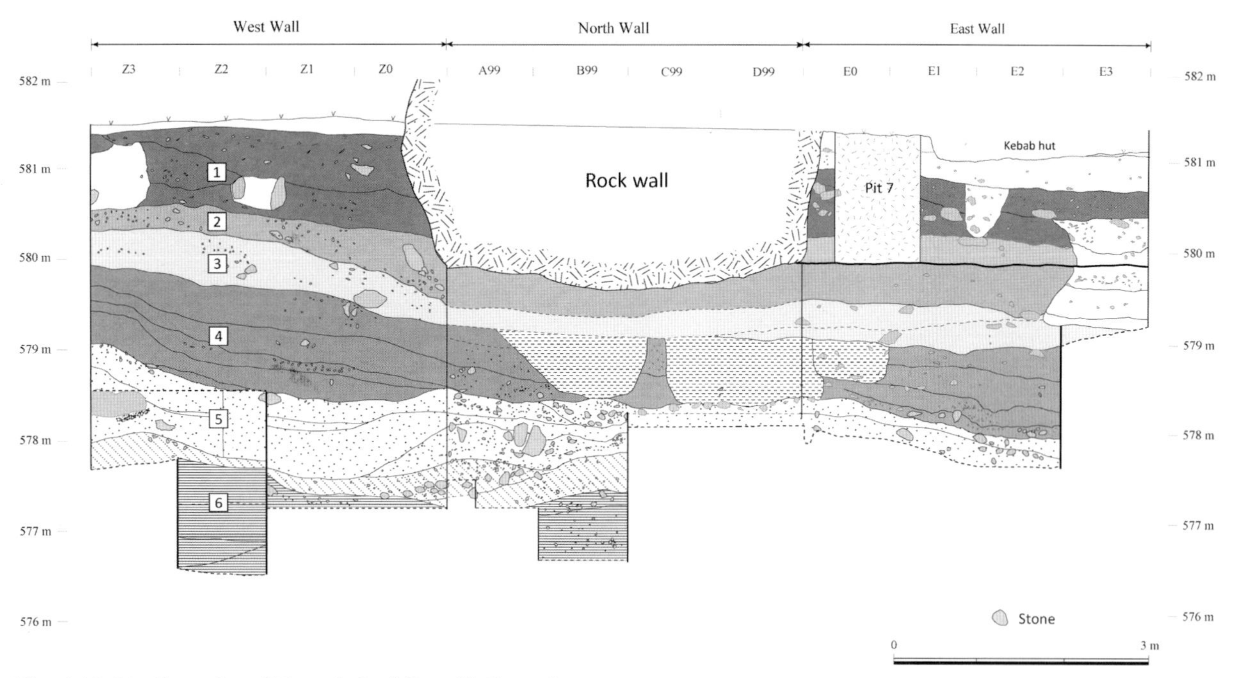

Fig. 4.16 Stratigraphy of Trench 9 of Damjili Cave 1.

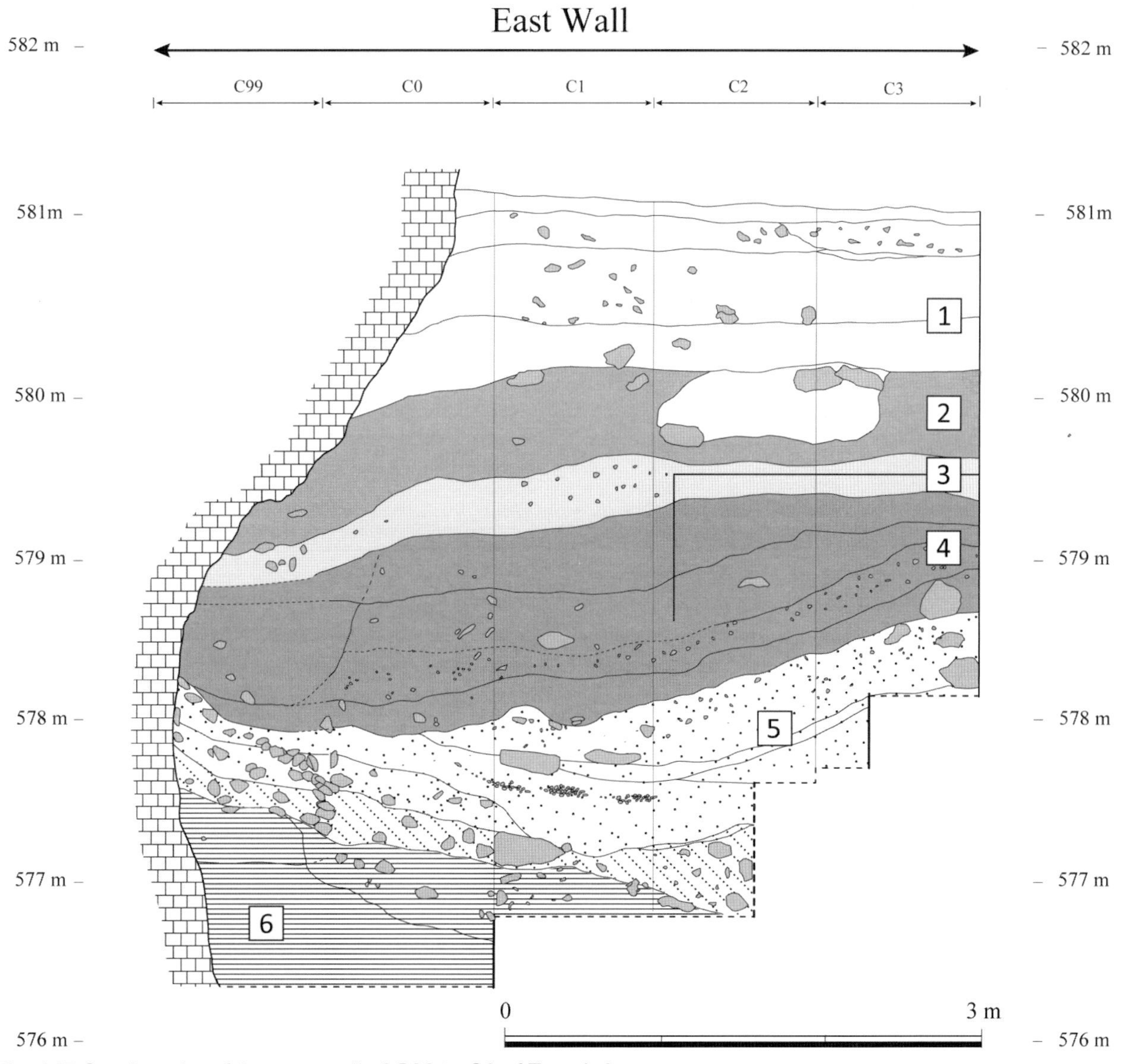

Fig. 4.17 Stratigraphy of the east wall of C99 to C3 of Trench 9.

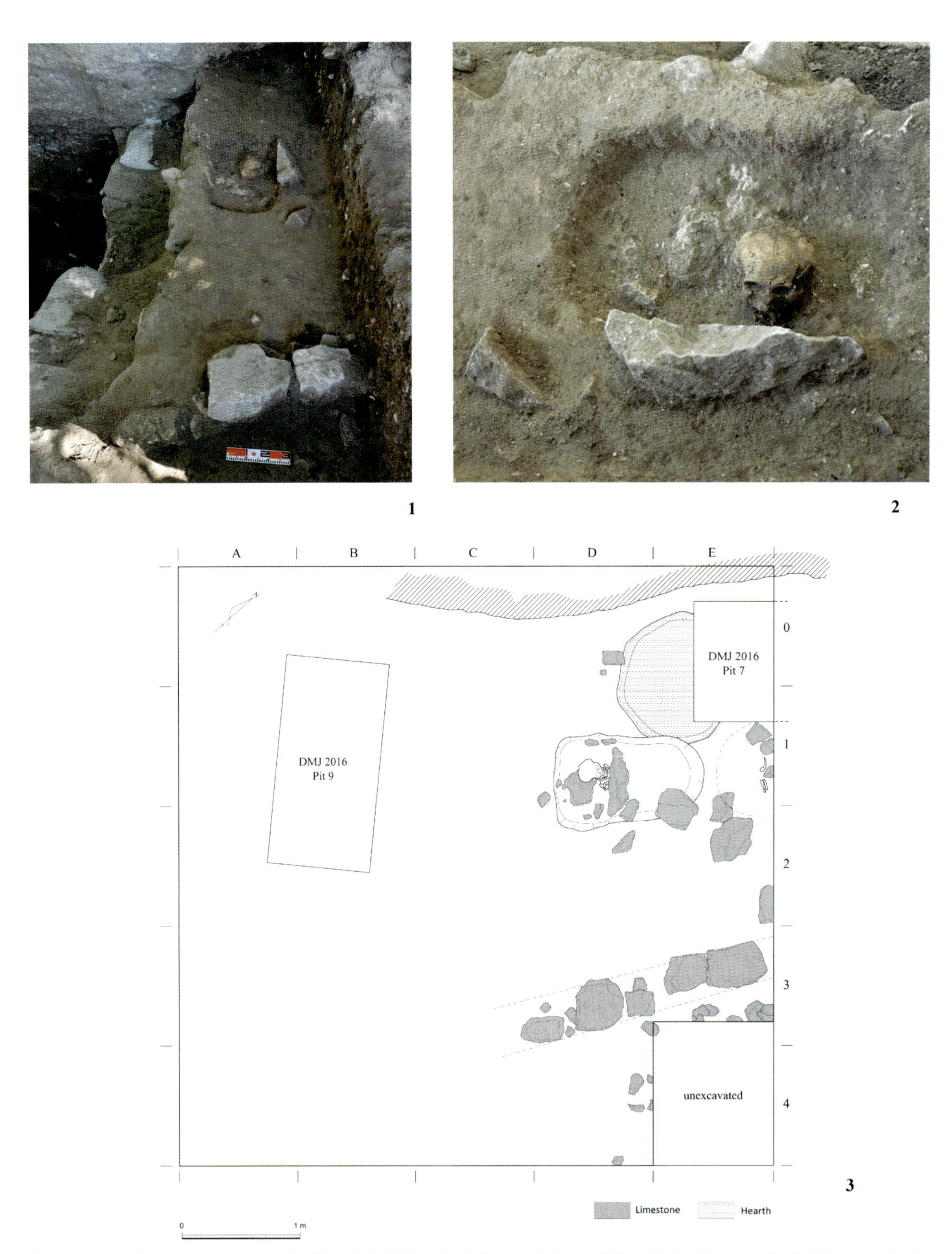

Fig. 4.18 Medieval constructions in Trench 9, Unit 1.1. 1: General view of Unit 1.1 looking north; 2: A human burial of Unit 1.1 looking west; 3: Plan of the features in Unit 1.1.

and storage facilities. These are interesting data for interpreting the differential use of the cave areas of the Damjili Complex in the medieval period.

Structures of the earlier phase of Unit 1 (Unit 1.2) are also characterized by well-built stone walls similarly aligned but in a different location from the those of Unit 1.1 (Fig. 4.19: 1). No human burials were recovered from this phase. A fire place was recovered nearby the main stone wall (Fig. 4.19: 2, 3).

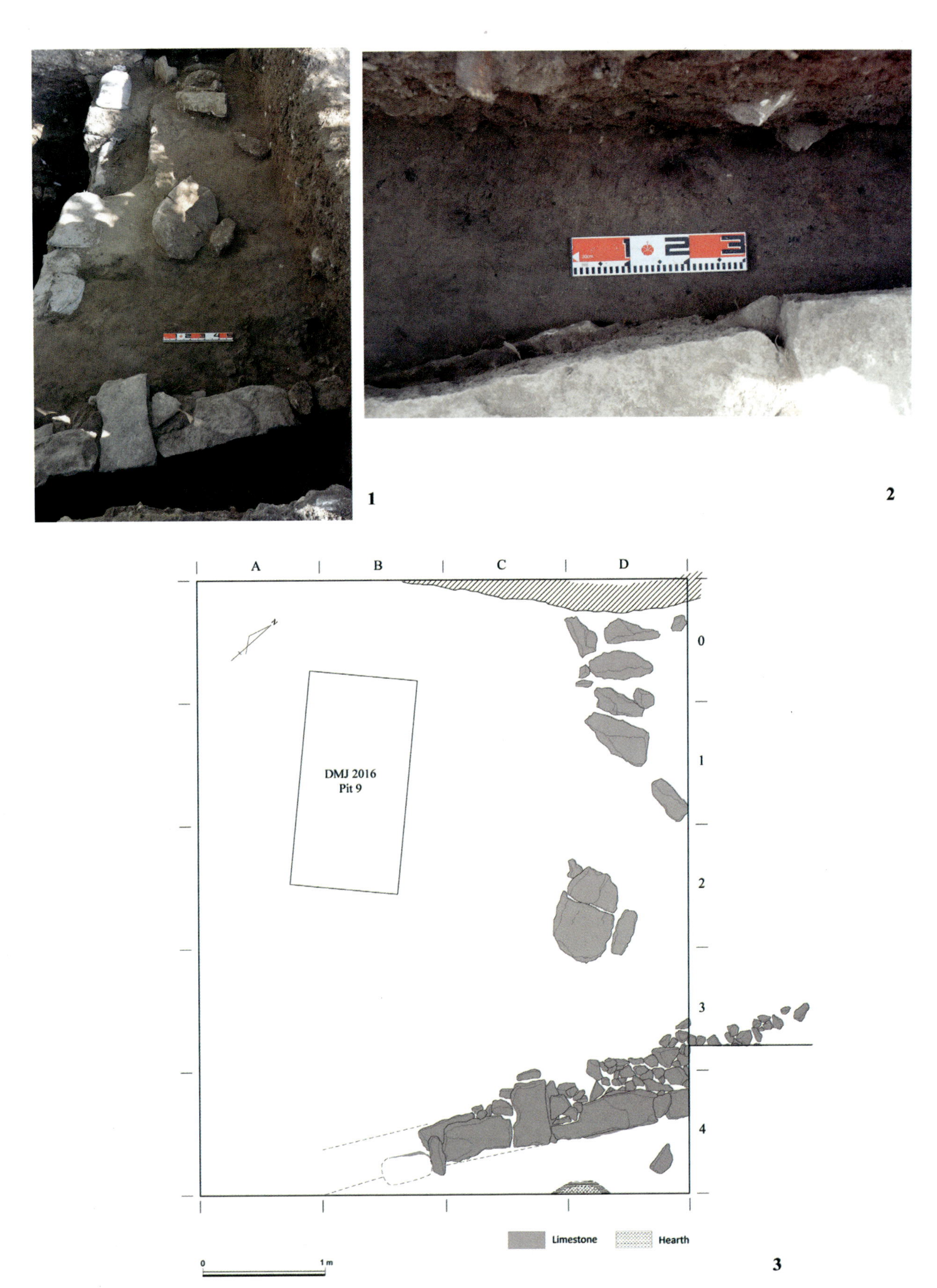

Fig. 4.19 Medieval constructions in Trench 9, Unit 1.2. 1: General view of Unit 1.2 looking north; 2: A fire place south of the main stone wall; 3: Plan of the features in Unit 1.2.

Unit 2
The underlying archaeological unit is characterized by relatively loose and homogeneous sediments containing dark gray ash. Its thickness is approximately 40 cm. Although a few fireplaces and ash distributions were noted (Fig. 4.20), no definitive construction was recovered. Nevertheless, recovered ceramics and other remains indicate Bronze Age dates. This estimate is supported by radiocarbon dating (Chapter 5 of this volume).

Unit 3
This unit, approximately 30–50 cm thick, is defined as a lithological unit of compact grayish-brown sediment with a small amount of limestone rubble (Fig. 4.16). The accumulation of ash and a few pits are the only recognizable features. The artifactual remains were sparse. Radiocarbon dates indicate a Chalcolithic period from the late 5th to early fourth millennium BC (Chapter 5 of this volume). However, the archaeological materials from this unit are too limited to understand their nature. Considering the absence of this unit from Trench 7 altogether (Nishiaki et al. 2019), the occupation of the site during the Chalcolithic period seems to have been sparse in comparison with other periods.

Unit 4
Units 4 and 5 are of primary importance for our research on the Mesolithic-Neolithic cultural transition. Unit 4 consists of reddish-brown soil and dark gray ash layers; it is approximately 1–1.2 m thick in total. The excavations suggest that this unit could be divided into four subunits of Units 4.1 to 4.4 (Figs 4.16, 4.17; Nishiaki et al. 2022). From a sedimentological perspective, they can be grouped into two phases. The sediments of the early phase (Units 4.4 and 4.3) are characterized by a significant amount of black ash, whereas the latter (Units 4.2 and 4.1) consist of compact brown sediments. The blackish color of the earlier phase is likely due to the more common fireplaces associated with many black ashes and charcoal remains (see below).

Unit 4.1 is represented by a sparse distribution of limestone blocks, mostly in the northeastern part of the trench (Fig. 4.21). A small amount of charcoal remains was also recovered (Square D0). It is unclear whether the limestone blocks represent the remains of any standing structure. There is no evidence of the association of mudbricks or clay with these stones.

The subunit of Unit 4.2 is characterized by a denser distribution of limestone blocks in the northeastern part of the trench (Fig. 4.22). We also identified two fireplaces containing numerous charcoal remains in Squares A0 and D0. The distribution of the limestone blocks appears to have a semicircular shape (Fig. 4.22: 2), accommodating the Square D0 fireplace. However, it is necessary to evaluate this spatial configuration after expanding the excavation area to the east.

Unit 4.3 showed a wider distribution of limestone blocks than the above subunits (Fig. 4.23). At least two fire places were recovered in Squares A3–B3 and C1–D1 (Fig. 4.23: 1, 3). The most interesting is that of Squares C1–D1 (Fig. 4.23: 2). It was situated in the middle of a circular stone-walled construction with a diameter of approximately 3 to 4 m. This fireplace is an oval pit measuring approximately 70 × 50 cm and about 25 cm deep, filled with limestone rubble showing evidence of firing, and it is 10–15 cm in diameter. Reddish-brown burned soil and scattered charcoal were recovered. These features resemble the "cobble-filled pits" commonly known at Neolithic settlements across Southwest Asia (Nishiaki in press). Upper Mesopotamia pits are classified into two types: shallow pits with gently sloped walls (*cuvettes creusées*) and deep pits with steep walls (*fosses-foyers*), both filled with burned cobbles and ashes (Molist 1985). The fire place from Unit 4.3 is similar to the former type. Parallels have also been discovered at Hacı Elamxanlı Tepe (Nishiaki et al. 2021). A fire-place was also discovered within the encircled stone structure of Unit 4.2. However, it was not a cobble-filled pit type but a simple shallow hearth with some stones (Fig. 4.20).

The lowest Neolithic level, Unit 4.4, is also characterized by a circular distribution of limestone blocks (Fig. 4.24: 1, 3). Their distribution pattern is similar to that noted for Units 4.2 and 4.3. The limestone blocks appear to have formed a circular wall encircling the fireplace in Square D1. The fireplace was of the same type as in Unit 4.3. It was slightly larger in diameter, up to 1 m, and 40 cm deep (Fig. 4.24: 2).

Unit 5
The stratigraphic unit below the Neolithic layer is about 1-m thick, situated about 3.2 to 3.5 m below the ground surface. It consists of three distinct sedimentological layers: a reddish-brown, coarse-grained soil unit containing abundant limestone rubble (Unit 5.1), a blackish-gray layer with ashes (Unit 5.2), and a very hard blackish-gray layer with clayish matrix (Figs. 4.16; 4.17). The upper layer inclined downward to the cave wall, while the earlier ones inclined downward to the cave mouth. This stratigraphic pattern differs markedly from that of the Unit 4 Neolithic layers, which accumulated rather horizontally. Moreover, the top of Unit 5.1

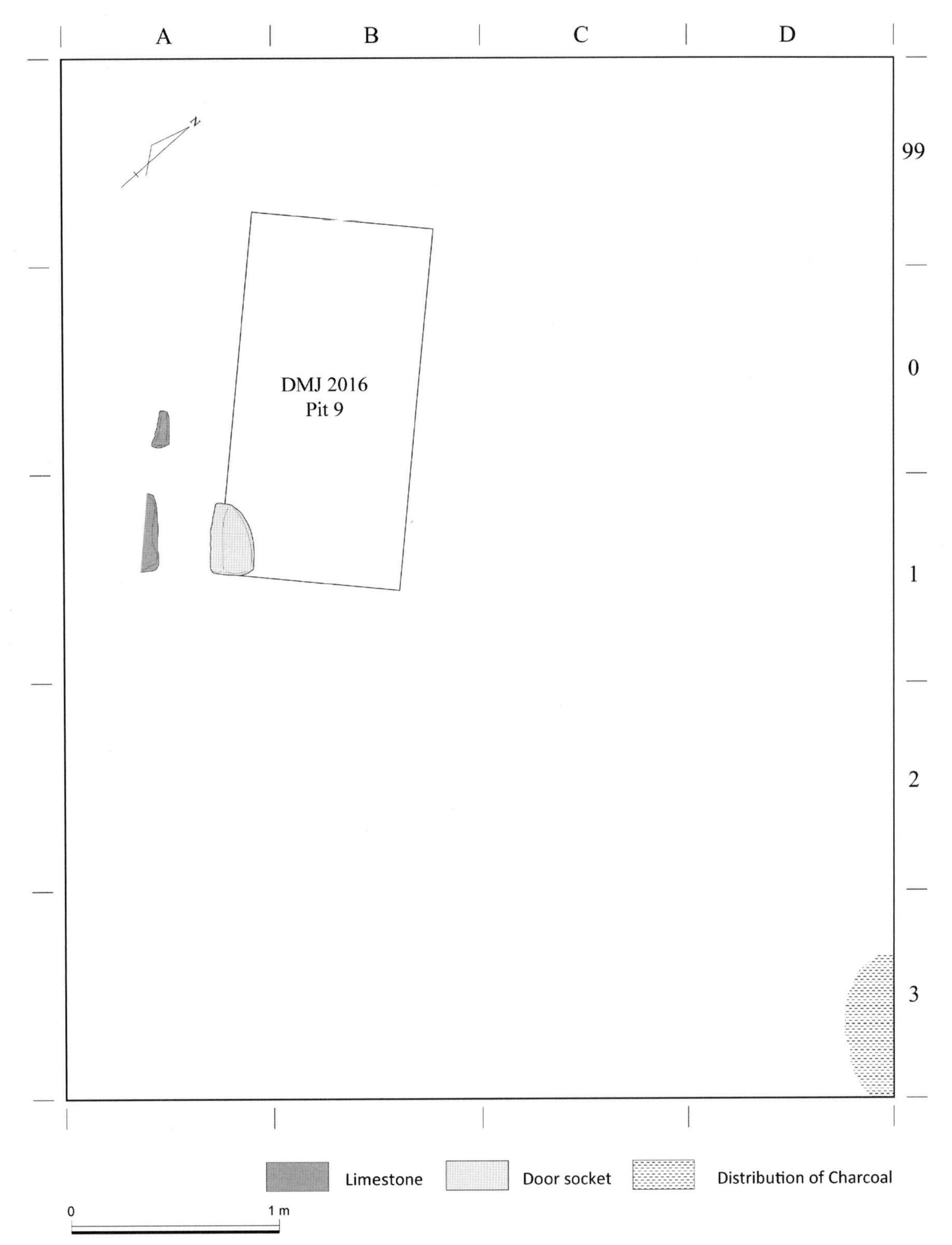

Fig. 4.20 Bronze Age plan, Unit 2.

has been truncated by erosion (Nishiaki et al. 2022). Overall, there seems to have been a stratigraphic discontinuity between Units 4.4 and 5.1, and hence, between the Neolithic and Mesolithic layers.

Unit 5.1 is characterized by a patchy distribution of large and small limestone blocks and a fireplace. Unlike Unit 4, the limestone distribution does not show a meaningful pattern suggesting the existence of layer any construction (Fig. 4.25: 1, 3). In addition, the fireplace (Fig. 4.25: 2, 3) is situated away from the stone blocks in Square A3 at the southwest corner of the excavation area. It is not a cobble-filled pit as we saw in the Neolithic layers.

Below Unit 5.1, there is a thin ashy layer that marks the boundary of the underlying Unit 5.2 (Fig. 4.17). The characteristic features of Unit 5.2 are also the distribution of limestone blocks, which do not indicate spatial patterning, and a few fireplaces

1

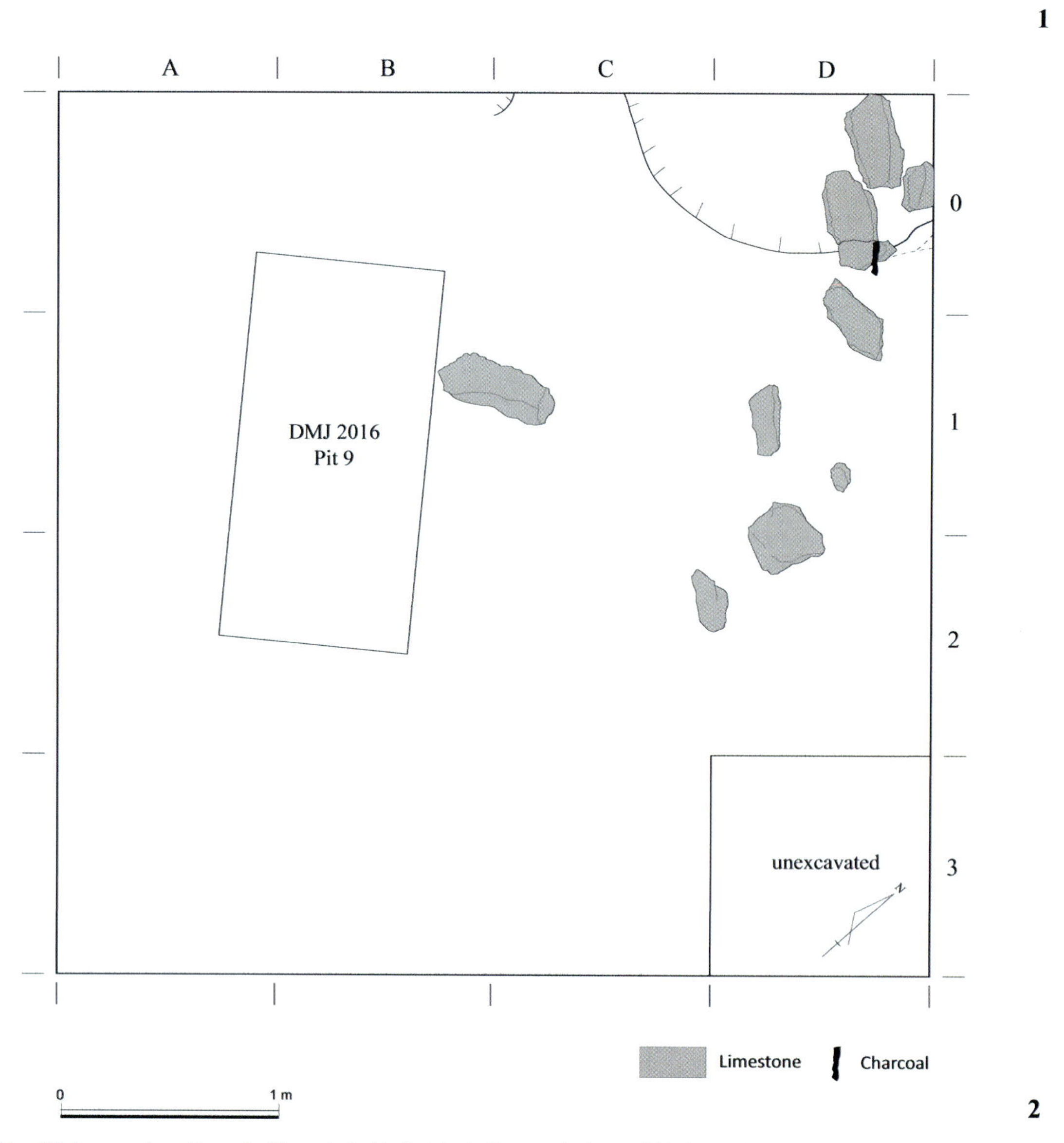

2

Fig. 4.21 Neolithic constructions in Trench 9, Unit 4.1. 1: General view of Unit 4.1 looking north; 2: Plan of the features in Unit 4.1.

1

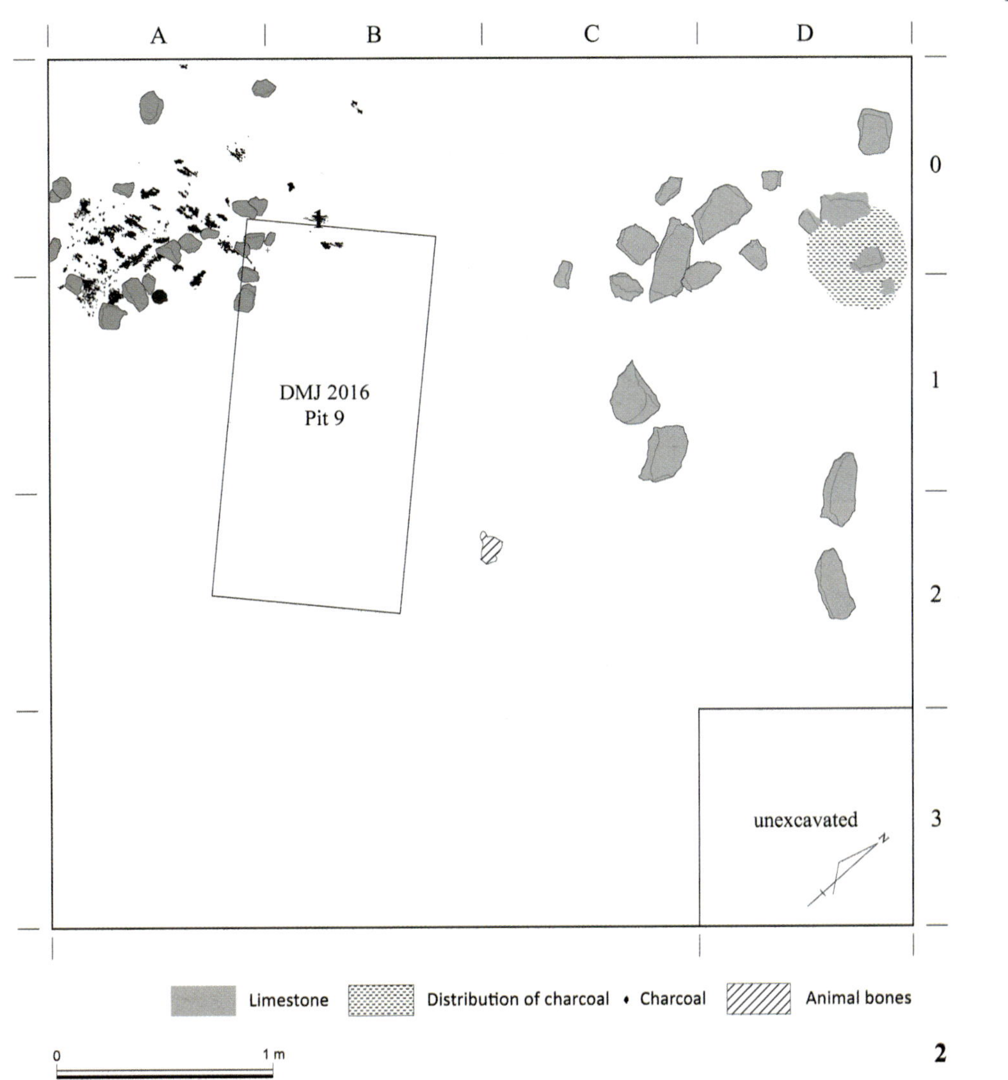

2

Fig. 4.22 Neolithic constructions in Trench 9, Unit 4.2. 1: General view of Unit 4.2 looking north; 2: Plan of the features in Unit 4.2.

Fig. 4.23 Neolithic constructions in Trench 9, Unit 4.3. 1: General view of Unit 4.3 looking north; 2: Cobble-filled pit in Square C1–D1 looking north; 3: Plan of the features in Unit 4.3.

(Figs. 4.26; 4.27). Interestingly, one of the fireplaces is located, as in Unit 5.1, at the southwest corner of the excavation area or Square B3, suggesting occupational continuity between Units 5.2 and 5.1 (Figs. 4.25: 1; 4.26). Additionally, there was another fireplace (Square B3; Figs. 4.26: 2; 4.27). It was covered by large limestone, but not the cobble-filled one discovered in the earlier levels of the Neolithic.

The lowest part of Unit 5, or Unit 5.3, consists of a blackish-gray silt layer and hard clay sediments containing yellowish-brown clay blocks and limestone cobbles (Fig. 4.28). Its sedimentological

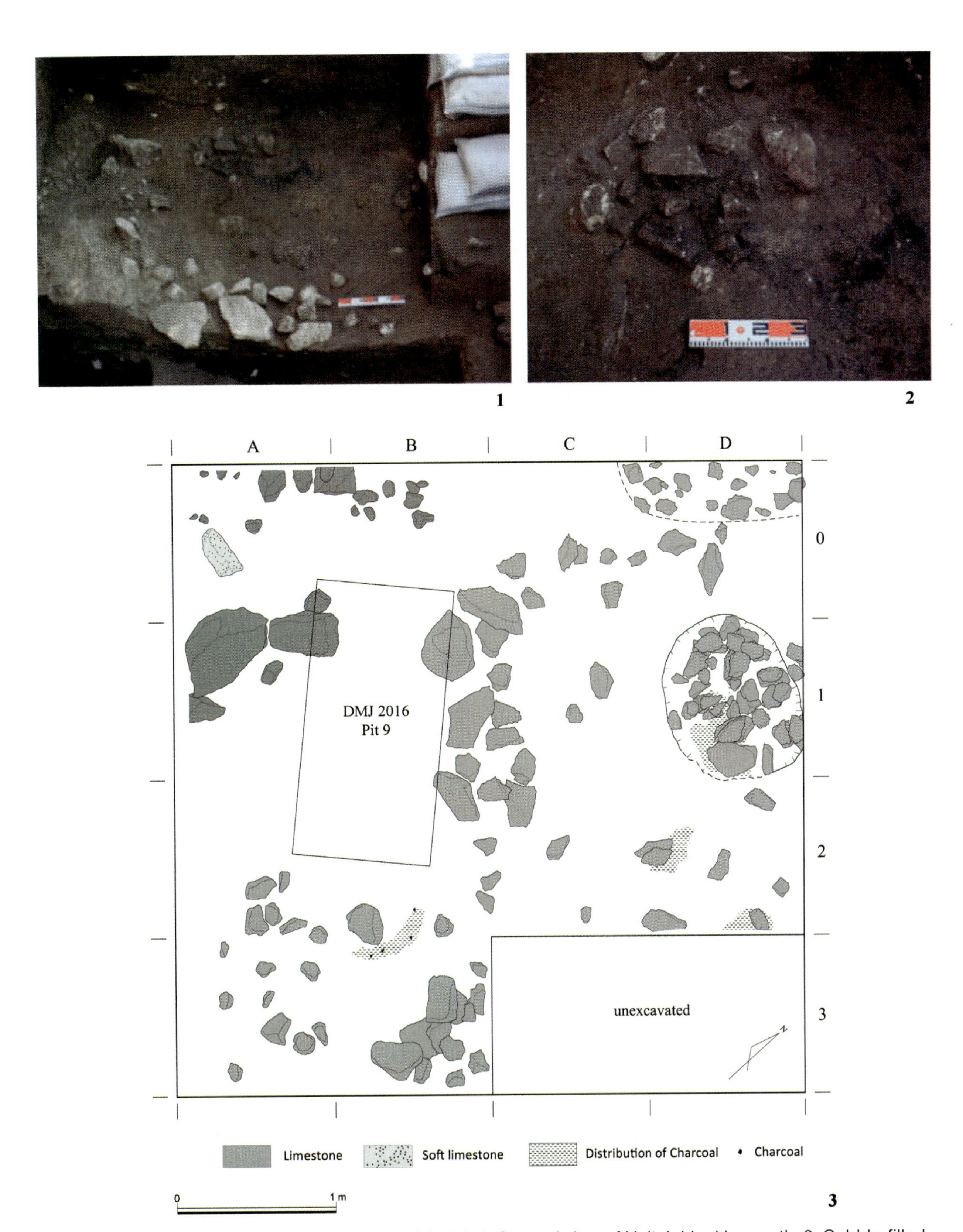

Fig. 4.24 Neolithic constructions in Trench 9, Unit 4.4. 1: General view of Unit 4.4 looking north; 2: Cobble-filled pit in Square D1 looking north; 3: Plan of the features in Unit 4.4.

characteristics indicate their origin from water activities, most likely waterfall actions (Chapter 3 of this volume). There were no *in situ* living floors in this unit. However, a small amount of Mesolithic and Paleolithic artifacts were recovered (Chapter 7 of this volume).

Unit 6

Unit 6 was investigated in a restricted area of 1 × 4 m in Squares B99–B2 (Fig. 4.1). This unit also consists of very hard yellowish- or greenish-brown clay layers.

To identify whether there is an *in situ* layer con-

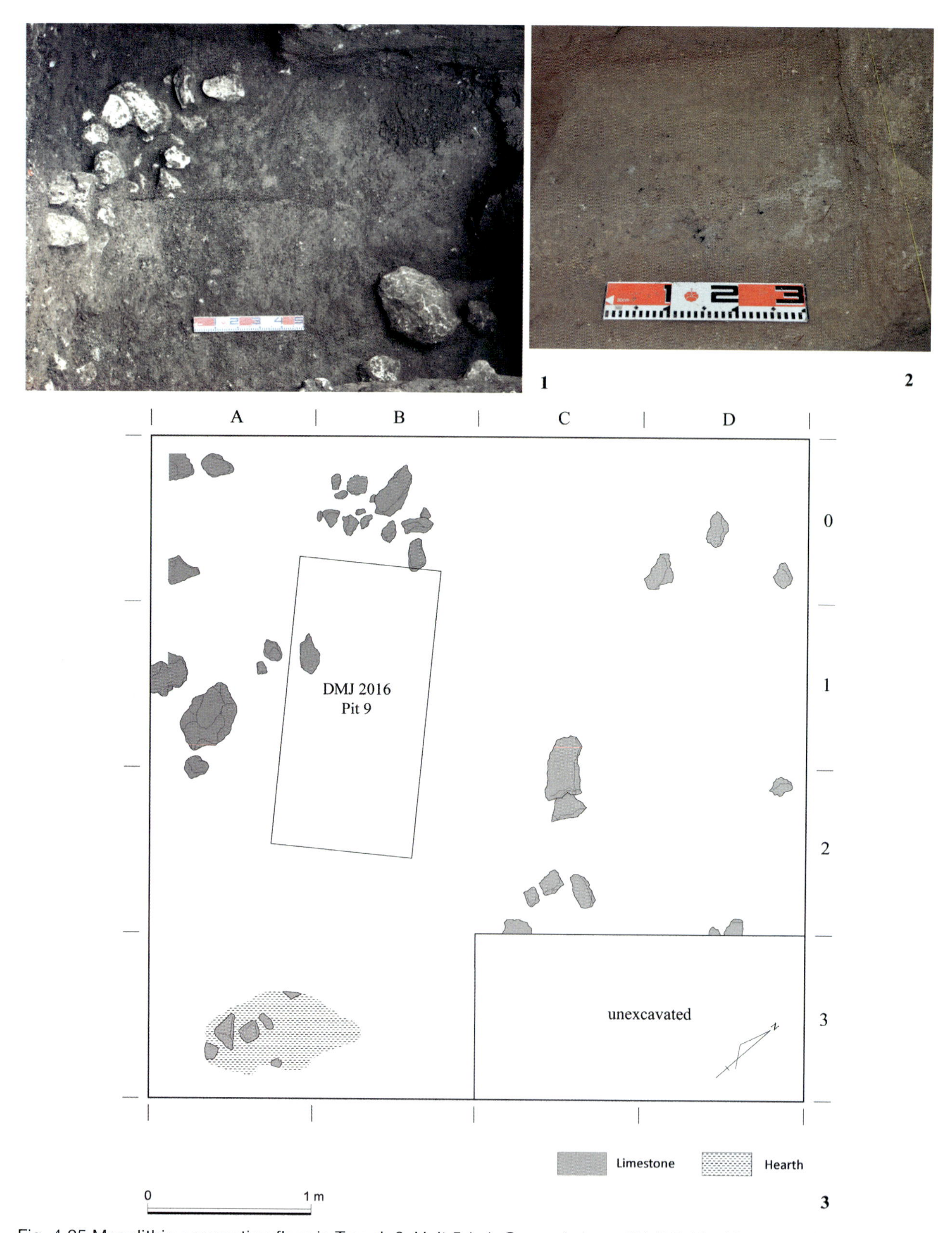

Fig. 4.25 Mesolithic occupation floor in Trench 9, Unit 5.1. 1: General view of Unit 5.1 looking east; 2: Fireplace in Square A3 looking south; 3: Plan of the features in Unit 5.1.

taining Paleolithic remains we made deeper soundings to penetrate the hard clay layer downward in Squares B0 (Fig. 4.28: 1) and B2 (Fig. 4.28: 2). The deepest point reached approximately 1.5 m deep from the top of the Unit 6 deposits (Fig. 4.28: 2). Although the top part of this unit yielded a very small amount of lithic artifacts, including Paleolithic ones best interpreted to have derived from mixed contexts (Chapter 7 of the present volume), the lower parts were completely sterile in the archaeological terms.

1

2

Fig. 4.26 Mesolithic occupation floor in Trench 9, Unit 5.2. 1: General view of Unit 5.2 looking north; 2: Fireplace in Square B3 looking south.

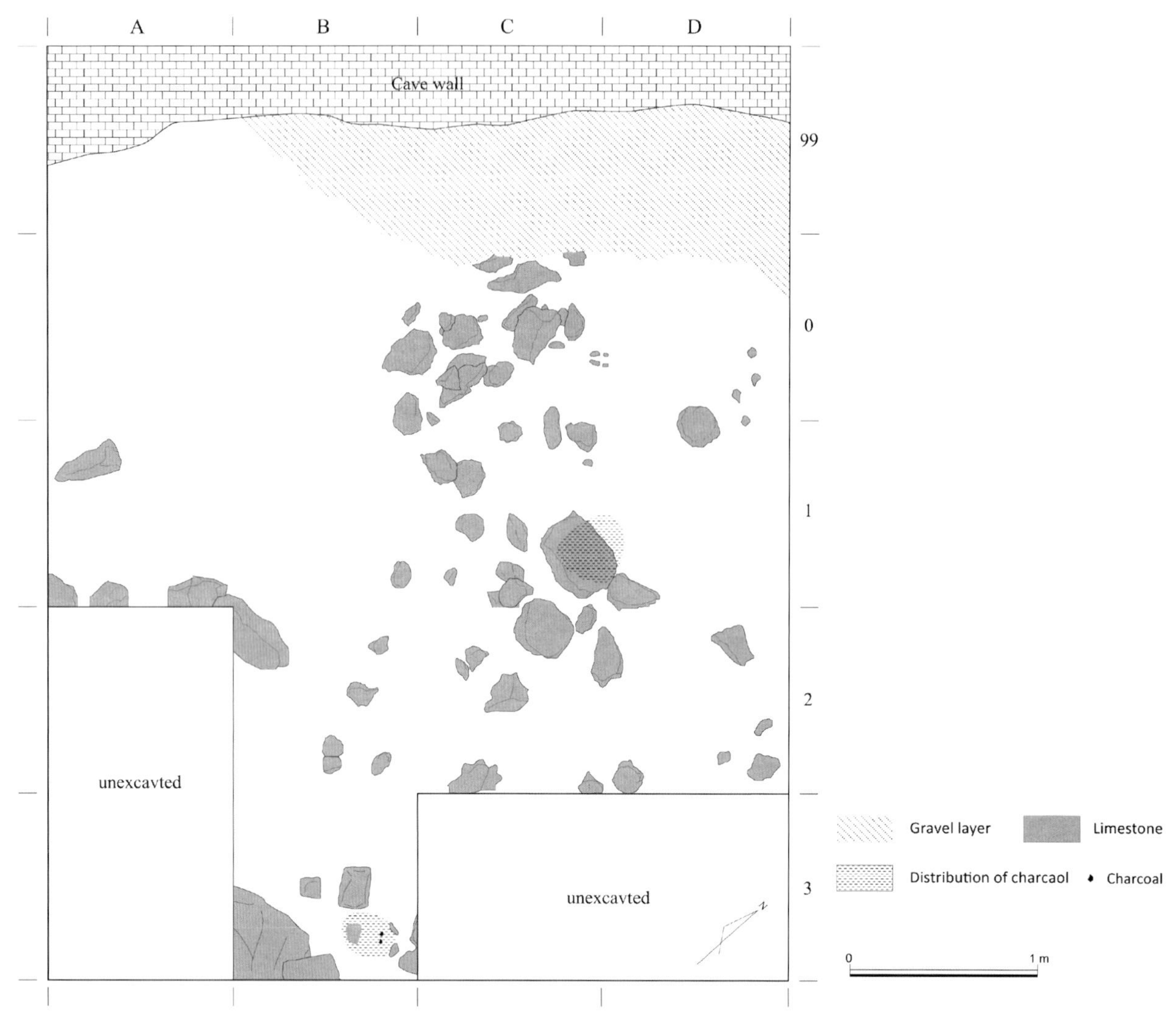

Fig. 4.27 Plan of the Mesolithic occupation floor in Trench 9, Unit 5.2.

4.4 Summary

Damjili Cave was excavated by S. N. Zamyatnin and M. Huseynov in the 1950s, who stated that archaeological remains from different periods, ranging from the Middle Paleolithic to the Neolithic, were recovered in mixed contexts (Appendix of this volume; Huseynov 2010: 224). Our investigations between 2016 and 2022 confirm that this statement is applicable to the area closer to the waterfall basin at the west end of the cave complex. Two of our sounding trenches opened in the western area (Trenches 8 and 10) were found to be almost sterile, except for the medieval and later remains, largely filled with sandy sediments containing considerable amounts of limestone gravel. It is suggested that the sediments in the western area were exposed to water action, resulting in unfavorable conditions for the preservation of prehistoric sediments.

On the other hand, the renewed excavations revealed that the eastern area of Damjili Cave, accommodated well-stratified cultural deposits. Although the Paleolithic deposits are found to be disturbed by water action, as in the 1950s, the deposits from the Mesolithic to the later periods were well preserved. Particularly important is that the sequence contains a period of transition from the Mesolithic (Unit 5) to Neolithic (Unit 4), a central subject of the present study. The stratified data provide us with a unique opportunity in the South Caucasus to document the Mesolithic-Neolithic transitional processes, results of whose analysis will be addressed in the other chapters of this volume.

We also investigated Damjili Cave 2, the rockshelter located to the east of Cave 1. The results showed that the sheltered area was used sporadically in medieval and later times for specific purposes such as temporary camps and storage places. The deposits containing abundant fresh limestone rubble strongly suggest that Damjili Cave 2 is a recent formation unlikely to have accommodated prehistoric human habitations.

1

2

Fig. 4.28 Deep soundings in Trench 9, Unit 6. 1: Square B0 looking north; 2: Square B2 looking west.

References

Huseynov, M. (2010) *The Lower Paleolithic of Azerbaijan*. Baku: National Academy of Sciences of Azerbaijan (in Russian with English summary).

Molist, M. (1985) Les structures de combustion de Cafer Höyük, Malatya, Turquie: étude préliminaire après trois campagnes. *Cahiers de l'Euphrate* 4: 32–52.

Nishiaki, Y., M. Tao, S. Kadowaki, M. Abe, and H. Tano (2001) Excavations in Sector A of Tell Kosak Shamali: The stratigraphy and architectures. In: *Tell Kosak Shamali: The Archaeological Investigations on the Upper Euphrates, Syria. Vol. 1: Chalcolithic Architecture and the Earlier Prehistoric Remains*, edited by Y. Nishiaki and T. Matsutani, pp. 49–113. UMUT Monograph 1. Oxford: Oxbow Books.

Nishiaki, Y., A. Zeynalov, M. Mansrov, C. Akashi, S. Arai, K. Shimogama, and F. Guliyev (2019) The Mesolithic-Neolithic interface in the Southern Caucasus: 2016–2017 Excavations at Damjili Cave, West Azerbaijan. *Archaeological Research in Asia* 19: 100140.

Nishiaki, Y. and F. Guliyev (2020) *Göytepe: The Neolithic Excavations in the Middle Kura Valley, Azerbaijan*. Oxford: Archaeopress.

Nishiaki, Y., F. Guliyev, and S. Kadowaki (2021) *Hacı Elamxanlı Tepe: The Archaeological Investigations of an Early Neolithic Settlement in West Azerbaijan*. Berlin: ex oriente.

Nishiaki, Y., A. Zeynalov, M. Mansrov, and F. Guliyev (2022) Radiocarbon chronology of the Mesolithic-Neolithic sequence at Damjili Cave, Azerbaijan, Southern Caucasus. *Radiocarbon* 64(2): 309–322.

Nishiaki, Y. (in press) Neolithic cooking in Upper Mesopotamia: A preliminary study of cobble-filled pits from Tell Seker al-Aheimar, Northeast Syria. In: *Neolithic in Syria*, edited by F. Borrell, H. Alarashi, and E. Healey. Berlin: ex oriente.

Zamyatnin, S. N. (1958) Scientific research on the Stone Age of Azerbaijan, Autumn 1953. *Bulletin of the Institute of History* 13: 5–18 (in Azerbaijani).

Chapter 5

Radiocarbon chronology for the cultural sequence of Damjili Cave

Yoshihiro Nishiaki

5.1 Introduction

Damjili Cave is one of the rare archaeological sites in the South Caucasus, known to date, to have produced stratified archaeological records during the Late Pleistocene and Holocene. Although investigations in the 1950s merely recovered disturbed cultural deposits (Appendix of this volume), our resumed work has revealed *in situ* deposits from the Mesolithic to medieval periods; however, no primary Paleolithic deposits have been identified. The Mesolithic to medieval records are associated with sufficient radiocarbon dates, providing an excellent opportunity to establish a Holocene cultural chronology of this part of South Caucasus. As described earlier (Chapter 4 of this volume), we opened ten sounding trenches at the Damjili Cave complex in 2016, among which Trench 9 was subjected to more extensive excavations in later seasons. This chapter reports the radiocarbon dates obtained during the 2016–2022 seasons at Damjili Cave.

5.2 Samples for radiocarbon dating

The basic stratigraphy of Damjili Cave follows the sequence established in Trench 9, which consists of Units 1–6. As the lowest Unit 6 did not yield any primary archaeological findings, our radiocarbon dating was based on samples from Units 1–5.

All the samples from Trench 9 were charcoal remains collected from either fireplaces or ash distributions in primary contexts (Table 5.1). On the other hand, samples from the soundings at neighboring Trenches 7 and 8 were as follows: Trench 7 partially overlapped Trench 9 at the east end, thus enabling its stratigraphy to be correlated with the basic stratigraphy of Trench 9. Charcoal samples were obtained from layers equivalent to Units 2 and 4.2 of Trench 9 (Nishiaki et al. 2019: fig. 3). In contrast, Trench 8 was six meters away to the west from Trench 9, without a direct connection. However, we assigned two human bone samples (#1 and #2 in Table 5.1) recovered from Tombs 1 and 2 of Trench 8 (Chapter 4 of this volume: fig. 4.10) to Unit 1 of Trench 9. The reason was: first, the comparability of the sedimentological features containing those tombs and those of the Trench 9 Unit 1 layers, and second, the fact that a few comparable tombs were also discovered in Unit 1 of Trench 9 (Chapter 4 of this volume: fig. 4.14).

5.3 Radiocarbon chronology for Damjili Cave

Most radiocarbon dates from the Damjili Cave were published in Nishiaki et al. (2019, 2022). This chapter adds to the corpus three unpublished dates recently made available (#3, #4, and #26 in Table 5.1). A Bayesian statistical analysis based on 27 dates defines the chronology of Trench 9 of Damjili Cave.

Unit 1: Historic period (5th to 10th centuries AD)

The dates for Unit 1 suggest the existence of two subunits: 5–6th and 8–10th centuries AD (Table 5.1; Fig. 5.1). We also identified two architectural phases for this unit (Chapter 4 of this volume). Considering the proximity of both the stratigraphy and plan, the two architectural phases are rather likely to represent a single phase of a younger possible subunit. Thus, a safer assumption, at this stage of the research, would be that this unit represents a historic period, mainly the medieval period (Chapter 16 of this volume).

Unit 2: Bronze Age (c. 2800–2200 BC)

The two dates available for this period (#5 and #6 in Table 5.1) cover the third millennium BC, corresponding to the Early Bronze Age in the South Caucasus (Chapter 16 of this volume). This suggests a relatively large occupation gap between Units 2 and 1 and between Units 3 and 2 (see below).

Unit 3: Chalcolithic period (c. 4500–3700 BC)

This unit, dating from the late fifth to early fourth millennium BC, was poorly documented in our excavations. There have not been discovered any structural remains. Furthermore, the corresponding deposits were identified only in Trench 9 and not in Trench 7 (Nishiaki et al. 2019). Nevertheless,

Table 5.1 Radiocarbon dates for the Mesolithic (Unit 5) to Historic (Unit 1) units of Damjili Cave and the results of a Bayesian stratigraphic analysis. * In the column for contexts, "Pit" denotes one of the trenches opened in the 2016 season.

					Unmodelled (BC/AD)		Modelled (BC/AD)			
#	Contexts*	Units	Lab no.	uncal BP	68.3% probability	95.4% probability	68.3% probability	95.4% probability	Median	Agreement Index
	Boundary End						848–1071	781–1430	976	
1	DMJ16-Pit 8. Grave 2	Unit 1	TKA-17147	1160 ± 25	776–952	772–977	775–943	772–971		98.2
2	DMJ16-Pit 8. Grave 1	Unit 1	TKA-17146	1165 ± 25	776–948	772–975	775–894	771–958		100.5
3	DMJ22-B2-2022.1	Unit 1	TKA-26454	1401 ± 20	610–658	605–661	611–658	605–662		97
4	DMJ22-B2-2022.1	Unit 1	TKA-26453	1583 ± 20	435–538	427–545	452–542	429–548		98.4
	Boundary Unit 2/1						150–534	1238–559	323	
5	DMJ16-Pit 7. 7	Unit 2	TKA-17144	3790 ± 25	2284–2147	2296–2138	2284–2147	2297–2138		99
6	DMJ16-Pit 7. 8	Unit 2	IAAA-160715	4130 ± 30	2857–2629	2871–2581	2856–2627	2870–2581		99.5
	Boundary Unit 3/2						3644–3101	3657–2747	3303	
7	DMJ16-Pit 9. 5	Unit 3	IAAA-170936	4,860 ± 30	3698–3543	3708–3532	3700–3544	3709–3533		98
8	DMJ16-Pit 9. 6	Unit 3	IAAA-170937	5,390 ± 30	4327–4177	4336–4065	4327–4177	4336–4066		99.6
9	DMJ16-Pit 9. 7	Unit 3	TKA-17148	5695 ± 25	4547–4461	4603–4455	4547–4461	4601–4454		100.5
	Boundary Unit 4-1/3						5335–4542	5348–4523	4940	
10	DMJ16-Pit 9. 10	Unit 4-1	TKA-17149	6365 ± 25	5368–5313	5467–5223	5367–5313	5466–5223		99.8
	Boundary Unit 4-2/4-1						5661–5470	5676–5331	5556	
11	DMJ16-Pit 7. 10	Unit 4-2	IAAA-160716	6740 ± 30	5704–5624	5716–5571	5661–5625	5697–5568		110.8
	Boundary Unit 4-3/4-2						5697–5646	5721–5626	5671	
12	DMJ16-Pit 9. 13.1	Unit 4-3	IAAA-160717	6790 ± 30	5718–5661	5727–5633	5720–5676	5731–5652		107.6
13	DMJ16-Pit 9. 13.2	Unit 4-3	IAAA-160718	6810 ± 30	5723–5668	5736–5636	5721–5677	5735–5656		109.5
	Boundary Unit 4-4/4-3						5757–5691	5808–5668	5727	
14	DMJ18-A0-9	Unit 4-4	IAAA-180673	6925 ± 30	5834–5750	5886–5728	5840–5756	5887–5734		99.2
15	DMJ18-A0-12.1	Unit 4-4	IAAA-180674	6925 ± 30	5834–5750	5886–5728	5840–5756	5887–5734		99.2
16	DMJ18-A0-12.2	Unit 4-4	IAAA-180675	7095 ± 30	6014–5921	6028–5895	6011–5921	6025–5897		101.8
17	DMJ18-A0-13	Unit 4-4	IAAA-180676	7170 ± 30	6063–6015	6075–5987	6030–5991	6053–5926		80.9
	Boundary Unit 5-1/4-4						6054–6015	6069–5994	6033	
18	DMJ16-Pit 9. 14	Unit 5-1	TKA-17150	7170 ± 35	6066–6012	6079–5932	6067–6036	6076–6016		111
19	DMJ16-Pit 9. 15	Unit 5-1	TKA-17151	7195 ± 30	6069–6025	6160–5988	6096–6036	6079–6019		113.6
	Boundary Unit 5-2/5-1						6101–6042	6164–6025	6075	
20	DMJ17-B3-13.2	Uniyt 5-2	IAAA-170939	7270 ± 30	6216–6073	6223–6067	6220–6136	6224–6074		102.5
21	DMJ17-B3-13.1	Uniyt 5-2	IAAA-170938	7350 ± 30	6240–6090	6337–6079	6241–6095	6341–6081		98.7
22	DMJ16-Pit 9. 17.2	Uniyt 5-2	IAAA-160720	7360 ± 30	6331–6090	6361–6082	6332–6095	6360–6085		99.3
23	DMJ16-Pit 9. 17.1	Uniyt 5-2	IAAA-160719	7400 ± 30	6364–6229	6385–6102	6363–6229	6385–6106		100.3
24	DMJ16-Pit 9. 21.1	Uniyt 5-2	IAAA-160721	7490 ± 30	6420–6266	6428–6249	6419–6266	6428–6249		99.7
25	DMJ16-Pit 9. 21.2	Uniyt 5-2	IAAA-160722	7500 ± 30	6425–6269	6434–6252	6425–6269	6434–6251		99.7
26	DMJ22-C1.12	Uniyt 5-2	TKA-26451	7609 ± 25	6464–6436	6497–6419	6462–6436	6476–6421		101.7
27	DMJ18-B0.11	Uniyt 5-2	IAAA-180672	7650 ± 30	6504–6444	6571–6433	6481–6441	6562–6427		113
	Boundary Start						6542–6456	6652–6441	6506	

the radiocarbon dates from Trench 9 point to a Chalcolithic age. The excavated area of the cave was probably only sparsely occupied during this period.

Unit 4: Neolithic period (c. 6000–5300 BC)

This unit represents one of the most important parts of the cultural sequence of an excavation area. It is divided into four subunits, each containing living floors (Chapter 4 of this volume). All radiocarbon dates for these subunits indicate early–mid sixth millennium BC, a period typical of the Neolithic in the South Caucasus (Nishiaki et al. 2022). Though there is again a chronological gap from the overlying Unit 3, interestingly, no significant occupational

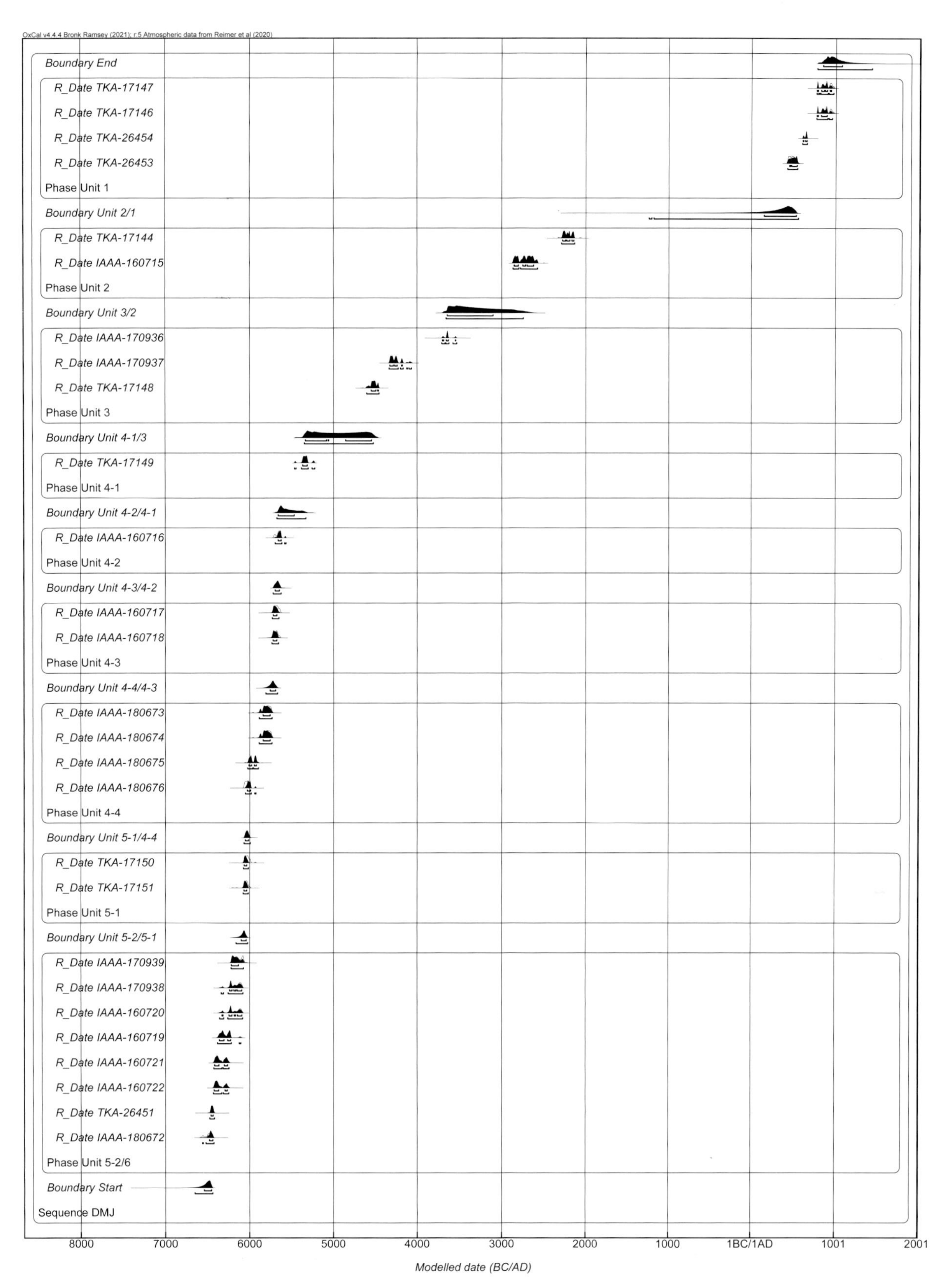

Fig. 5.1 A Bayesian modelling of radiocarbon dates for Units 1 to 5 of Damjili Cave.

hiatus is identifiable between the subunits of Unit 4. Neolithic occupations appear to have occurred continuously during the first three quarters of the sixth millennium BC.

Unit 5: Mesolithic period (c. 6500–6000 BC)

Unit 5 is dated from the late seventh millennium BC, immediately preceding the Neolithic era of Unit 4. Sedimentological analysis suggests a possible stratigraphic gap between Units 4 and 5 (Chapter 4 of this volume). However, the gap is hardly identifiable by means of archaeological radiocarbon dating: it is estimated to be as short as a few decades, if at all (Fig. 5.1; Table 5.1). The two subunits of Unit 5 are also dated to be similar to each other, suggesting that Mesolithic occupations continued over this period.

5.4 Conclusions

Our radiocarbon dating of Trench 9 demonstrates that human occupations occurred intermittently in the Mesolithic, Neolithic, Chalcolithic, Bronze, and medieval periods. The discovery of this impressive stratified sequence, at a single site, is the first in the archaeology of the South Caucasus and will serve as an invaluable resource for many different fields of study.

Our research is primarily concerned with the sequence of the Mesolithic (Unit 5) to Neolithic (Unit 4). The continuous occupations at Damjili Cave from the Mesolithic period of the late seventh millennium BC are remarkable. Notably, radiocarbon dating alone shows chronological continuity. The question of whether a "cultural" continuity can be identified is a matter of intensive research (Chapter 17 of this volume).

Another important insight obtained from our data concerns the end of the Neolithic occupations in the region. As at other sites with a sufficient number of radiocarbon dates, the Neolithic occupations at Damjili Cave were abandoned at approximately 5300 BC (Fig. 5.1). Although the cause(s) for this shared phenomenon has not been specified, Damjili Cave attests that it took place not only at mound settlements in the plains, but also at a cave site in the mountain foothills.

Given the absence of a stratified site, such as Damjili Cave, chronological data from the Chalcolithic, Bronze Age, and medieval periods are also worthy of further analysis. Detailed studies of well-dated archaeological records from each period will undoubtedly make significant contributions to the study of the Holocene cultural developments in the region.

References

Nishiaki, Y., A. Zeynalov, M. Mansrov, C. Akashi, S. Arai, K. Shimogama, and F. Guliyev (2019) The Mesolithic-Neolithic interface in the Southern Caucasus: 2016–2017 Excavations at Damjili Cave, West Azerbaijan. *Archaeological Research in Asia* 19: 100140.

Nishiaki, Y., A. Zeynalov, M. Munsrov, and F. Guliyev (2022) Radiocarbon chronology of the Mesolithic-Neolithic sequence at Damjili Cave, Azerbaijan, Southern Caucasus. *Radiocarbon* 64(2): 309–322.

Chapter 6

Luminescence dating of cultural deposits at Damjili Cave

Toru Tamura and Yoshihiro Nishiaki

6.1 Introduction

Damjili Cave, situated in West Azerbaijan, is a limestone rockshelter that contained cultural deposits from the Middle Paleolithic to Medieval Ages. Although the most layers have been sufficiently dated by radiocarbon dating (Nishiaki et al. 2019, 2022; also see Chapter 5 of this volume), a stratigraphic question has remained unsolved. It is about the origin of Middle Paleolithic artifacts. The excavations of the Trench 9 area yielded a certain amount of Middle Paleolithic lithics from the basal part of the cultural deposits and the top of the sterile yellowish gray clay sediments (see Chapters 4 and 7 of this volume). The lithic artifacts exhibit Levallois technology, which confirmed the discovery of comparable artifacts in the 1950s (Appendix of this volume), potentially providing valuable information of the *Homo neanderthalensis* occupations in the South Caucasus. However, our radiocarbon dating has not dated the layers yielding Middle Paleolithic artifacts. As a matter of fact, those layers are likely beyond the limit of radiocarbon dating. Accordingly, we attempted to date the related deposits by means of an independent method of radiocarbon dating, that is luminescence dating.

Luminescence dating is a method that provides sedimentary ages by determining the accumulation of natural radiation dosed on sediment grains through intensity of luminescence emitted from grains, referred to as equivalent dose (D_e), and dividing it by the dose rate. Over the past 25 years, it has become a reliable method through key methodological developments, such as the quartz optically-stimulated luminescence (OSL) single aliquot regenerative (SAR) method (Murray and Wintle 2000) and the feldspar post-infrared infrared-stimulated luminescence (pIRIR) method (Thomsen et al. 2008), to be applied to archeological studies especially in cases with insufficient datable materials for radiocarbon dating and beyond the limit of radiocarbon dating application, 40–50 ka (e.g., Duller et al. 2008; Bailiff 2019).

This chapter reports results of luminescence dating of nine sediment samples based on three different signals to provide independent evidence for the chronology of sediment accumulation identified in the trench at Damjili Cave.

6.2 Samples

The main trench (Trench 9) was excavated at N 41° 8′ 32″ and E 45° 13′ 59″ with the ground elevation of 581 m above the present sea level (Fig. 6.1). It is situated at an extensive limestone rockshelter facing the south, where the sediment is considered as having been derived from the local limestone and aeolian dusts. The trench was rectangular and dug to a depth of approximately 4.5 m with four vertical walls referred to as North, East, South, and West sections (Fig. 6.1).

Six stratigraphic units, 1 to 6, were identified in the trench with laterally undulating boundaries (Fig. 6.1; also see Chapter 4 of this volume). Unit 1 consists of light brownish gray sediments with abundant limestone rubble and is considered as a medieval layer. Unit 2 is characterized by relatively homogenous sediment containing dark gray ashes and Bronze Age artifacts. Unit 3 is grayish brown sediment with a small amount of limestone rubble. Radiocarbon dates and pottery obtained from Unit 3 suggest an age of the Chalcolithic period. Unit 4 consists of reddish brown soil and dark gray ash layers, containing artifacts that suggest a Neolithic age of the sixth millennium BC. Unit 5 is composed of reddish brown, coarse-grained soil and blackish gray ashes. Lithic artifacts as well as the associated radiocarbon dates suggest that the unit was formed during the seventh millennium BC, corresponding to the late Mesolithic age. Unit 6 is yellow-brownish gray sediment with numerous weathered limestone cobbles.

The Middle Paleolithic artifacts in question were mostly recovered in the lowest part of Unit 5 (Unit 5.3) and the top of Unit 6 (Chapter 7 of this volume). They are small in number and discovered together with Mesolithic artifacts. Actually, our stratigraphic observations have dated Unit 5 to the Mesolithic and Unit 6 to archaeologically sterile (Chapter 4 of this volume). Although these suggest a secondary context of the origin of the Middle Paleolithic artifacts, alternatively, there is also a possibility that Mesolithic artifacts intruded into the original layer of the Middle Paleolithic. Our luminescence dating aims to provide independent evidence for the chronology of the layers containing the Middle Paleolithic artifacts.

Sediment samples for luminescence dating were taken by hammering light-tight plastic tubes 15 cm long into the trench walls. A total of nine samples were collected for analysis. Two uppermost samples represent Unit 4 and the lowermost one was collected from Unit 6 (Fig. 6.1; Table 6.1). Other six samples were obtained from Unit 5, especially from its lower part.

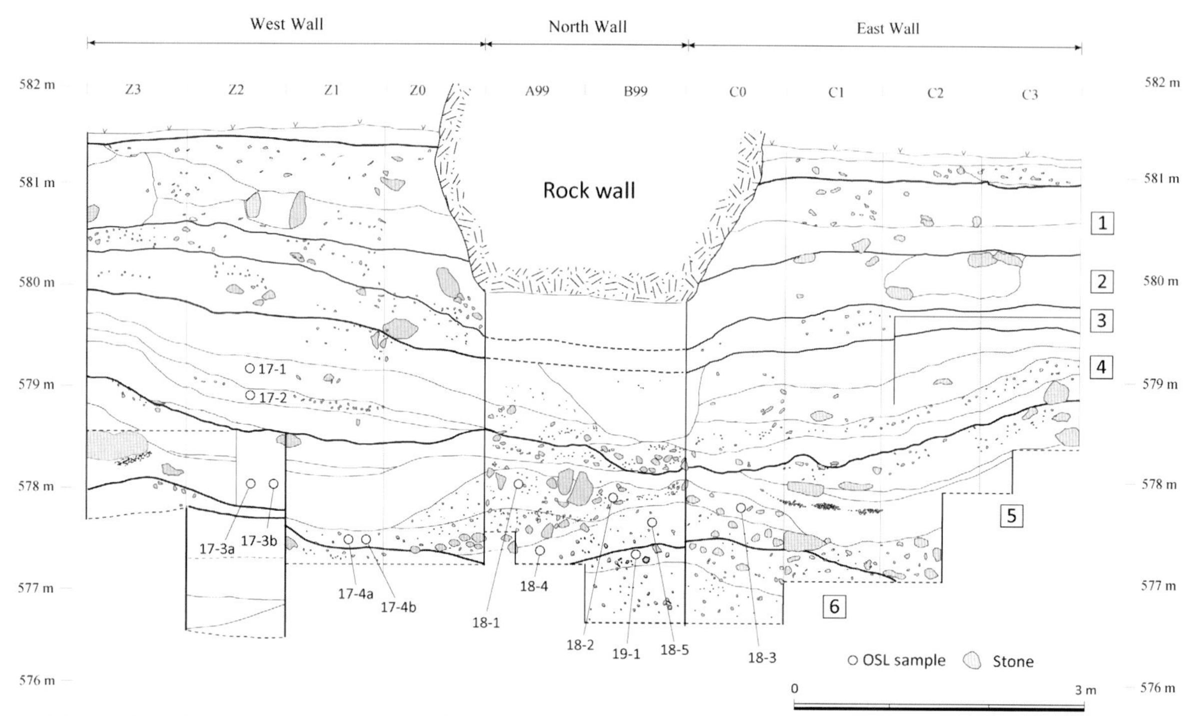

Fig. 6.1 Stratigraphic section identified in the main trench at Damjili Cave, showing sample locations.

Table 6.1 Details of sample for luminescence dating and estimate of comic dose rate.

Sample ID	Lab code	Depth (m)	Unit	Section	U (ppm)	Th (ppm)	K (%)	Rb (ppm)	Water content (%)	Cosmic dose rate (Gy/ka)
DJ17-1	gsj17340	2.3	4.2	West A2	0.92	2.43	0.74	29	17.4	0.082 ± 0.008
DJ17-2	gsj17342	2.6	4.3	West A2	1.1	3.45	0.95	39	19.7	0.079 ± 0.008
DJ17-3	gsj17344	3.5	5.2	West A2	1.4	5.51	1.5	63	25.2	0.071 ± 0.007
DJ17-4	gsj17346	4.0	5.3	West A1	1.8	7.56	3.3	85	20.9	0.067 ± 0.007
DJ18-2	gsj18250	3.7	5.3	North B0	2.6	8.9	3.3	76	24.1	0.070 ± 0.007
DJ18-3	gsj18251	3.7	5.3	East B0	1.7	7.8	3.1	81	18.8	0.069 ± 0.007
DJ18-4	gsj18252	4.2	5.3	North A0	1.8	7.7	2.3	86	27.4	0.066 ± 0.007
DJ18-5	gsj18253	3.9	5.3	North B0	1.8	8.3	2.0	84	24.4	0.068 ± 0.007
DJ19-1	gsj19420	4.1	6	North B0	2.1	8.5	2.4	85	24.3	0.066 ± 0.007

6.3 Methods

6.3.1 Sample preparation

Luminescence dating was carried out at the Geological Survey of Japan. Samples were prepared under controlled red light to avoid affecting the luminescence signals. Sediment within 5 cm from tube ends was removed and used for measurements of water content and dosimetry. The remaining samples were processed for luminescence measurements. The samples were treated with hydrochloric acid and hydrogen peroxide to remove carbonate and organic matter. Samples were then processed with a settling cylinder to extract polymineral fine grains of 4–11 μm diameter. Quartz fine grains were further purified by processing a fraction of the polymineral fine grains with hexafluorosilicic acid. Extracted quartz and polymineral grains were mounted on stainless-steel discs to form large (6 mm in diameter) aliquots for luminescence measurements.

6.3.2 Luminescence measurements

Luminescence measurements were performed with a TL-DA-20 Risø TL/OSL reader equipped with blue and infrared (IR) LEDs for stimulation and a $^{90}Sr/^{90}Y$ beta source for laboratory irradiation. For stimulations of quartz grains by blue LED emitted optically-stimulated luminescence (OSL) through a Hoya U-340 was measured with a photomultiplier tube. Luminescence from polymineral grains stimulated by IR LED through a combination of Schott BG3, BG39, and GG400 filters was measured with a photomultiplier tube to exclusively receive signals emitted from K-feldspar grains (Huntley et al. 1991).

The single-aliquot regenerative-dose (SAR) protocol was applied to quartz grains (Table 6.2; Murray and Wintle 2000). IR test of quartz grains indicated that there is no significant feldspar contamination. Following dose-recovery and preheat plateau tests on sample DJ17-3, optimal preheat temperatures of 200 °C was chosen with a cutheat of 160 °C. To determine equivalent dose (D_e), four regeneration points were measured including 0 Gy and a replicate of the first regeneration point. Data from aliquots were rejected if recycling ratios were beyond 1.0 ± 0.1. Six replicates per sample were measured.

The modified SAR protocol of post-IR IRSL (infrared-stimulated luminescence) measured at 150 °C ($pIRIR_{150}$) after a prior IRSL at 50 °C (IR_{50}) was applied to polymineral grains (Table 6.3; Thomsen et al. 2008; Reimann and Tsukamoto 2012). Preheat temperatures were fixed at 180 °C for 60 s. Four regeneration points and a replicate of the first point were measured and data with recycling ratios outside 1.0 ± 0.1 were rejected. Fading tests were also performed on aliquots after the SAR protocol to determine fading rates (expressed as g_{2days}-value) following the method of Auclair et al. (2003). Six replicates per sample were measured.

6.3.3 Dose rate and age determination

The environmental dose rate was determined using the DRAC program of Durcan et al. (2015) based on contributions of both natural radionuclides in sediments and cosmic rays. Concentrations of potassium (K) were measured by inductively coupled plasma optical emission spectrometry and those of uranium (U), thorium (Th), and rubidium (Rb) by inductively coupled plasma mass spectrometry; these results

Table 6.2 Summary of the quartz OSL single-aliquot regenerative dose protocol used in this study.

Step	ProCess	Signal
1	Preheat at 200°C for 10 s	
2	Blue OSL stimulation at 125°C for 20 s	Lx
3	Test dose	
4	Cutheat at 160°C	
5	Blue OSL stimulation at 125°C for 20 s	Tx
6	Dose and return to step 1	

Table 6.3 Summary of the IR_{50} and $pIRIR_{150}$ single-aliquot regenerative dose protocol used in this study.

Step	Process	Signal
1	Preheat at 180°C for 60 s	
2	IR stimulation at 50°C for 100 s	Lx for IR_{50}
3	IR stimulation at 150°C for 100 s	Lx for $pIRIR_{150}$
4	Test dose	
5	Preheat at 180°C for 60 s	
6	IR stimulation at 50°C for 100 s	Tx for IR_{50}
7	IR stimulation at 150°C for 100 s	Tx for $pIRIR_{150}$
8	Dose and return to step 1	

were converted to dose rates by applying the conversion factors of Guérin et al. (2011). The attenuation factors used for beta and alpha rays were based on Mejdahl (1979) and Bell (1980), respectively. We used an a-value of 0.038 ± 0.002 and 0.086 ± 0.0038 for OSL of quartz grains and IR_{50} and $pIRIR_{150}$ of polymineral grains, respectively (Rees-Jones 1995). As the samples were collected nearby the rock-shelter, cosmic dose rates were assumed as a half of those calculated for the open-air site according to Prescott and Hutton (1994). The final D_e value was determined by applying the Central Age Model (Galbraith et al. 1999) for individual samples. D_e values were then divided by the environmental dose rate to obtain age estimates (Table 6.4). All ages are expressed relative to 2020 CE. For IR_{50} and $pIRIR_{150}$ ages, fading correction was carried out with the g_{2days}-value based on Huntley and Lamothe (2001) and using the R Luminescence Package (Kreutzer et al. 2012; Fuchs et al. 2015).

6.4 Results and discussion

6.4.1 Luminescence properties

Quartz OSL is characterized by high signal to noise ratios and dominated by the fast component (Fig. 6.2: A). As a result, dose-response curves are well-defined with a single saturated exponential curve. The recycling ratio of individual aliquot is largely within 1.0 ± 0.1 with only two aliquots rejected out of 54 aliquots measured (Table 6.4), indicating the SAR

Table 6.4 Estimate of total dose rates and quartz OSL ages. n = the number of aliquots used for determining the equivalent dose, D_e = equivalent dose.

Sample ID	Lab code	Depth (m)	Unit	Total dose rate (Gy/ka)	n	De (Gy)	Age (ka)
DJ17-1	gsj17340	2.3	4.2	1.18 ± 0.05	6	13.4 ± 0.2	**11.2 ± 0.5**
DJ17-2	gsj17342	2.6	4.3	1.45 ± 0.06	6	12.7 ± 0.2	**8.6 ± 0.4**
DJ17-3	gsj17344	3.5	5.2	2.04 ± 0.09	6	18.1 ± 0.4	**8.8 ± 0.4**
DJ17-4	gsj17346	4.0	5.3	3.85 ± 0.18	5	21.6 ± 0.3	**5.5 ± 0.3**
DJ18-2	gsj18250	3.7	5.3	4.06 ± 0.18	6	26.3 ± 1.3	**6.4 ± 0.4**
DJ18-3	gsj18251	3.7	5.3	3.76 ± 0.17	6	24.4 ± 0.9	**6.4 ± 0.4**
DJ18-4	gsj18252	4.2	5.3	2.89 ± 0.12	5	21.3 ± 0.6	**7.8 ± 0.6**
DJ18-5	gsj18253	3.9	5.3	2.82 ± 0.12	6	21.4 ± 0.9	**7.5 ± 0.5**
DJ19-1	gsj19420	4.1	6	3.20 ± 0.14	6	26.1 ± 0.5	**8.1 ± 0.4**

protocol works properly. The mean D_es derived from quartz OSL are between 21.3 and 26.3 Gy for the lower six samples while the upper three samples yield lower mean D_es, roughly being consistent with the stratigraphy (Table 6.4; Fig. 6.2: B).

Bright polymineral IR_{50} is observed and results in well-defined dose-response curves with a single saturated exponential curve (Fig. 6.2: B). No aliquot shows a recycling ratio outside 1.0 ± 0.1 and thus all of 54 aliquots were used for the D_e determination. The mean D_es derived from IR_{50} are between 8.7 and 19.8 Gy, illustrating a similar vertical trend to those from quartz OSL (Table 6.5).

Polymineral $pIRIR_{150}$ is relatively dim with lower signal to noise ratios (Fig. 6.2: C). Nevertheless, dose-response curves are defined well with a single saturated exponential curve (Fig. 6.2: C). Only an aliquot shows a recycling ratio outside 1.0 ± 0.1 and was rejected. The mean D_es derived from $pIRIR_{150}$ are between 12.1 and 27.2 Gy, characterized by a similar vertical trend to those from quartz OSL and polymineral IR_{50} (Table 6.6).

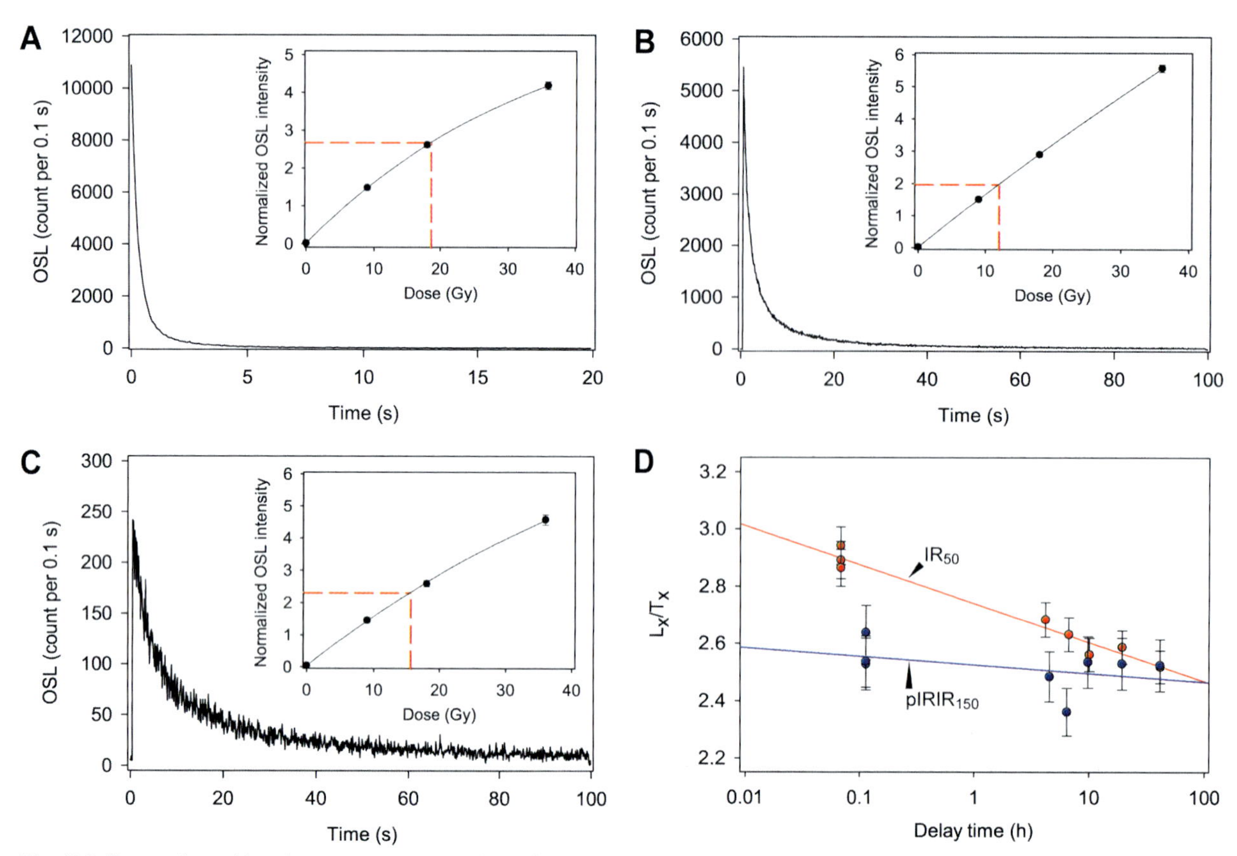

Fig. 6.2 Examples of luminescence properties of sample DJ17-3. Decay curves and dose-response curves (inset) of A) Quartz OSL, B) polymineral IR_{50}, and C) polymineral $pIRIR_{150}$. D) Fading test results of IR_{50} and $pIRIR_{150}$ for a single aliquot. Regression lines provide fading rates.

In fading tests of IR_{50} and $pIRIR_{150}$, a clear decrease of luminescence intensity with the time since irradiation, or delay time was observed and fitted with a linear regression in logarithmic time scale (Fig. 6.2: D). The mean g_{2days}-value calculated for individual sample is characterized by a modest range (Tables 6.5 and 6.6), and the variation is considered random rather than systematic changes. The average g_{2days}-values of IR_{50} and $pIRIR_{150}$ are 5.37 ± 0.10 and 0.82 ± 0.23 %/decade, respectively, and were used for the fading correction.

6.4.2 Dose rates

The total dose rates for quartz OSL are characterized by a broad range, 1.18–4.06 Gy/ka, with those for the upper three samples clearly lower than the rest (Table 6.4; Fig. 6.3: B). The broad range is attributed to large variations in radionuclide contents, which appear to reflect variable contents of sand-sized limestone clasts (Table 6.1). The upper three samples contain more sand- and silt-sized limestone clasts. For IR_{50} and $pIRIR_{150}$, the total dose rates are estimated slightly higher than those for quartz OSL owing to the irradiation from internal K and a higher a-value, and also show a broad range, 1.35–4.53 Gy/ka, with those for the upper three samples clearly lower than the rest.

6.4.3 Luminescence ages

All luminescence ages estimated here fall within the Holocene period, ranging from 5.0 ± 0.3 to 11.2 ± 0.5 ka (Tables 6.4–6.6). Age estimates from different signals for each sample are generally consistent with each other while stratigraphic inconsistency is identified among the samples. Ages of samples DJ17-2, 17-3, 18-4 and 18-5 are consistent with the radiocarbon dates if considering uncertainties, while that of DJ17-1 and those of DJ17-4, 18-2 and 18-3 appear to be overestimated and underestimated, respectively.

The consistent multiple age estimates for each sample supports the reliability of D_e estimates. Each sample is dated with three independent signals, quartz OSL, and polymineral IR_{50} and $pIRIR_{150}$. For each sample, resultant three age estimates are

Table 6.5 Estimate of total dose rates and polymineral IR_{50} ages. n = the number of aliquots used for determining the equivalent dose, D_e = equivalent dose.

Sample ID	Lab code	Depth (m)	Unit	Total dose rate (Gy/ka)	n	De (Gy)	Uncorrected age (ka)	g2days-value (%/decade)	Corrected age (ka)
DJ17-1	gsj17340	2.3	4.2	1.35 ± 0.06	6	8.8 ± 0.1	6.4 ± 0.3	5.4 ± 0.3	**10.7 ± 0.5**
DJ17-2	gsj17342	2.6	4.3	1.66 ± 0.07	6	8.7 ± 0.1	5.2 ± 0.2	5.7 ± 0.1	**8.6 ± 0.4**
DJ17-3	gsj17344	3.5	5.2	2.31 ± 0.10	6	12.1 ± 0.1	5.2 ± 0.2	5.2 ± 0.2	**8.6 ± 0.4**
DJ17-4	gsj17346	4.0	5.3	4.24 ± 0.19	6	15.0 ± 0.2	3.5 ± 0.2	4.7 ± 0.2	**5.7 ± 0.3**
DJ18-2	gsj18250	3.7	5.3	4.53 ± 0.19	6	16.9 ± 0.2	3.7 ± 0.2	5.5 ± 0.3	**6.0 ± 0.3**
DJ18-3	gsj18251	3.7	5.3	4.15 ± 0.18	6	16.8 ± 0.2	4.1 ± 0.2	6.2 ± 0.1	**6.6 ± 0.3**
DJ18-4	gsj18252	4.2	5.3	3.25 ± 0.14	6	13.5 ± 0.1	4.1 ± 0.2	5.8 ± 0.2	**6.7 ± 0.3**
DJ18-5	gsj18253	3.9	5.3	3.21 ± 0.13	6	14.5 ± 0.2	4.5 ± 0.2	5.4 ± 0.2	**7.4 ± 0.4**
DJ19-1	gsj19420	4.1	6	3.62 ± 0.15	6	19.8 ± 0.6	5.5 ± 0.3	4.3 ± 0.3	**9.0 ± 0.5**

Table 6.6 Estimate of total dose rates and polymineral $pIRIR_{150}$ ages. n = the number of aliquots used for determining the equivalent dose, D_e = equivalent dose.

Sample ID	Lab code	Depth (m)	Unit	Total dose rate (Gy/ka)	n	De (Gy)	Uncorrected age (ka)	g2days-value (%/decade)	Corrected age (ka)
DJ17-1	gsj17340	2.3	4.2	1.35 ± 0.06	6	12.2 ± 0.3	8.9 ± 0.4	1.6 ± 0.2	**9.5 ± 0.5**
DJ17-2	gsj17342	2.6	4.3	1.66 ± 0.07	6	12.1 ± 0.3	7.2 ± 0.4	1.1 ± 0.3	**7.7 ± 0.4**
DJ17-3	gsj17344	3.5	5.2	2.31 ± 0.10	6	16.3 ± 0.4	6.9 ± 0.3	1.5 ± 0.6	**7.4 ± 0.4**
DJ17-4	gsj17346	4.0	5.3	4.24 ± 0.19	6	20.1 ± 0.5	4.7 ± 0.2	1.0 ± 1.0	**5.0 ± 0.3**
DJ18-2	gsj18250	3.7	5.3	4.53 ± 0.19	6	25.4 ± 0.6	5.6 ± 0.3	0.8 ± 0.4	**5.9 ± 0.3**
DJ18-3	gsj18251	3.7	5.3	4.15 ± 0.18	6	24.8 ± 1.3	6.0 ± 0.4	1.2 ± 0.6	**6.3 ± 0.5**
DJ18-4	gsj18252	4.2	5.3	3.25 ± 0.14	6	19.0 ± 0.8	5.8 ± 0.3	0.8 ± 0.6	**6.2 ± 0.4**
DJ18-5	gsj18253	3.9	5.3	3.21 ± 0.13	5	21.2 ± 0.7	6.6 ± 0.3	0.1 ± 0.8	**7.0 ± 0.4**
DJ19-1	gsj19420	4.1	6	3.62 ± 0.15	6	27.2 ± 1.1	7.5 ± 0.4	-0.6 ± 1.0	**8.0 ± 0.5**

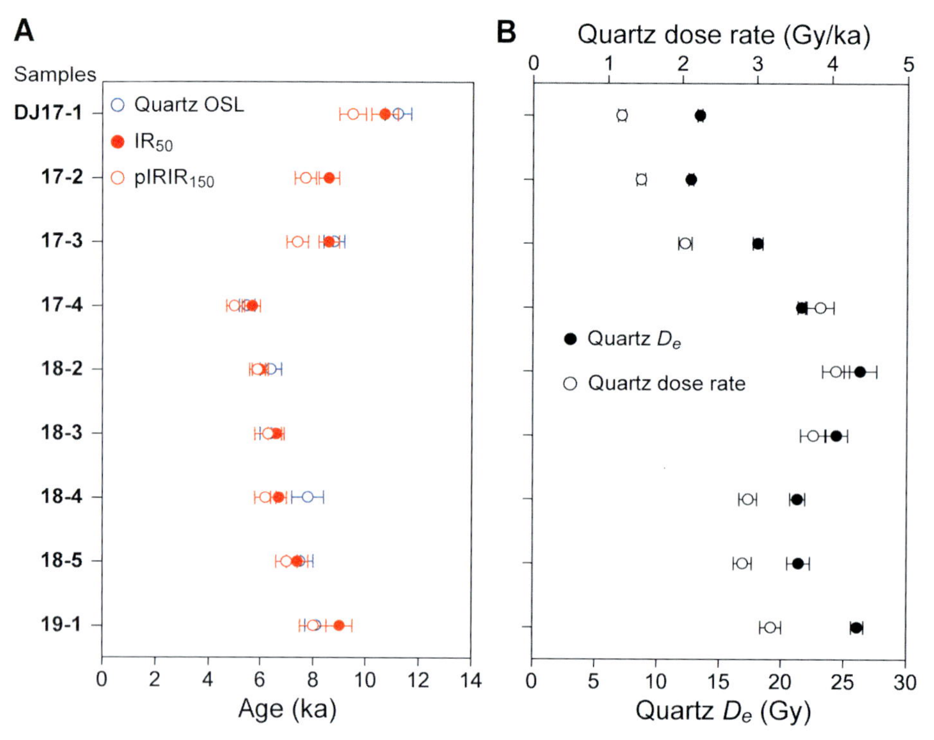

Fig. 6.3 Vertical variations in A) all luminescence age estimates and B) D_e calculated from and total dose rate for quartz OSL. Plots are arranged in the stratigraphic order.

roughly identical with each other if considering uncertainties even though these signals have different properties. Different signals are bleached by sunlight at different rates; the quartz OSL and polymineral $pIRIR_{150}$ are the fastest and slowest, respectively (e.g., Ishii et al. 2021). Incomplete bleaching causes inconsistency between age estimates of different signals. The consistency of the age estimates for each sample is thus likely to represent that the samples were bleached completely before the deposition (Murray et al. 2012), as also suggested by their aeolian origin. Athermal instability is also variable for different luminescence signals. While quartz OSL is supposed to be stable, IR_{50} and $pIRIR_{150}$ suffer from athermal instability, referred to as anomalous fading, with different rates. Thus, fading correction is required for appropriate age estimates from polymineral IR_{50} and $pIRIR_{150}$. While the correction process possibly introduces some uncertainties, it is considered as appropriate here because of the consistency between the two age estimates. For each sample, there is only a slight (< 15 %) difference in the total dose rate between the quartz and polymineral fine grains (Tables 6.4–6.6). Therefore, the consistent age estimates from multiple signals likely owes to reliable D_e estimates.

The clear stratigraphic inconsistency of the luminescence ages in contrast to the reliable D_e estimates infer problems in the dose-rate determination. Two clear luminescence age reversals are identified in the cultural deposits at Damjili Cave; one between samples DJ17-3 and DJ17-4 and another between DJ17-1 and DJ17-2. In comparisons with the radiocarbon dates, these reversals are likely attributed to underestimated ages of samples DJ17-4, DJ18-2 and DJ18-3 and an overestimated age of sample DJ17-1, respectively. As the D_e estimates are reliable, these inaccuracies in age estimates reflect problems in the dose-rate determination. At Damjili Cave, the dose-rate determination can be inaccurate owing to the migration of radionuclides and/or spatial heterogeneity of sediment (e.g., Duller 2008). In wet environments where water percolates through time, such as Damjili Cave, uranium is mobile together with water, which then causes the dose rate to change through time. This makes it difficult to accurately estimate the average dose rate over the burial period from the condition at the time of sampling. With abundant limestone rubble nearby, the actual gamma dose rate could be lower than that determined only from the sediment without taking into the heterogeneity. Thus, the dose rate of the lower part of Unit 5, where samples DJ17-4, DJ18-2, and DJ18-3 were collected, are possibly overestimated. Furthermore, sand- and silt-sized limestone clasts, which are abundantly contained in sample DJ17-1, could lower the apparent beta dose rate than actual (e.g., Cunningham et al. 2019), accounting for the underestimation of the total dose rate of sample DJ17-1. Therefore, it is considered that properties of the sediment succession at Damjili Cave possibly cause inaccuracy of the dose-rate determination that leads to inconsistency of the final luminescence age estimates.

In spite of the stratigraphic inconsistency, the luminescence ages determined here generally supports the Holocene age of the sediment succession observed at Damjili Cave. Sample DJ19-1 is the only sample collected from Unit 6 and supposed as the oldest among nine samples analyzed. In contrast to the presence of lithic artifact with Middle Paleolithic Levallois products, all of quartz OSL, polymineral IR_{50} and $pIRIR_{150}$ ages are roughly consistent with the radiocarbon age that indicates the unit corresponds to the Mesolithic. Sample DJ19-1 also suffers from the dose rate issues. However, for sample DJ19-1 the beta dose from potassium, which is irrelevant to the issues, is estimated c. 1.5 Gy/ka. Thus, the total dose rate of sample DJ19-1 should be larger than 1.5 Gy/ka, revealing the quartz OSL age is < 20 ka at least, considering that the quartz D_e, 26.1 Gy, is divided by 1.5 Gy/ka. Therefore, the luminescence ages of sample DJ19-1 indicate that the Middle Paleolithic artifacts found in Unit 6 are secondary materials and are likely to support the Holocene age of the cultural deposits at Damjili Cave.

6.5 Conclusion

This chapter reports luminescence ages of cultural deposits at Damjili Cave based on measurements of three different signals, quartz OSL, polymineral IR_{50} and $pIRIR_{150}$. The major aim is to investigate the radiometric age of the stratigraphic layers of Units 5.3 and the top part of Unit 6, which yielded Middle Paleolithic artifacts together with Mesolithic artifacts. The resultant age estimates for these deposits should be interpreted cautiously owing to problems in dose-rate determination. However, our analysis suggests a date to those layers to the Mesolithic period, or a period at least younger than 20 ka, which does not match the expected dating from the Middle Paleolithic artifacts. In other words, the present study concludes that the Middle Paleolithic artifacts from the excavated trench are stratigraphically secondary materials. This conclusion is evidently based on a small set of experiments using a limited amount of samples. More dating will further contribute to evaluating this interpretation. At the same time, research of the even lower part of Unit 6 deposits and even other trenches will be also useful to understand the origins of the Middle Paleolithic artifacts of Damjili Cave.

References

Auclair, M., M. Lamothe, and S. Huot (2003) Measurement of anomalous fading for feldspar IRSL using SAR. *Radiation Measurements* 37: 487–492.

Bailiff, I. K. (2019) Applications in archaeological contexts. In: *Handbook of Luminescence Dating*, edited by M. D. Bateman, pp. 321–349. Caithness: Whittles Publishing.

Bell, W. T. (1980) Alpha dose attenuation in quartz grains for thermoluminescence dating. *Ancient TL* 12: 4–8.

Cunningham, A. C., T. Tamura, and S. J. Armitage (2019) Applications to coastal and marine environments. In: *Handbook of Luminescence Dating*, edited by M. D. Bateman, pp. 259–292. Caithness: Whittles Publishing.

Duller, G. A. T. (2008) *Luminescence Dating: Guidelines on Using Luminescence Dating in Archaeology*. Swindon: English Heritage.

Durcan, J. A., G. E. King, and G. A. Duller (2015) DRAC: dose rate and age calculator for trapped charge dating. *Quaternary Geochronology* 28: 54–61.

Fuchs, M. C., S. Kreutzer, C. Burow, M. Dietze, M. Fischer, C. Schmidt, and M. Fuchs (2015) Data processing in luminescence dating analysis: an exemplary workflow using the R package 'Luminescence.' *Quaternary International* 362: 8–13.

Galbraith, R. F., R. G. Roberts, G. M. Laslett, H. Yoshida, and J. M. Olley (1999) Optical dating of single and multiple grains of quartz from jinmium rock shelter, northern Australia: part i, experimental design and statistical models. *Archaeometry* 41: 339–364.

Guérin, G., N. Mercier, and G. Adamiec (2011) Dose-rate conversion factors: update. *Ancient TL* 29: 5–8.

Huntley, D. J. and M. Lamothe (2001) Ubiquity of anomalous fading in K-feldspars and the measurement and correction for it in optical dating. *Canadian Journal of Earth Sciences* 38: 1093–1106.

Huntley, D. J., D. I. Godfrey-Smith, and E. H. Haskell (1991) Lightinduced emission spectra from some quartz and feldspars. *Nuclear Tracks and Radiation Measurements* 18: 127–131.

Ishii, Y., T. Tamura, D. S. Collins, and B. Ben (2021) Applicability of OSL dating to fine-grained fluvial deposits in the Mekong River floodplain, Cambodia. *Geochronometria* 48: 351–363.

Kreutzer, S., C. Schmidt, M. C. Fuchs, M. Dietze, M. Fischer, and M. Fuchs (2012) Introducing an R package for luminescence dating analysis. *Ancient TL* 30: 1–8.

Mejdahl, V. (1979) Thermoluminescence dating: beta-dose attenuation in quartz grains. *Archaeometry* 21: 61–72.

Murray, A. S. and A. G. Wintle (2000) Luminescence dating of quartz using an improved single-aliquot regenerative-dose protocol. *Radiation Measurements* 32: 57–73.

Murray, A. S., K. J. Thomsen, N. Masuda, J. P. Buylaert, and M. Jain (2012) Identifying well-bleached quartz using the different bleaching rates of quartz and feldspar luminescence signals. *Radiation Measurements* 47: 688–695.

Nishiaki, Y., A. Zeynalov, M. Mansrov, C. Akashi, S. Arai, K. Shimogama, and F. Guliyev (2019) The Mesolithic-Neolithic Interface in the Southern Caucasus: 2016–2017 Excavations at Damjili Cave, West Azerbaijan. *Archaeological Research in Asia* 19: 100140.

Nishiaki, Y., A. Zeynalov, M. Munsrov, and F. Guliyev (2022) Radiocarbon chronology of the Mesolithic-Neolithic sequence at Damjili Cave, Azerbaijan, Southern Caucasus. *Radiocarbon* 64(2): 309–322.

Prescott, J. R. and J. T. Hutton (1994) Cosmic ray contributions to dose rates for luminescence and ESR dating: large depths and long-term time variations. *Radiation Measurements* 23: 497–500.

Rees-Jones, J. (1995) Optical dating of young sediments using fine-grain quartz. *Ancient TL* 13: 9–14.

Reimann, T. and S. Tsukamoto (2012) Dating the recent past (< 500 years) by post-IR IRSL feldspar–examples from the north sea and Baltic sea coast. *Quaternary Geochronology* 10: 180–187.

Thomsen, K. J., A. S. Murray, M. Jain, and L. Bøtter-Jensen (2008) Laboratory fading rates of various luminescence signals from feldspar-rich sediment extracts. *Radiation Measurements* 43: 1474–1486.

Part II:

Artifacts and Subsistence Remains recovered at Damjili Cave in 2016–2022

Chapter 7

Middle Paleolithic lithic artifacts from Damjili Cave

Yoshihiro Nishiaki

7.1 Introduction

The preliminary reports of the 1950s investigations at Damjili Cave address "the discovery of partially destroyed layers of the Middle and Late Paleolithic as well as of Mesolithic and Neolithic" (Huseynov 2010: 224, *sic*), and "of the discovered stone material, only an insignificant number (some 20 items) dates back to the Mousterian period, something that does not allow describing them statistically. Pointers (eight specimens) were made of short triangular flint flakes. Racloir-shaped tools (12 specimens) are made of elongated, wide, and more massive flakes. Typologically, they are subdivided into rounded, elongated, pointed, denticulate, and notched ones" (Huseynov 2010: 237, *sic*).

In other words, the 1950s investigations yielded a small collection of Middle Paleolithic lithic artifacts from disturbed deposits. Our excavations of Trench 9 in 2016–2022, situated approximately 40 m to the east of the 1950's main trench, produced a similar picture: a few dozen Middle Paleolithic lithic artifacts were discovered in a disturbed context. However, we were able to identify the stratigraphic context of the Paleolithic findings in more detail. They were recovered mainly from the lowest Mesolithic (Unit 5.3) and the top of the basal sterile layers (Unit 6), together with Mesolithic artifacts (Table 7.1; Fig. 7.1).

Recovered from disturbed deposits, the Middle Paleolithic artifacts referred to in this chapter were selected materials: lithic artifacts showing the "Middle Paleolithic technology," which mostly in-

Table 7.1 Middle Paleolithic artifacts from the 2016–2022 excavations at Damjili Cave.

Stratigraphic unit		Mixed	Unit 1	Unit 2	Unit 3	Unit 4	Unit 5	Unit 6	Total
Obsidian (n=5)									
Tools	Levallois flake						1		1
	Levallois flake, rettouched					1			1
	Levallois point, retouched					1			1
	Mousterian point, on Levallois flake					1			1
	Side scraper convex, on Levallois flake		1						1
Flint (n=23)									
Debitage	Levallois flake, debordant						1		1
	Flakes					2	1		3
Tools	Levallois flakes			1		4	3		8
	Levallois flake, rettouched					1			1
	Levallois point						1		1
	Levallois point, retouched					1			1
	Levallois point, pseudo					1			1
	Levallois blade						1		1
	End scraper, on Levalois blade						1		1
	Side scraper convex, on Levallois flake						1		1
	Notched flake							1	1
	Denticulated flake							1	1
	Retouched flakes					1		1	2
Total (n=28)		0	1	1	0	13	10	3	28

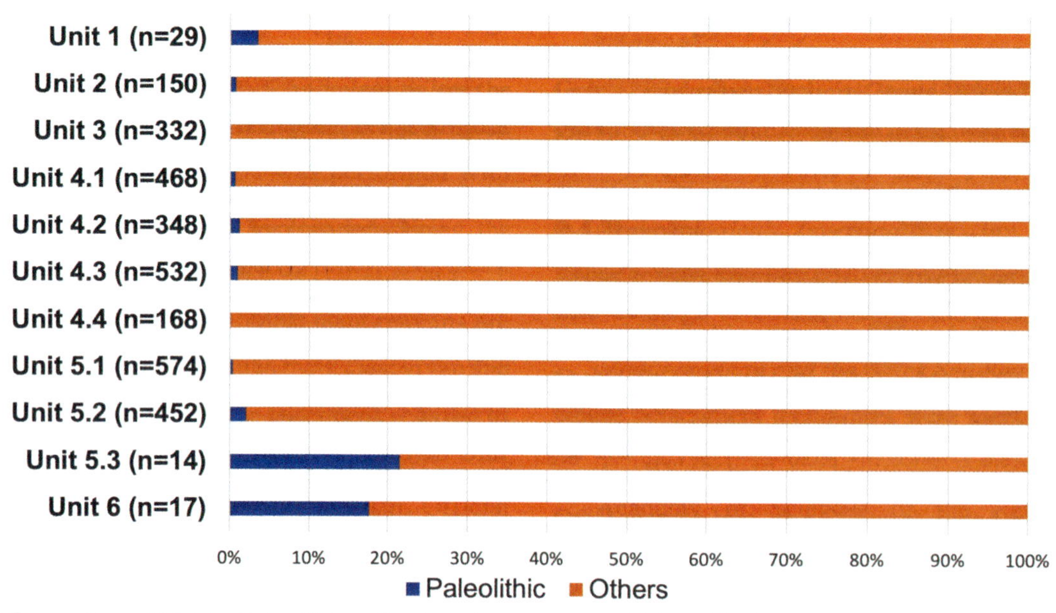

Fig. 7.1 Proportion of Middle Paleolithic artifacts in the recovered lithic artifacts assemblages by periods.

volves Levallois and related blank production technology, were selected on a techno-typological basis. Given that non-diagnostic tools and debitage are likely to be omitted, the count should be regarded as the minimum for the Middle Paleolithic collection of Damjili Cave.

7.2 Middle Paleolithic lithic artifacts from Damjili Cave

We identified at least 28 lithic artifacts with the Middle Paleolithic features. Despite their recovery from disturbed stratigraphic contexts, surprisingly, the surface conditions of those artifacts displayed little post-depositional damage such as heavy wear, scratches, and edge breakage. Instead, many of them showed rather fresh conditions. These Paleolithic artifacts were probably redeposited over a short time period on a geological scale.

The raw material for these artifacts consists of 14 "flint," seven andesite, five obsidian, and two tuff specimens. The "flint" artifacts are likely to be further divisible to specific categories such as chalcedony, rhyolite, and even siliceous tuff. Precise petrological identification of non-obsidian lithic raw materials remains incomplete. Based on this, we describe the assemblage as obsidian or non-obsidian pieces, as follows:

Obsidian specimens

Five obsidian artifacts were typologically identified in the Middle Paleolithic (Table 7.1; Fig. 7.2). They were fabricated on relatively thick blanks produced using Levallois and its related technologies. They exhibited dorsal scars in either convergent unidirectional (Fig. 7.2: 1, 2) or multidirectional directions (Fig. 7.2: 3), or at a limited part uncovered with cortex (Fig. 7.2: 4). Their platforms are roughly faceted (Fig. 7.2: 1, 2, 4) or dihedral (Fig. 7.2: 3). Careful faceting, often called the *Chapeau de Gendarme* type (Bordes 1961), is never identified.

Non-obsidian specimens

The non-obsidian assemblage consisted of 23 lithic artifacts, including those on 15 Levallois blanks (Table 7.1; Figs. 7.3; 7.4). Their composition, namely 11 flakes (Figs. 7.3: 1–4, 9; 7.4: 1–3, 6), two points (Figs. 7.3: 5; 7.4: 4), and three blades (Figs. 7.3: 6–8, 10; 7.4: 5, 7, 8), indicate the common use of flake-oriented core reduction technology at Damjili Cave. Other artifacts also exhibited technological features related to Levallois, such as the production of pseudo-Levallois points (Fig. 7.3: 11) and robust Levallois core preparation flakes (*éclat dèbordant*) (Figs. 7.3: 1; 7.4: 1).

Their dorsal scar patterns, reflecting core reduction technology, exhibited the same features as those noted for obsidian Levallois products; they were predominantly unidirectional or convergent (Figs. 7.3: 2–5, 7–10; 7.4: 2–4, 6–8). Multidirectional preparatory scars on the cores were only occasionally identified (Figs. 7.3: 1, 6; 7.4: 1, 5). Their butts indicate a limited use of faceting; when present, faceting was carried out only roughly (e.g., Fig. 7.3: 2, 7–9). Furthermore, faceted butts do not exhibit a distinct shape such as the *Chapeau de Gendarme* type.

In terms of typology, the retouched tools in the present assemblage consisted of lightly retouched pieces only (Fig. 7.3: 3, 7–10). They do not include heavily retouched side-scarpers as often recovered at Middle Paleolithic sites in the region (Zeynalov 2016).

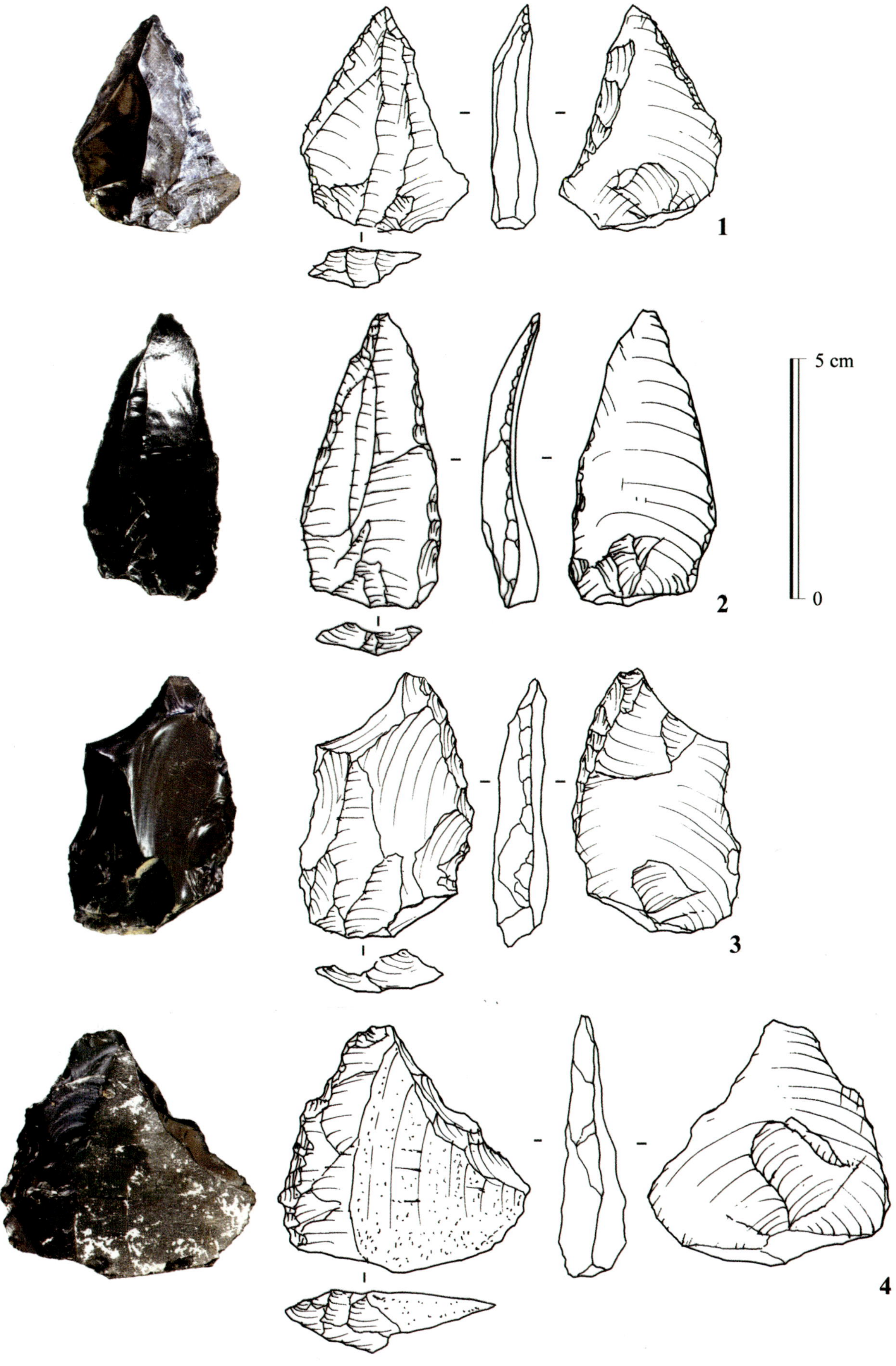

Fig. 7.2 Middle Paleolithic obsidian artifacts from the 2016–2022 excavations at Damjili Cave. 1, 2: Mousterian points; 3, 4: Side scrapers.

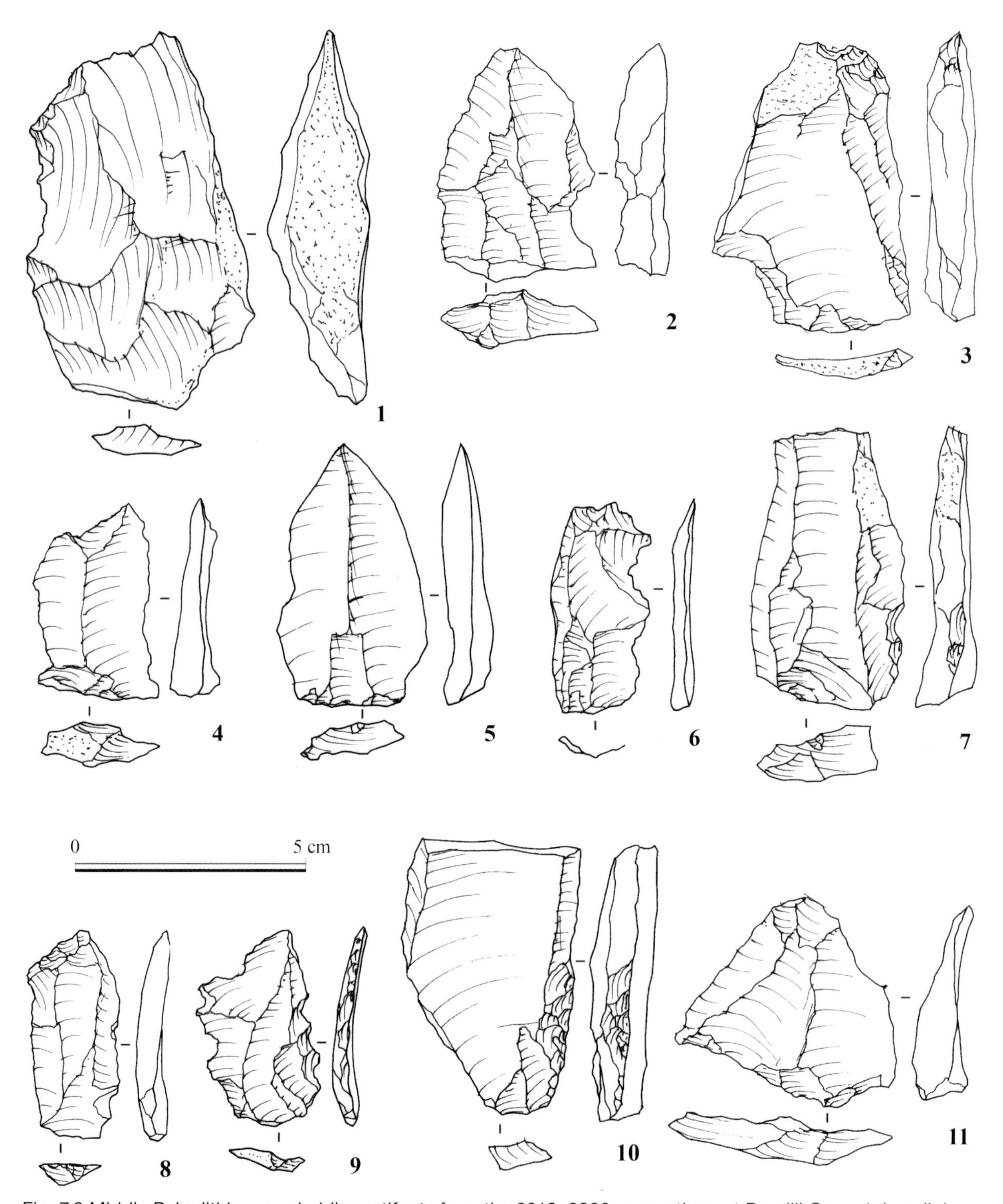

Fig. 7.3 Middle Paleolithic non-obsidian artifacts from the 2016–2022 excavations at Damjili Cave. 1: Levallois debordant flake; 2, 4: Levallois flakes; 3: Retouched non-Levallois flake; 5: Atypical Levallois point; 6, 7: Levallois blades; 8, 10: Retouched Levallois blades; 9: Denticulate on Levallois flake; 11: Pseudo Levallois point.

7.3 Discussion

Overall, the techno-typological features of the Middle Paleolithic obsidian and non-obsidian assemblages from our investigations at Damjili Cave resemble each other, therefore, likely representing a single industry. This observation is based on a non-stratified collection that should not be used for a deeper analysis. However, considering that only a small number of Middle Paleolithic sites have been excavated in Azerbaijan, general comments on their cultural affiliation should be useful for the future study.

The techno-typological aspects of our collection resemble those recovered in the 1950s (Huseynov 2010; also see Appendix of this volume). My examination of the 1950s collection at the Institute of Archaeology, Ethnography, and Anthropology

Fig. 7.4 Middle Paleolithic non-obsidian artifacts from the 2016–2022 excavations at Damjili Cave. 1: fig. 7.3: 1; 2: fig. 7.3: 2; 3: fig. 7.3: 3; 4: fig. 7.3: 5; 5: fig. 7.3: 6; 6: fig. 7.3: 9; 7: fig. 7.3: 7; 8: fig. 7.3: 10. 1–7: flint; 8: andesite.

confirms that both collections exhibited a use of the Levallois technology and the common manufacturing of scrapers and points (Fig. 7.5). This similarity suggests that comparable Middle Paleolithic remains were widely distributed from the western to the eastern ends of the cave.

A comparison with the Middle Paleolithic of Dash Salahli Cave, located in the southwest foothills of Avey Mountain, will be useful to understand the nature of the Damjili assemblage. Dash Salahli Cave was excavated in 1958 by M. Huseynov, who discovered 326 Middle Paleolithic artifacts from a 30–35 cm thick layer embedded between Holocene sediments and the bedrock (Huseynov 2010: 237). According to the preliminary accounts, they consisted of triangular, sub-triangular, blade flakes, and retouched tools made on those blanks. The general features definitely indicate Middle Paleolithic. However, they differ from the Damjili assemblage in the higher frequency of blade blanks and tools, which are rare at Damjili (Table 7.1). This respect should be in mind when making detailed techno-typological comparison between Dash Salahli whose details have not been published, and Damjili Caves in the future.

In the meantime, it is interesting that Huseynov (2010: 237) notes the scarcity of cores in comparison to the abundance of flakes and tools in Dash Salahli. He interprets this fact to represent that the core reduction took place outside and the tool manufacturing within the cave, thereby surmising that the Dash Salahli Cave was a "permanent" residential place of the Middle Paleolithic hunters (Huseynov 2010: 178). However, the same evidence can also be interpreted in other ways. For example, it may indicate that cores and finished tools were brought to the site, and after removing the necessary blanks for producing tools, the remaining cores were taken to the next camp. The latter interpretation considers the site as a temporary camp within a mobile residential system.

Regardless, further discussion on this issue is essential for us to interpret the Middle Paleolithic evidence at Damjili Cave. Based on a small non-stratified collection, I would acknowledge a temporary rather than "permanent" nature of the Damjili Cave human occupations. A supporting observation is the diverse collection of raw materials used to produce these artifacts, including obsidian, "flint," and andesite (Fig. 7.4). As mentioned earlier, "flint" is very

Fig. 7.5 Selected Middle Paleolithic andesite artifacst from the 1950's excavations at Damjili Cave. 1: Retouched Levallois blade; 2–5: Side-scrapers. 1–5: andesite.

likely to be divisible to more specific rock categories. This great diversity of raw materials for a small sample size suggests that the lithic assemblage in question could reflect the history of the occupants' previous journeys, exploiting different local raw materials on the way to the next in their mobile residential system (Kuhn 1995).

Recent developments in Middle Paleolithic research in Armenia and Georgia, provide a useful framework for reference regarding the chronological position of the Damjili Cave assemblage. This suggests that the Middle Paleolithic in the South Caucasus is divisible into at least two phases (Gasparyan et al. 2014). The earlier phase is characterized by a blade-rich industry that produces numerous blade tools and elongated points. The representative assemblage was from Djurchula, Georgia, and has been radiometrically dated to 250 ka (Mercier et al. 2010). Dating at Hovk 1, Armenia, indicates that a comparable industry associated with numerous elongated blanks may have survived until 100 ka (Pinhasi et al. 2011). On the other hand, the later phase, occurring mainly in MIS4 and early MIS3 is characterized by a flake-based industry that produces many short, retouched points and scrapers (Ghukasyan et al. 2011). The late Middle Paleolithic industry in Azerbaijan is also known at Gazma Cave, Nakhchivan (Zeynalov 2016). The Damjili Cave data share basic techno-typological features with these assemblages but lack a particular type of artifact, called the Yerevan points, popularly found at late Middle Paleolithic sites in the southern Caucasus (Liagre et al. 2006). Its absence from Damjili Cave could be due to its small sample size or chronological difference. Another possible cause is regional variability; the Middle Paleolithic on the northern side of the South Caucasus has rarely been well documented (Golovanova and Doronichev 2003). The stratified Middle Paleolithic assemblages from Azykh and Taglar Caves, Karabakh (Huseynov 2010) would be worthy of further study to establish a framework to identify the chronological and regional diversity of the Middle Paleolithic industries in the Lesser Caucasus.

7.4 Conclusions

Our provisional interpretation is that Damjili Cave was repeatedly occupied as a short-term camp in the late Middle Paleolithic. We also suggest that Middle Paleolithic deposits were disturbed before or during the early Mesolithic period. This is implied by the observed nature of the present collection and the lack of related assemblages obtained through proper stratigraphic excavations in the study region. To further discuss the Damjili Paleolithic lithics, it is essential to increase the lithic collections obtained through controlled excavations. In this regard, the Gazakh region seems promising for Paleolithic research. Our research suggests a widespread distribution of Middle Paleolithic sites in this region. For example, at the Neolithic site of Göytepe excavated since 2008 (Nishiaki and Guliyev 2020), a few Middle Paleolithic artifacts were excavated (Fig. 7.6: 1, 2); they were probably brought in by the Neolithic people as exotic materials. In addition, there is a rich Paleolithic scatter of Mousterian artifacts from Yataq Yeri (Fig. 7.6: 3–8; see Chapter 2 of this volume). Middle Paleolithic artifacts comprise an important portion, including an obsidian broad-based Levallois point (Fig. 7.6: 8).

Fig. 7.6 Middle Paleolithic artifacts from the surrounding region of Damjili Cave. 1, 2: Göytepe; 3–8: Yataq Yeri. 1, 3–7: flint; 2, 8: obsidian.

Further research on the Middle Paleolithic in this region will undoubtedly contribute to a better understanding of the cultural dynamics of the era before the emergence of modern humans in the southern Caucasus. For this, systematic artifact sampling and radiometric dating is at most required.

References

Bordes, F. (1961) *Typologie du Paléolithique Ancien et Moyen*. Bordeaux: Publications de l'Institut de Préhistoire de l'Université de Bordeaux.

Gasparyan, B., C. P. Egeland, D. S. Adler, R. Pinhasi, P. Glauberman, and H. Haydosyan (2014) The Middle Paleolithic occupation of Armenia: summarizing old and new data. *Stone Age of Armenia*, edited by B. Gasparyan and M. Arimura, pp. 65–105. Kanazawa: Kanazawa University.

Ghukasyan, R., D. Colonge, S. Nahapetyan, V. Ollivier, B. Gasparyan, H. Monchot, and C. Chataigner (2011) Kalavan-2 (North of Lake Sevan, Armenia): A new Late Middle Paleolithic site in the Lesser Caucasus. *Archaeology, Ethnology and Anthropology of Eurasia* 38(4): 39–51.

Golovanova, L. V. and V. B. Doronichev (2003). The Middle Paleolithic of the Caucasus. *Journal of World Prehistory* 17(1): 71–140.

Huseynov, M. (2010) *The Lower Paleolithic of Azerbaijan*. Baku: National Academy of Sciences of Azerbaijan (in Azerbaijani with English summary).

Kuhn, S. (1995) *Mousterian Lithic Technology: An Ecological Perspective*. Princeton: Princeton University Press.

Liagre, J., B. Gasparyan, V. Ollivier, and S. Nahapetyan (2006) Angeghakot 1 (Armenia) and the identification of the Mousterian cultural facies of « Yerevan points » type in the Southern Caucasus. *Paléorient* 32(1):1 5–18.

Mercier, N., H. Valladas, L. Meignen, J.-L. Joron, and N. Tushabramishvili (2010) Dating the Early Middle Palaeolithic laminar industry from Djruchula Cave, Republic of Georgia. *Paléorient* 36(2): 163–173.

Nishiaki, Y. and F. Guliyev (2020) *Göytepe: The Neolithic Excavations in the Middle Kura Valley, Azerbaijan*. Oxford: Archaeopress.

Pinhasi, R., B. Gasparian, S. Nahapetyan, G. Bar-Oz, L. Weissbrod, A. A. Bruch, R. Hovsepyan, and K. Wilkinson (2011) Middle Palaeolithic human occupation of the high-altitude region of Hovk-1, Armenia. *Quaternary International* 30: 3846–3857.

Zeynalov, A. (2016) *Neanderthal Man's Last Refuge*. Baku: Institute of Archaeology and Ethnography, Azerbaijan.

Chapter 8

Flaked stone artifacts of the Mesolithic to Bronze Age periods at Damjili Cave, the 2016–2022 seasons

Yoshihiro Nishiaki

8.1 Introduction

This chapter reports the flaked stone assemblages recovered from our investigations at Damjili Cave between 2016 and 2022. All those are from the main trench of Cave 1, which encompassed Trenches 7 and 9 defined in the sounding season of 2016. Other trenches at Cave 1 and all those at Cave 2 did not reveal prehistoric deposits (Chapter 4 of this volume). The Cave 1 sequence consists of six stratigraphic units from the Mesolithic (Unit 5) to the Medieval Ages (Unit 1) situated above the basal sterile layer (Unit 6). The excavations did yield Paleolithic artifacts as well. However, those artifacts, recovered only from secondary contexts are described separately (Chapter 7 of this volume).

8.2 Material and Method of Description

The collection consists of 3147 flaked stone artifacts (Table 8.1). Two aspects should be noted in the following description. One is that non-contextual specimens from top soil, disturbed layers, section scrapings, and other secondary contexts are grouped to "mixed" assemblages, which are referred to only when necessary. The second note is concerned with the raw material breakdown. As mentioned in Chapter 7 for the Middle Paleolithic artifacts, the definition of "flint" is rather arbitrary at this stage of our research. It can include other siliceous rocks such as hornfels and tuff. Accordingly, readers should regard the "flint" in the following techno-typological descriptions as representing a group of "non-obsidian" raw materials (Tables 8.2–8.5).

Techno-typological features of the obsidian and non-obsidian assemblages of Damjili Cave differ from each other. For example, the much more common production of blade blanks is noted on obsidian and heavy tools on non-obsidian raw materials. However, the same classification scheme was employed for the techno-typological descriptions. The assemblages were divided to four major categories: cores, core-management pieces, debitage, and retouched tools, each of which was then subdivided to specific types as follows.

8.2.1 Cores

The cores were classified principally according to the major target blank removals: blade and flake cores. When whose target blank forms are unclear due to heavy reduction, those cores were assigned to the exhausted type. The cores with blade removal scars in the present collection, or blade cores, always exhibit the use of a single-striking platform. They consist of two main types showing the use of pressure (e.g., Fig. 8.4: 3) and percussion debitage

Table 8.1 Raw materials used for the lithic assemblages from Damjili Cave.

Periods	Topsoil/ Mixed	Historic	Bronze Age	Chalcolithic	Neolithic				Mesolithic			Sterile	
		Unit 1	Unit 2	Unit 3	Unit 4.1	Unit 4.2	Unit 4.3	Unit 4.4	Unit 5.1	Unit 5.2	Unit 5.3	Unit 6	Total
Obsidian	36	25	118	263	369	279	373	132	261	52		2	1910
Flint	23	2	19	34	55	34	93	24	232	368	13	14	911
Tuff	1	2	4	22	10	11	29	6	24	17	1	1	128
Andesite	1		7	9	33	22	27	4	54	12			169
Jasper	2						1			1			4
Chalcedony	0		2	2	1	1	7	2	3	2			20
Limestone				2		1	2						5
Total	63	29	150	332	468	348	532	168	574	452	14	17	3147

Table 8.2 General inventory of the obsidian lithic assemblages from Damjili Cave.

Periods	Topsoil/ Mixed	Historic	Bronze Age	Chalcolithic	Neolithic				Mesolithic			Sterile	
		Unit 1	Unit 2	Unit 3	Unit 4.1	Unit 4.2	Unit 4.3	Unit 4.4	Unit 5.1	Unit 5.2	Unit 5.3	Unit 6	Total
Cores													
Blade core, pressure, single platform					1								1
Blade cores, percussion, single platform		1		1		1							3
Flake core, single platform													0
Flake core, multi-platform													0
Flake cores, on-flake	1		1		1								3
Exhausted cores			2	2	1				1				6
Core management pieces													
Core-edge elements	1		4	8	9	6	2	3	5	1			39
Core preparatory flake			1										1
Crested blades				1			1						2
Core fronts		1	1	2		1	1						6
Core bottoms		1				1				1			3
Core tablets	1				1	3	2	1					8
Blanks and other debitage													
Cortex flakes				2	1			3	1	1			8
Part-cortical flakes			6	6	7	2	9	4	1	1			36
Flakes	5	10	31	62	87	46	49	16	42	10		1	359
Part-cortex blades						1	1	4					6
Blades	8		7	15	18	55	95	21	52	6			277
Chips and fragments	11	3	25	106	142	77	100	47	102	20		1	634
Retouched tools	9	8	40	58	100	84	113	33	56	12			513
Paleolithic	0	1			1	2			1				5
Total	36	25	118	263	369	279	373	132	261	52	0	2	1910

(e.g., Fig. 8.4: 2) respectively. The flake cores are, on the other hand, all reduced by percussion. They were subdivided on the basis of striking platform configuration, namely single-platform (e.g., Fig. 8.4: 1) and multiple-platform types (e.g., Fig. 8.11: 1). When made on a thin flake, they are described as cores-on-flake (e.g., Fig. 8.2: 1).

8.2.2 Core management pieces

This category comprises blades and flakes which retain a specific part of their parent cores, regarded to have been detached for preparing and/or maintaining core forms during the reduction. Core-edge elements represent flakes or blades that were detached from the edge of either the striking platform and the working surface or the working surface and the core back. They can be produced at any stage of the core reduction (e.g., Fig. 8.10: 3). Core preparatory pieces in the present study practically denote the flakes produced from striking platform preparation (e.g., Fig. 8.10: 2). Crested blades are those detached to produce elongated guide ridges for the following blade removals. The core fronts (e.g., Fig. 8.4: 4), core bottoms, core tablets (e.g., Fig. 8.3: 3) are from the working surface, the distal end, and the striking platform of the core respectively.

8.2.3 Debitage pieces

Debitage denotes pieces detached from cores, classifiable into flakes, blades, and chips and fragments. Flakes and blades are divided on the basis of the remaining amount of cortical parts on the blank surface. Cortex flakes retain cortex at $\geq 50\%$ and part-cortex flakes/blades at $50\% >$ and $> 0\%$. Blades

Table 8.3. Obsidian tool inventory of Damjili Cave.

Periods	Topsoil/ Mixed	Historic	Bronze Age	Chalcolithic	Neolithic				Mesolithic			Sterile	
		Unit 1	Unit 2	Unit 3	Unit 4.1	Unit 4.2	Unit 4.3	Unit 4.4	Unit 5.1	Unit 5.2	Unit 5.3	Unit 6	Total
Geometric													
Lunate	1						2	1	3				7
Oblique truncation									1				1
Trapeze	1			1	2	4	13		12	1			34
Triangle													0
Backed piece													
Chokh type									1				1
Backed blade				2		1		1					4
Backed flake					1								1
Pressure-retouched													
Damjili tool				1		1							2
Strangulated blade										1			1
Borer													
Borer on blade						2	1		2				5
Borer on flake				1		1							2
Burin													
Angle burin			2		2	7	10		1	3			25
Angle+transversal burin					2	4							6
Dihedral burin	1		2			1	1						5
Transversal burin			1	1		2	1			1			6
Angle burin on truncation						1			2				3
Burin spall			1		3	3	6	1	2				16
Splintered piece													
Splintered piece		2	5	14	29	9	8	2	2				71
Splintered piece spall			1	3	3	2							9
Scraper													
Endscraper on blade					1		2	1	2				6
Endscraper on flake				1		1			2	1			5
Endscraper, round, on flake				1									1
Sidescraper		1											
Sidescraper, massive				2		1		1	1				5
Raclette							1						1
Miscellanneous blade tool													
Truncated blade						1	1	1					3
Denticulated blade					5	3	5	2	2				17
Notched blade	1			1		5	26						33
Retouched blade	3		5	5	7	10	20	10	11	4			75
Nibbled blade	1	1	6	2	14	11		11	5	1			52
Miscellanneous flake tool													
Denticulated flake		1	1	3	1	3	3	1	1				14
Notched flake	1		1				1						3
Retouched flake		1	10	8	15	2	2		1				39
Nibbled flake		1	1	4	8	3							17
Others													
Retouch flake		1	1	2									4
Spall			2	3	4	4	6		2				21
Tool fragment			1	3	3	2	4	1	3				17
Total	9	8	40	58	100	84	113	33	56	12	0	0	512

Table 8.4. General inventory of the non-obsidian lithic assemblages from Damjili Cave.

Periods	Topsoil/ Mixed	Historic	Bronze Age	Chalcolithic	Neolithic				Mesolithic			Sterile	
		Unit 1	Unit 2	Unit 3	Unit 4.1	Unit 4.2	Unit 4.3	Unit 4.4	Unit 5.1	Unit 5.2	Unit 5.3	Unit 6	Total
Cores													
Blade core, pressure, single platform							1						1
Blade core, percussion, single platform		1											1
Flake cores, single platform								2	1	1			4
Flake cores, multi-platform		1							1	1			3
Flake cores, discoidal	1								1				2
Exhausted cores			1	1						1			3
Core management pieces													
Core-edge elements				2	2	2	6	1	5	5		1	24
Core preparatory flake					1								1
Crested blade													0
Core front										1			1
Core bottom													0
Platform tablet													0
Blanks and others debitage													
Cortex flakes	1	1	2		8	3	3	1	14	10	1		44
Part-cortical flakes	1		4	5	14	5	7	4	16	19	1	2	78
Flakes	5		9	35	42	38	58	15	108	95	3	2	410
Part-cortex blades							1	1		3			5
Blades	2	1	2	1	2	3	6		23	26	2		68
Chips and fragments	10		7	22	22	10	50	5	107	181	2	7	423
Chunks	1								1	2			4
Retouched tools	6		6	3	6	6	22	7	35	46	2	3	142
Paleolithic			1		2	2	5	0	1	9	3		23
Total	27	4	32	69	99	69	159	36	313	400	14	15	1237

with ≥ 50% cortex were included in cortex flakes. Chips and fragments are unclassifiable small and/or fragmentary pieces, including thermally fractured fragments.

8.2.4 Retouched tools

Our system divides retouched tools to standardized tools (geometrics, backed pieces, pressure retouched pieces, perforators, burins, splintered pieces, and scrapers), heavy tools (choppers, picks, and hammers), miscellaneous tools (denticulates, notches, retouched, nibbled pieces, and truncations), and others (retouch flakes, spalls, and tool fragments).

Geometrics

The major types of geometrics are classified according to their plan shape (Figs. 8.1; 8.5; 8.12). Lunates are defined here as bi-truncated microliths whose length is as large as twice or more the width. Their back can be either straight (Lunate A) or round (Lunate B). One may consider Lunate A as an elongated trapeze. However, Lunate A in the present collection is evidently narrower than ordinary trapezes. Therefore, we regard it a typologically different group from trapezes. Those relatively short, Length/Width < 2, are assigned to trapezes, which can also be divided according to the back shape: straight (Trapeze A) or round (Trapeze B). When the back is pointed, it was termed Triangle.

Backed pieces

These are defined as blades/bladelets having one

Table 8.5. Flint tool inventory of Damjili Cave.

Periods	Topsoil/ Mixed	Historic	Bronze Age	Chalcolithic	Neolithic				Mesolithic			Sterile	
		Unit 1	Unit 2	Unit 3	Unit 4.1	Unit 4.2	Unit 4.3	Unit 4.4	Unit 5.1	Unit 5.2	Unit 5.3	Unit 6	Total
Geometric													
Lunate													0
Oblique truncation													0
Trapeze										4			4
Triangle							2		4	3			9
Backed piece													0
Chokh type				1					1				2
Backed blade							1		1				2
Backed flake													0
Pressure retouched													
Damjili tool						1							1
Strangulated blade													0
Borer													
Borer on blade			1						1				2
Borer on flake													0
Burin													
Angle burin						1	1	1	2				5
Angle+transversal burin													0
Dihedral burin							1						1
Transversal burin													0
Angle burin on truncation													0
Burin spall	1					1							2
Splintered piece													
Splintered piece						1							1
Splintered piece spall													0
Scraper													
Endscraper on blade			1							1			2
Endscraper on flake	2						5	2	1	3	1		14
Endscraper, round, on flake													0
Sidescraper									2	1			3
Sidescraper, massive			1	1					3				5
Raclette													0
Heavy tool													
Chopper							1	1	3	1			6
Pick									1				1
Hammer							1						1
Miscellanneous blade tool													
Truncation													0
Denticulated blade					1			1	3	4			9
Nibbled blade			1				2		1	7			11
Notched blade						1			1				2
Retouched blade	1				2		4		4	3			14
Miscellanneous flake tool													
Denticulated flake	1						1			2			4
Nibbled flake								1	1	5			7
Notched flake						1	1			1		2	5
Retouched flake	1		1	1	2		2	1	6	9	1	1	25
Others													
Retouch flake			1										1
Spall													0
Tool fragment					1					1			2
Total	6	0	6	3	6	6	22	7	35	45	2	3	141

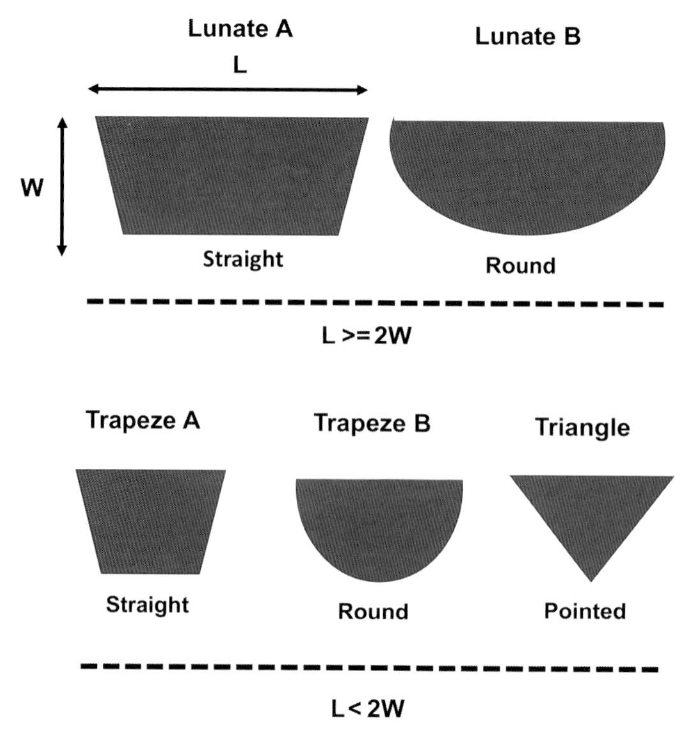

Fig. 8.1 Classification of lunates and trapezes.

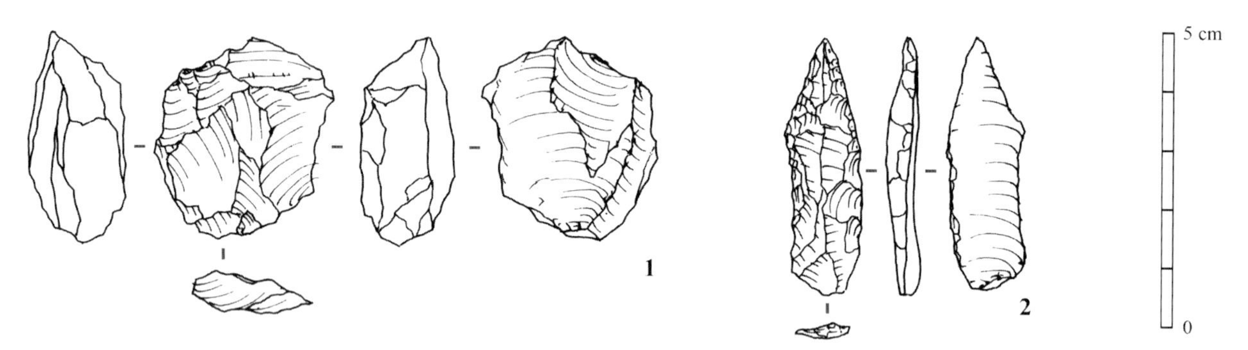

Fig. 8.2 Bronze Age lithic artifacts (Unit 2). 1: Core-on-flake; 2: Borer. 1: Obsidian; 2: Non-obsidian.

lateral edge backed with steep retouches. The most unique pieces include a group of tools with a teardrop shape (Fig. 8.13: 2, 3). Referring to the discovery of this type of artifacts at Chokh Rockshelter, Dagestan, they are called Chokh type (Nishiaki et al. 2019a).

The remaining pieces could have been made for cutting tools. At the same time, given their small size and that oblique truncation is also applied, they may include broken lunates or semi-finished geometrics.

Pressure retouched pieces

These represent a unique category of tools shaped by pressuring retouch, which consists of two types. One is tentatively referred to as "Damjili tools" (e.g., Figs. 8.3: 5; 8.6: 1; Nishiaki et al. 2019a). They exhibit a series of elaborate pressure retouches along the working edge. The black residues left parallel to the long axis of the blank and the presence of parallel striations suggest that those tools were used hafted for cutting some materials. Although they may be comparable to "Kmlo tools" popular in the Mesolithic of Armenian Highlands (Arimura et al. 2021), there are a series of dissimilarities in raw material use (obsidian vs. obsidian and flint), blank form (elongated flake vs. regular blade), retouch technology (steep vs. relatively flat), and tool form (hooked vs. less-hooked). In the meantime, these are reported as Damjili tools and their cultural contexts will be interpreted when more parallels are reported from related sites.

The second type represents strangulated blades, which have steep retouches to make bilateral notches (e.g., Fig. 8.13: 1). Again, this type resembles Kmlo tools or Cayönü tools. However, the irregular retouch forms do not allow us to assign them convincingly to this tool category.

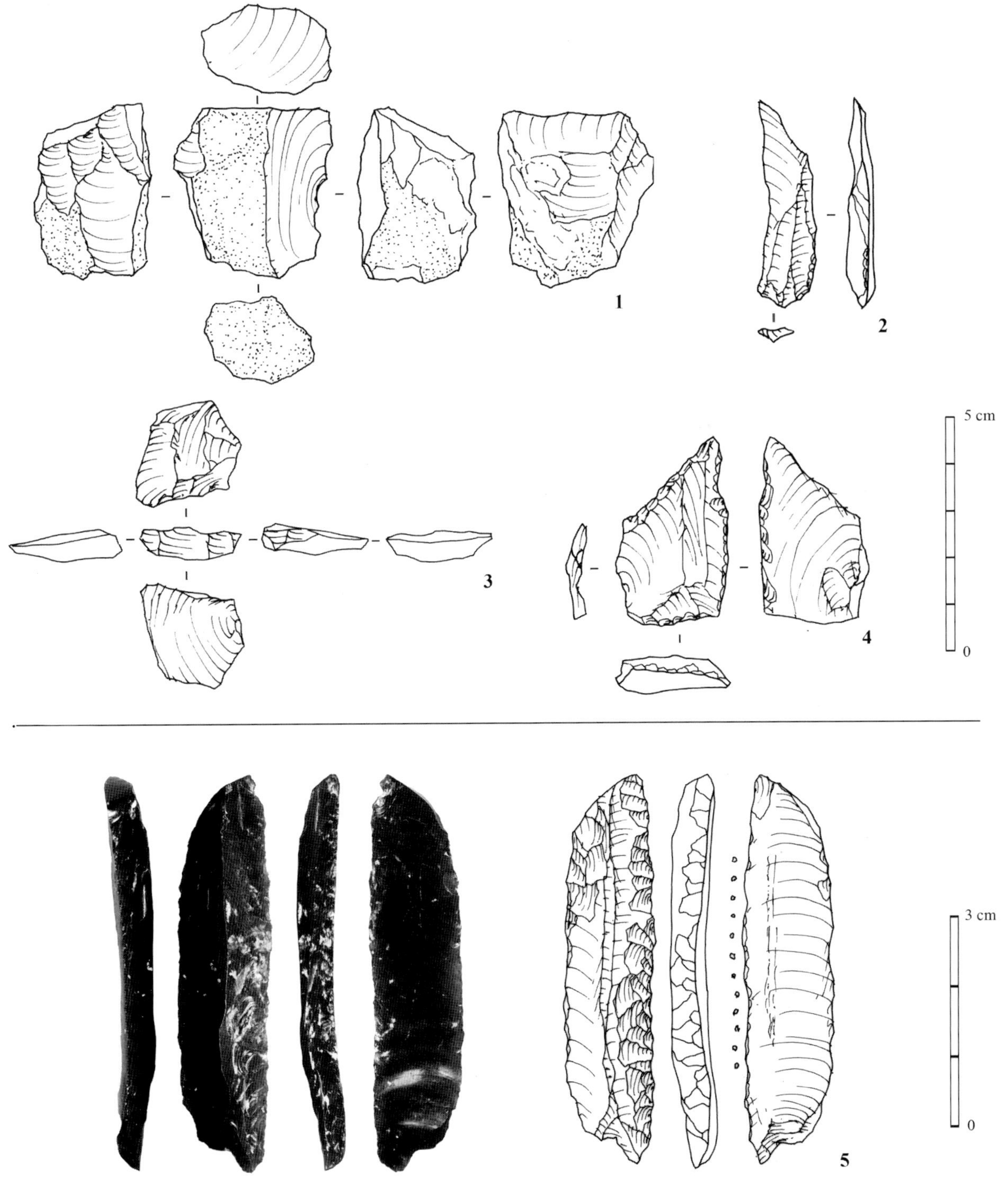

Fig. 8.3 Chalcolithic lithic artifacts (Unit 3). 1: Exhausted core; 2: Nibbled blade; 3: Core tablet; 4: Borer; 5: Damjili tool. 1: Flint; 2–5: Obsidian.

Borers

This tool category is characterized by having a pointed tip at one end of the blank for piercing. The tools are classified into those made on blade (Fig. 8.2: 2) and flake blanks (Fig. 8.3: 4).

Burins and splintered pieces

These two groups of tools were manufactured with a more or less similar retouching technology. Burins were made by blows delivered in perpendicular (Fig. 8.7: 1–3) and splintered pieces were struck in rather parallel to the blank surface (Fig. 8.7: 5–7). Our analysis of the specimens from Göytepe revealed a difficulty in typologically separating these tool groups (Nishiaki and Guliyev 2019). Splintered pieces can be described as even flat burins. However, splintered pieces generally do not exhibit specific retouch patterns, while the burins are easily divisible to several types: angle (e.g., Fig. 8.7: 2), dihedral (e.g., Fig. 8.7: 1), and angle on truncation types (e.g., Fig. 8.7: 3).

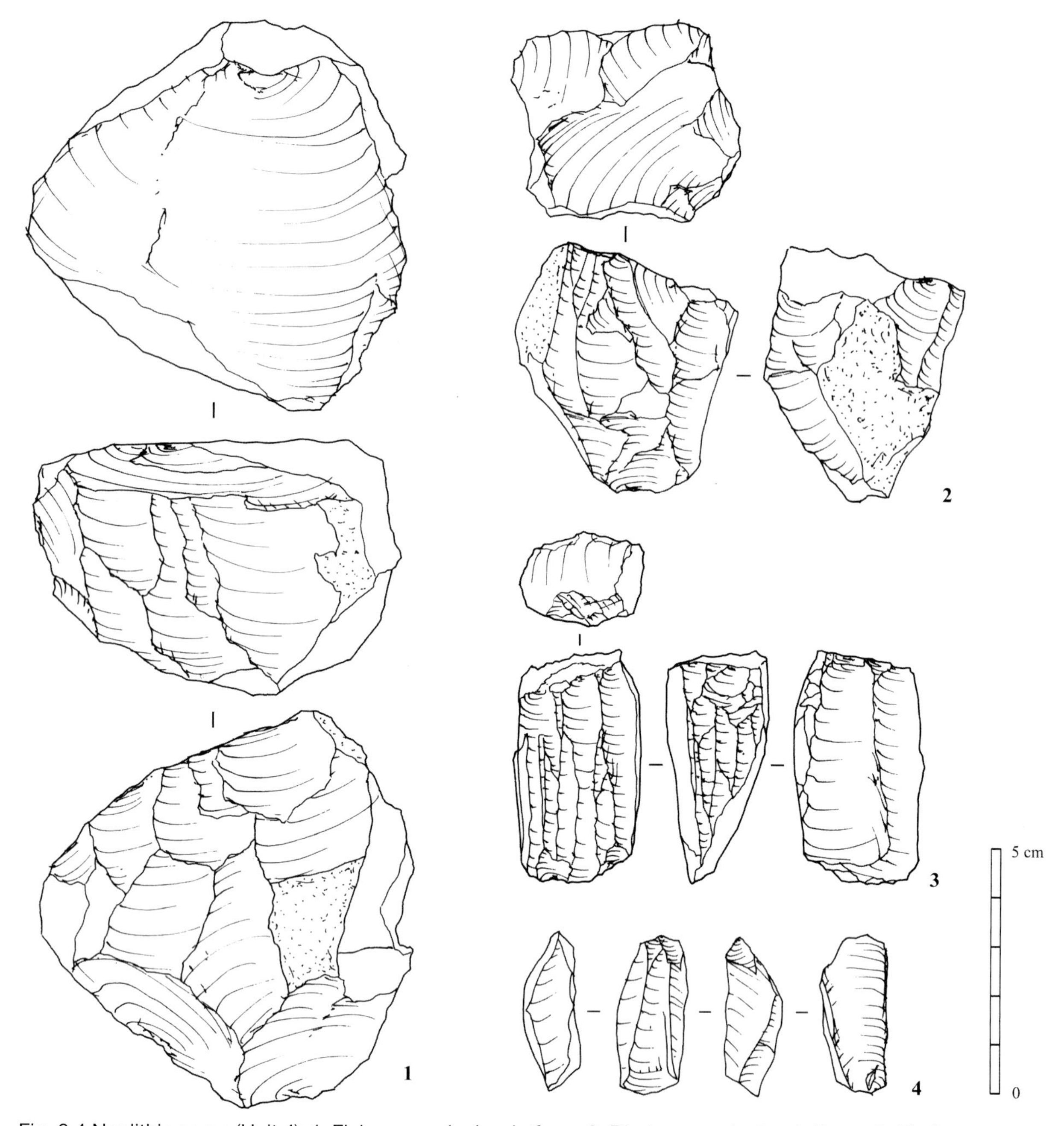

Fig. 8.4 Neolithic cores (Unit 4). 1: Flake core, single-platform; 2: Blade core, single-platform; 3: Blade core, single-platform, pressure; 4: Core front, detached from the bottom. 1, 2: Flint; 3, 4: Obsidian.

Scrapers

Scrapers also constitute the standardized retouched tools, which are divisible into endscrapers (e.g., Fig. 8.8), sidescrapers, massive scrapers (e.g., Fig. 8.9: 3, 4), and raclettes. Massive scrapers are defined as scrapers made on thick and large flake blanks. On the one hand, raclettes are scrapers made on thin flake blanks.

Core tools

The Damjili Cave lithic assemblages include tools made by *façonnage* processes: choppers and picks (Fig. 8.14: 3). The hammers seem to be recycled tools of used-up cores.

Miscellaneous tools

This group represents flake and blade tools fashioned with non-formalized retouches. According to the retouched edge-form, they were classified into denticulates, notches (e.g., Fig. 8.13: 9), retouched pieces (e.g., Fig. 8.9: 1, 2), and nibbled pieces (e.g., Fig. 8.3: 2). The nibbled pieces, displaying only light scars on the edge, could include those damaged through utilization. Blades with simple truncation at ends (e.g., Fig. 8.13: 8) are also included in this category.

Others

Retouched pieces too fragmentary to be identified

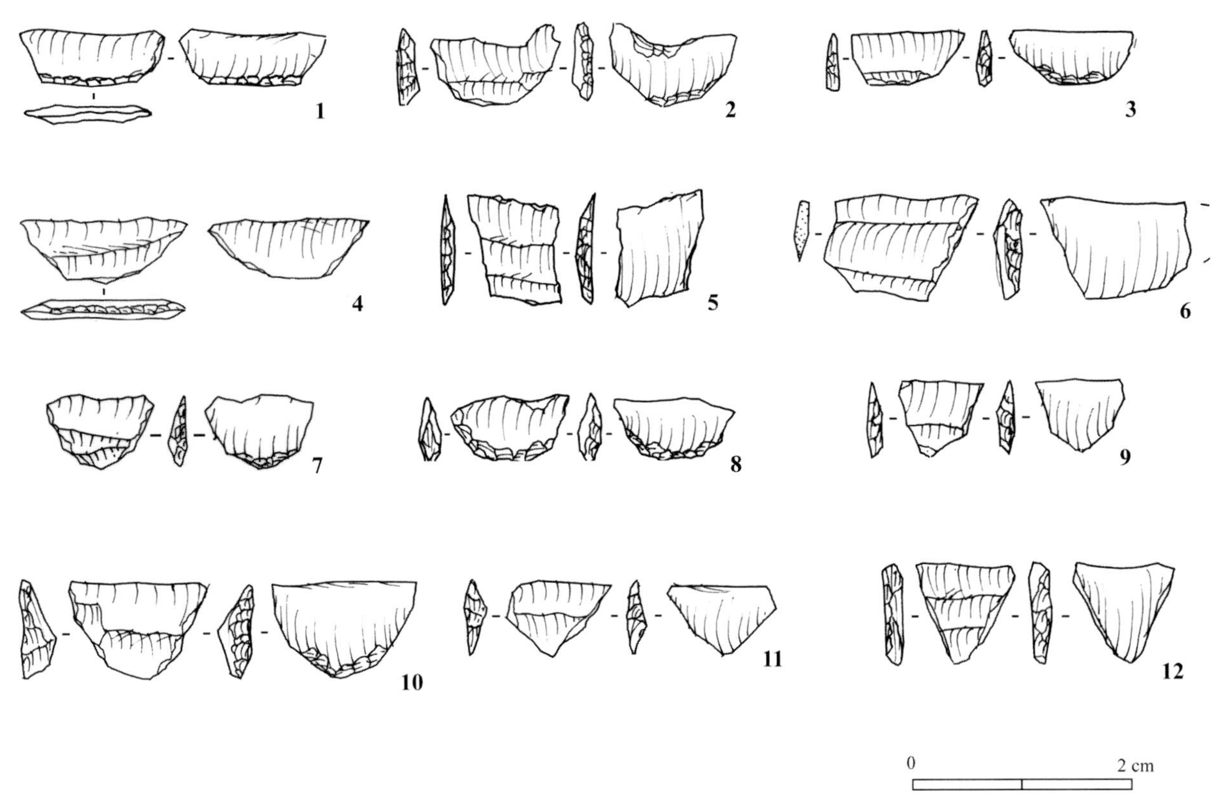

Fig. 8.5 Neolithic geometrics (Unit 4). 1: Lunate A; 2–4: Lunate B; 5, 6: Trapeze A; 7–10: Trapeze B; 11, 12: Triangle. 1–12: Obsidian.

as a specific tool type and small chips derived from retouching are included in this category.

8.3 Lithic assemblages

8.3.1 Historic period (Unit 1)

The latest cultural deposits of Damjili Cave belong to the medieval period, which is evidently no longer a stone age. The 29 artifacts from this unit, containing 25 obsidian pieces (25/29 or 86.2%), are most likely secondary materials derived from the underlying layers. They indeed include blade cores and their maintenance pieces typical of the earlier prehistoric periods (Tables 8.1–8.5).

8.3.2 Bronze Age (Unit 2)

The lithic assemblage from Unit 2 belongs to the Early Bronze Age of the third millennium BC. Archaeological sites of this period have been well known in this part of Azerbaijan mostly in the form of kurgans or surface scatters. Our survey at the foothills of Avey Mountain also documented comparable sites, which yielded a good amount of flaked stone artifacts as well as pottery sherds (Chapter 2 of this volume). The lithic remains from Damjili Cave Unit 2 consist of 118 obsidian and 32 non-obsidian artifacts (Table 8.1). The high frequency of obsidian pieces (118/150 or 78.7%) matches our observations at the surveyed Bronze Age sites.

Obsidian

The 118 obsidian artifacts consist of three cores, six core management pieces, 69 debitage pieces, and 40 retouched tools (Table 8.2). All obsidian cores, classified into exhausted and core-on-flake types (Fig. 8.2: 1), exhibit traces of flake production by percussion. The debitage composition also indicates an emphasis on flake production: blade products comprise only a small portion of the blanks (7/69 or 10.1%). The retouched tools are dominated miscellaneous pieces with non-systematic retouches (Table 8.3). Notable among standardized tools are burins (6/40 or 15.0%) and splintered pieces (6/40 or 15.0%) including their manufacturing spalls. Burins and splintered pieces are the main tool categories of the Neolithic period in the Southern Caucasus (Nishiaki and Guliyev 2019). Their common occurrence in the Bronze Age poses an intriguing question of the tool manufacturing tradition, which may have lasted over millennia.

Non-Obsidian

There are 32 non-obsidian artifacts, including one exhausted core, 24 debitage pieces, six tools, and a Paleolithic specimen (Table 8.4). The core reduction was evidently directed to flake production. The tools include a borer (Fig. 8.2: 2) and two scrapers. The remaining are miscellaneously retouched pieces (Table 8.5).

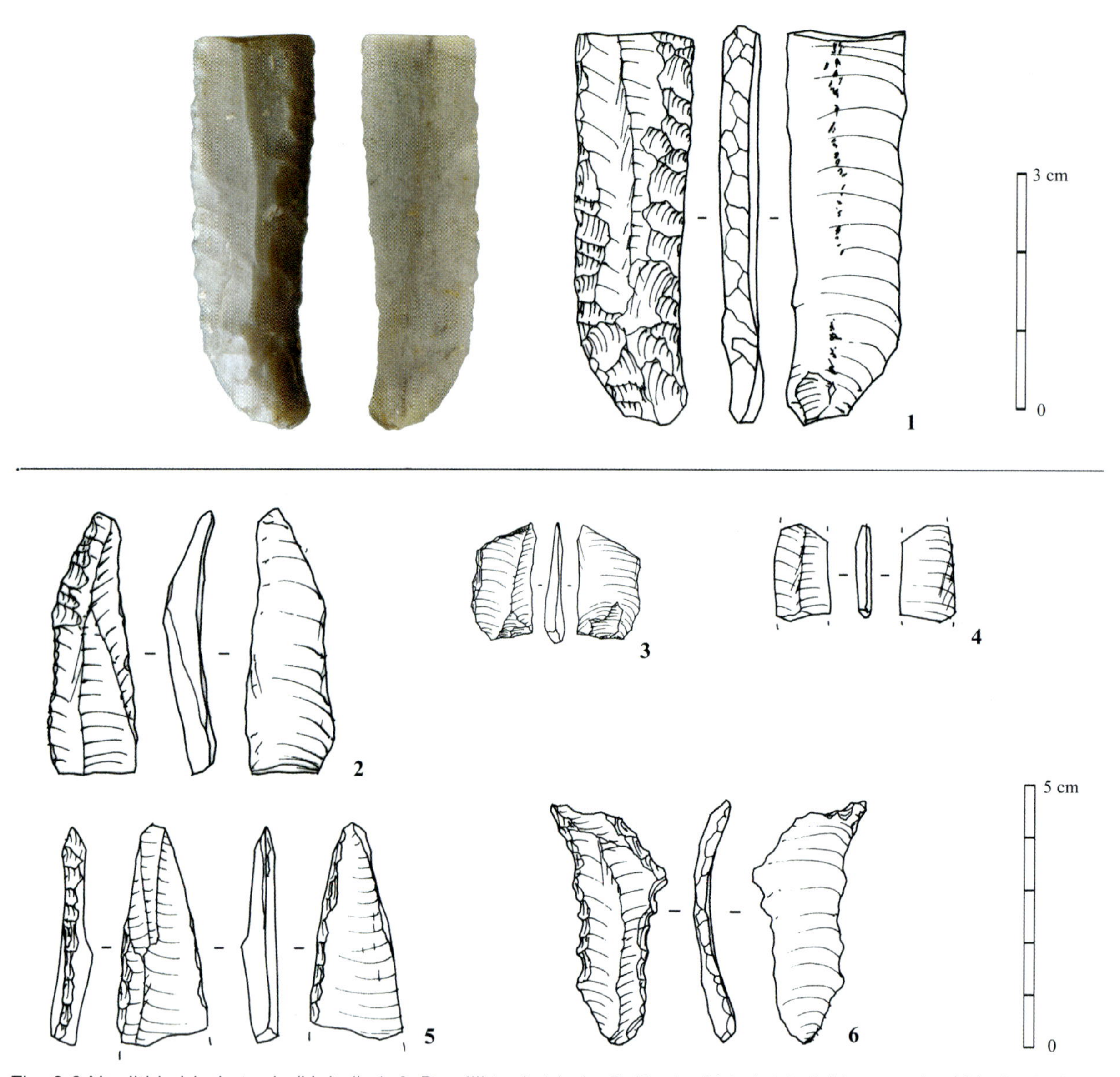

Fig. 8.6 Neolithic blade tools (Unit 4). 1, 2: Damjili tools blade; 3: Backed bladelet; 4: Unretouched blade; 5, 6: Borers. 1: Flint; 2–6: Obsidian.

8.3.3 Chalcolithic (Unit 3)

The Chalcolithic deposits of Damjili Cave are dated from the late fifth to early fourth millennium BC (Chapter 5 of this volume). They produced 332 flaked stone artifacts. As in the Bronze Age, the artifacts were mostly made of obsidian (263/332 or 79.2%; Table 8.1).

Obsidian

Most of the Chalcolithic obsidian artifacts belong to debitage pieces (191/263 or 72.6%; Table 8.2). Given this large amount of obsidian, it is worthy to note that it contains only three cores. Together with the occurrence of 11 core management pieces (Fig. 8.3: 3) and 58 retouched tools, this dataset collectively suggests a practice of intensive core reduction for obsidian raw material. As a matter of fact, two of the three obsidian cores are highly exhausted ones (Table 8.2).

The retouched tools include a range of standardized tools: a trapeze, a "Damjili tool" (Fig. 8.3: 5), two backed blades, a borer (Fig. 8.3: 4), a burin, two end-scrapers, two massive scrapers, and 14 splintered pieces (Table 8.3). The trapeze is probably an intrusion from the underlying Neolithic layer. The "Damjili tool" could also be an intrusive specimen, judging from its close typological similarities with the Neolithic Damjili tools and its distinct dissimilarity with other Chalcolithic tools that were predominantly made on flake blanks.

All four scrapers in the present assemblages are made on flake blanks. Also notable is the relatively common occurrence of splintered pieces, while burins, manufactured with a more or less comparable technology, are almost absent. This trend is comparable to that noted at the late phase of the Neolithic at Göytepe (Nishiaki 2020).

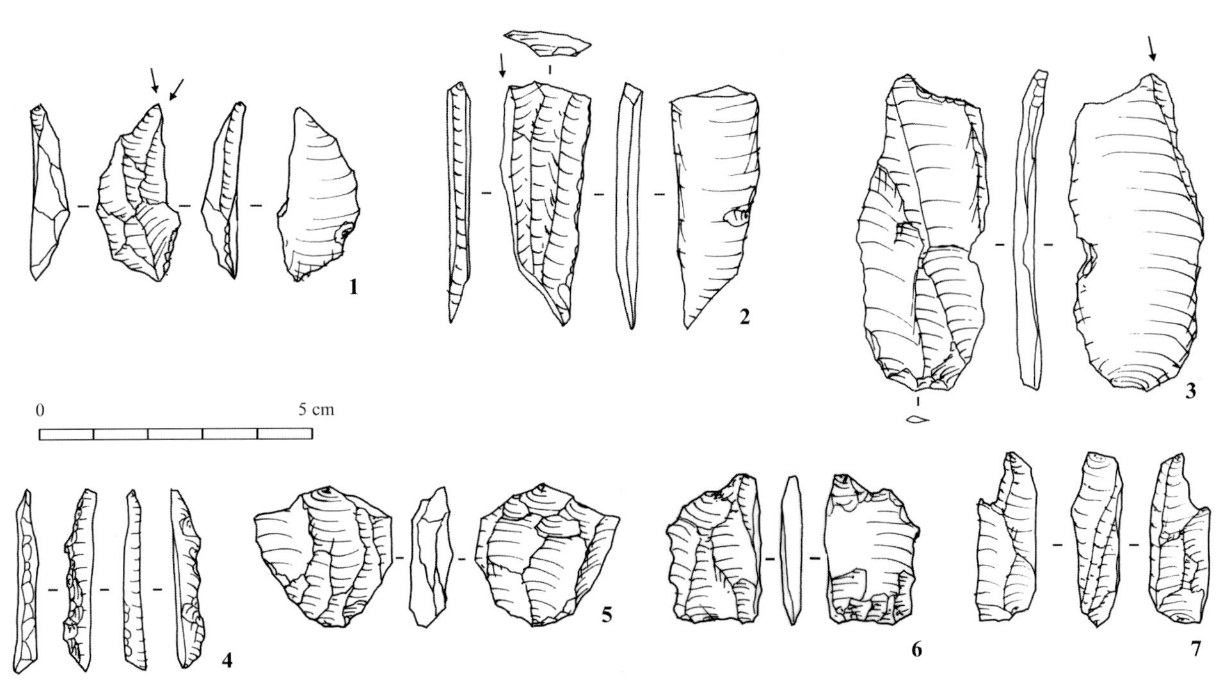

Fig. 8.7 Neolithic burins and splintered pieces (Unit 4). 1: Dihedral burin; 2: Angle burin; 3: Truncation burin; 4: Burin spall; 5–7: Splintered pieces. 1–7: Obsidian.

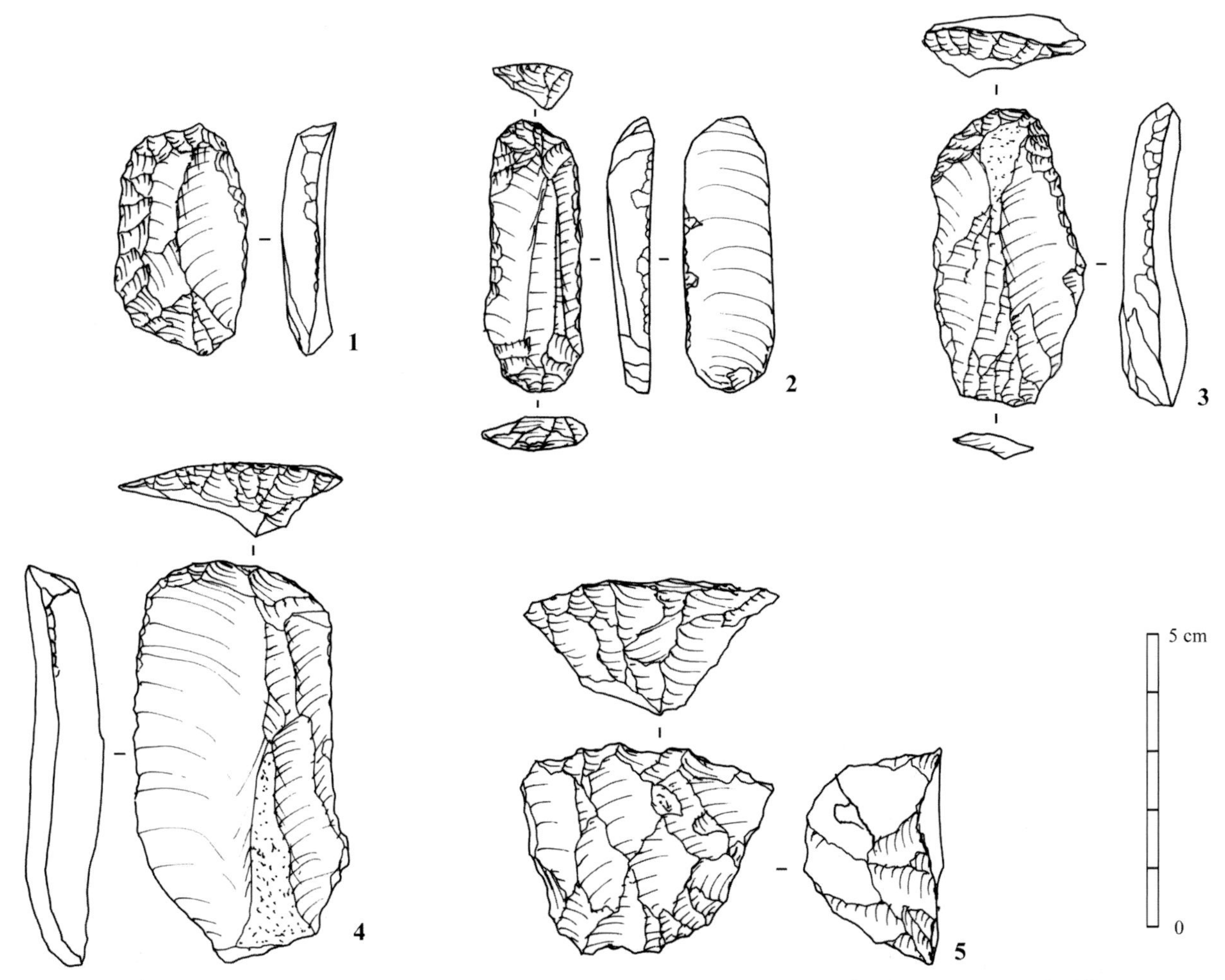

Fig. 8.8 Neolithic scrapers (Unit 4). 1–4: Endscrapers; 5: Round scraper. 1–4: Obsidian; 5: Non-obsidian.

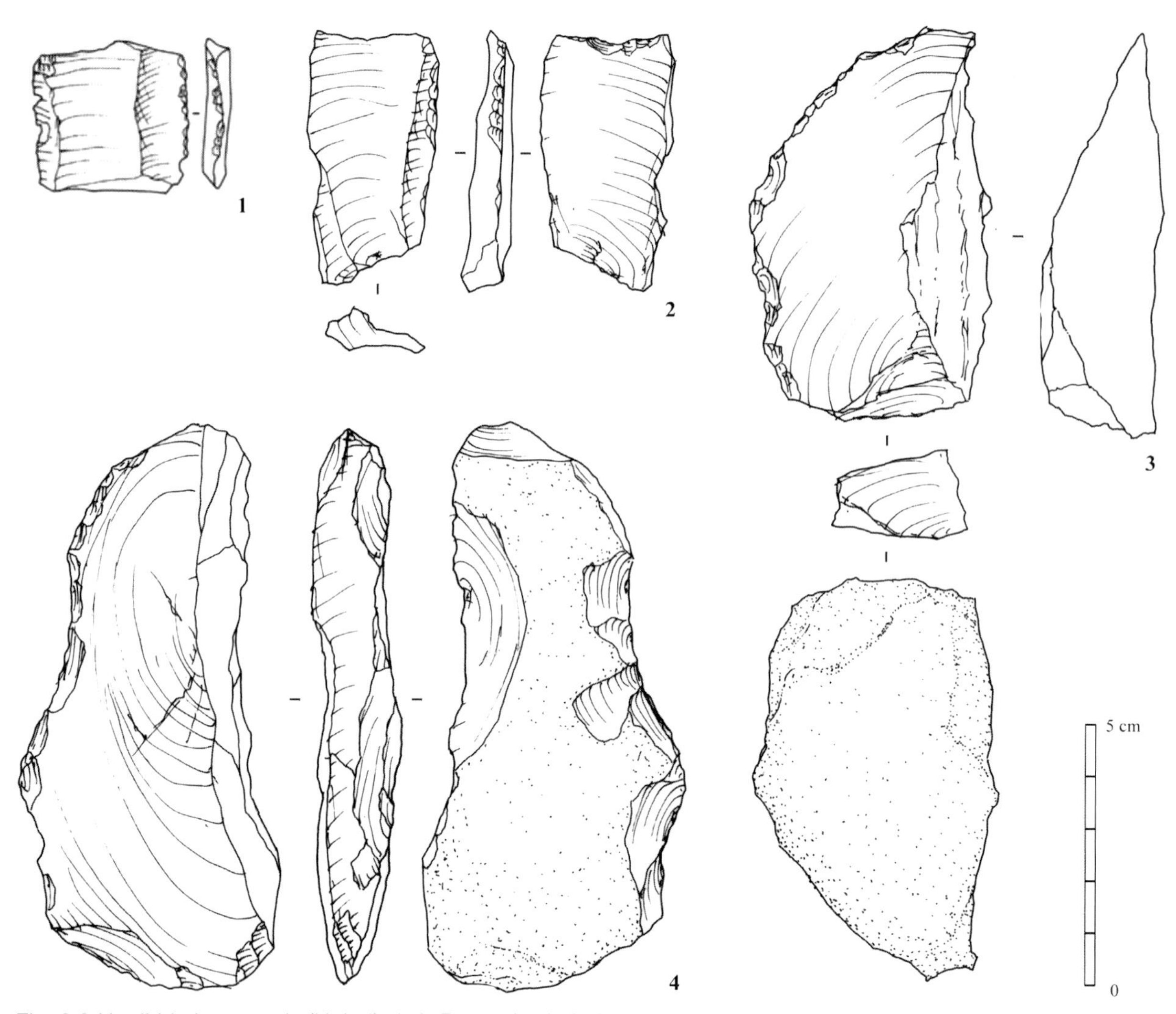

Fig. 8.9 Neolithic large tools (Unit 4). 1, 2: Retouched blades; 3, 4: Massive sidescrapers. 1: Obsidian; 2–4: Flint.

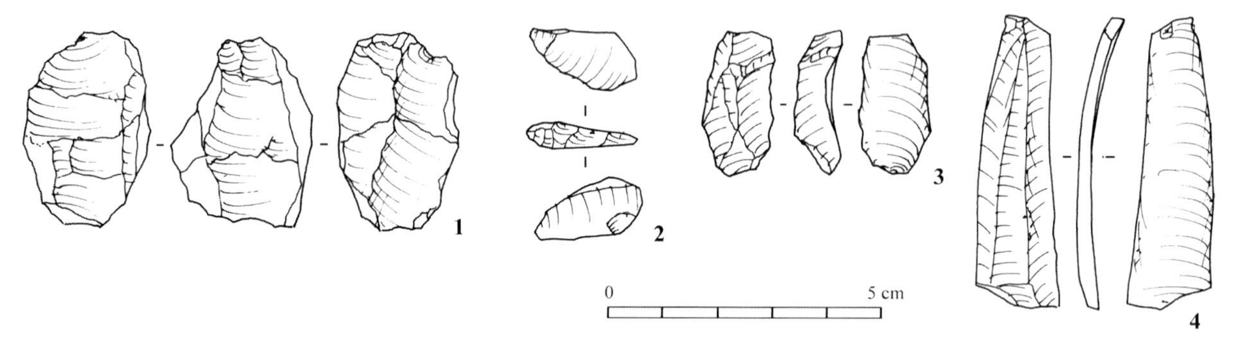

Fig. 8.10 Mesolithic core, core maintaining pieces, and debitage (Unit 5). 1: Blade core; 2: Core platform preparation flake; 3: Core edge blade; 4: Unretouched blade. 1–4: Obsidian.

Non-Obsidian

The rarity of cores characterizes the non-obsidian assemblage as well (Table. 8.4; 1/69 or 1.4%). Moreover, the single core from this unit is a non-characteristic exhausted piece on a small pebble (Fig. 8.3: 1). Retouched tools are also rare (Table 8.4; 3/69 or 4.3%). Therefore, the assemblage is dominated by debitage pieces, demonstrating a strong emphasis on flake production. The retouched tools include one Chokh type, which is very likely to be an intrusive piece from the earlier layers, and a sidescraper on a massive flake blank. The general techno-typological features of the non-obsidian assemblage are similar to those of the obsidian assemblage, indicating that these different raw materials were treated in a similar way.

8.3.4 Neolithic (Unit 4)

The Neolithic is the period when the eastern area

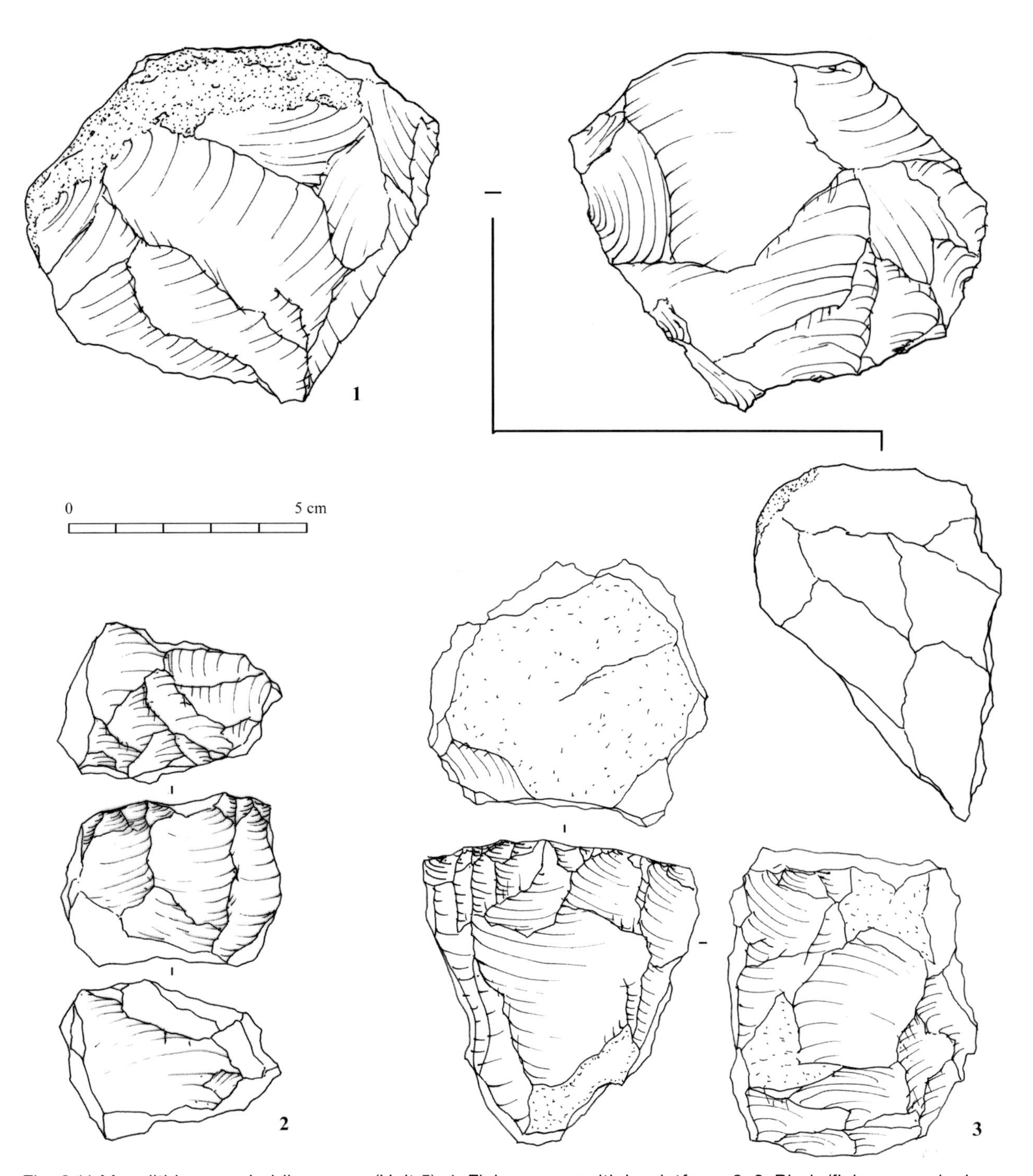

Fig. 8.11 Mesolithic non-obsidian cores (Unit 5). 1: Flake core, multiple-platform; 2, 3: Blade/flake core, single-platform.

of Damjili Cave was most densely occupied. Our investigations yielded the largest flaked stone assemblage in this period, consisting of 1153 obsidian (1153/1516 or 75.6%; Table 8.1) and 363 non-obsidian specimens. The assemblage is dated from the first three quarters of the sixth millennium BC (Chapter 5 of this volume), which have been divided into four subunits. It is described collectively in this section and its chronological changes according to the subunits are examined later.

Obsidian

The rare occurrence of obsidian cores noted in the Bronze and Chalcolithic units is observed even more strongly in the Neolithic unit (4/1153 or 0.3%; Table 8.2). The intensive reduction of obsidian cores is again suggested. The four cores include one blade core for pressure debitage (Fig. 8.4: 3). It is the only pressure blade core recovered at Damjili Cave, although the blades and bladelets of this cave sufficiently demonstrate a common use of pressure debitage. Conformingly, nearly a half of the obsidian debitage represent blade forms (195/419 or 46.5%).

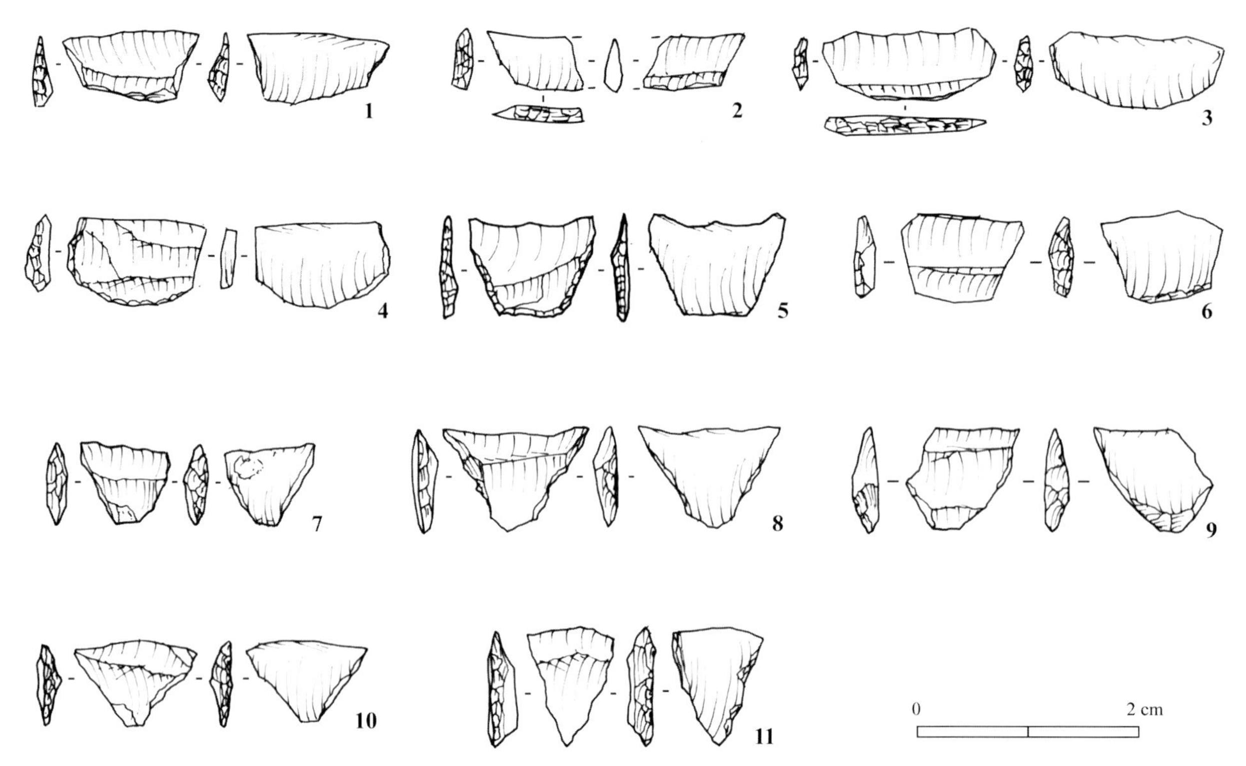

Fig. 8.12 Mesolithic geometrics (Unit 5). 1, 2: Lunates A; 3, 4: Linates B; 5, 6: Trapezes A; 7–9: Trapezes B; 10, 11: Triangles. 1, 5, 7, 8, 10, 11: Flint; 2–4, 6, 9: Obsidian.

The abundance of core management pieces (31 pieces) including carefully flaked striking-platform tablets and core refreshing flakes (Fig. 8.4: 4) suggests that obsidian cores were reduced to a considerable degree. It is also possible that the usable cores were taken to the next camp or settlement according to the Neolithic mobile system.

The retouched obsidian tools of this unit display the largest range of tool types at Damjili Cave (Table 8.3). They consist of 22 geometric tools, three backed pieces, two pressure-retouched pieces, four borers, 44 burins and their byproducts, five endscrapers, two massive sidescrapers, and 53 splintered pieces and their byproducts. The geometrics are comprised by one Type A lunates (Fig. 8.5: 1) and three Type B lunates (Fig. 8.5: 2–4), five Type A trapezes (Fig. 8.5: 5, 6) and 14 Type B trapezes (Fig. 8.5: 7–10), and two triangles (Fig. 8.5: 11, 12). The retouch exhibits a consistent retouch pattern: both ends were truncated by obverse retouches, the back was shaped by obverse (Fig. 8.5: 4), inverse (Fig. 8.5: 2, 3, 7, 10), or bifacial retouches (Fig. 8.5: 1, 8), or left unretouched (Fig. 8.5: 5, 6, 9). Those with a bifacially retouched back resemble Helwan lunates of the late Epipaleolithic.

Compared to the Chalcolithic obsidian tool assemblage, the Neolithic obsidian assemblage contain a larger number of tools made on blade blanks such as Damjili tools (Fig. 8.6: 1, 2), backed bladelets (Fig. 8.6: 3), and borers (Fig. 8.6: 5, 6). Also notable is a much higher frequency of burins and their byproducts in the Neolithic assemblages (Table 8.3; Fig. 8.7: 1–4). As mentioned elsewhere, the use of burin technology in the South Caucasus should be interpreted in relation to the employment of splintered piece technology (Fig. 8.7: 5–7; Nishiaki and Guliyev 2019). A difference from the Chalcolithic assemblage is also seen in the scraper typology: while scrapers were predominantly made on flake blanks in the Chalcolithic, the Neolithic scrapers were manufactured on as commonly as on blade blanks (4/8 or 50.0%; Fig. 8.8: 1–4).

Non-Obsidian

However, the 363 non-obsidian Neolithic flaked stone artifacts also contain only a few cores (3/363 or 0.8%). Non-obsidian cores were reduced to produce flake blanks. Blade blanks occur at only a small portion of the non-obsidian debitage assemblage (13/211 or 6.2%; Table 8.4), which sharply contrasts the very common production of blade blanks in the obsidian assemblage.

The non-obsidian Neolithic retouched pieces exhibit a more or less comparable range of tool types as noted for the obsidian assemblage. For example, the geometric tools occur in the non-obsidian assemblage (2/41 or 4.9%) at an almost similar frequency in the obsidian assemblage (22/330 or 6.7%). The burins' frequency is also comparable in obsidian (5/41 or 12.2%) and non-obsidian assemblages

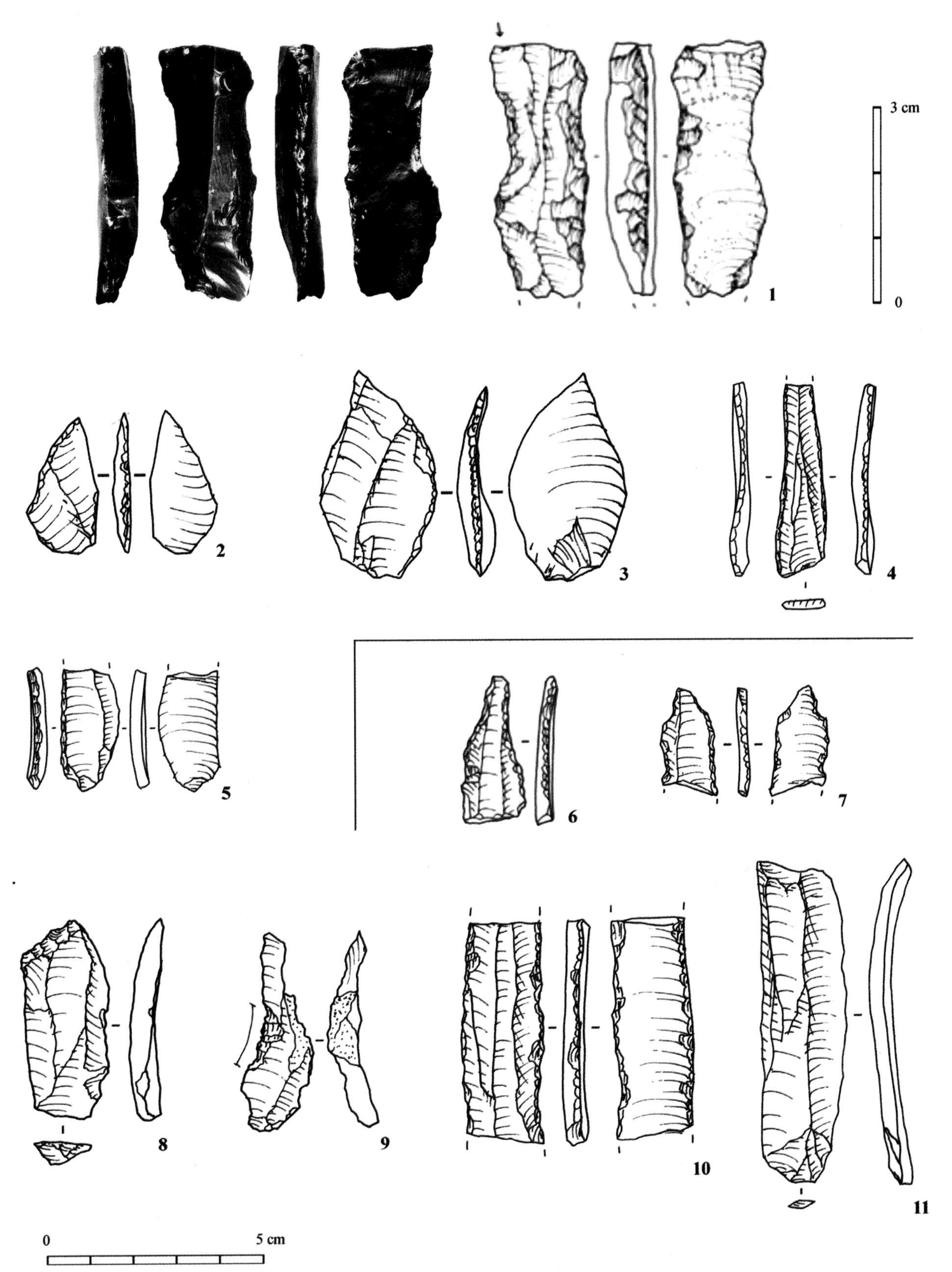

Fig. 8.13 Mesolithic blade tools (Unit 5). 1: Strangulated blade; 2, 3: Chokh points; 4, 5: Backed blades; 6, 7: Borers; 8: Truncated blade; 9: Notched blade; 10, 11: Retouched blades. 1, 4, 6, 7, 10, 11: Obsidian; 2, 3, 5, 8, 9: Flint.

(44/330 or 13.3%). On the other hand, there are tool groups that were almost exclusively made on specific raw materials. Splintered pieces were mostly manufactured on obsidian (53/330 or 16.1%) rather than on non-obsidian materials (1/41 or 2.4%). The difference indicates that the obsidian and non-obsidian raw materials were utilized differently from each other.

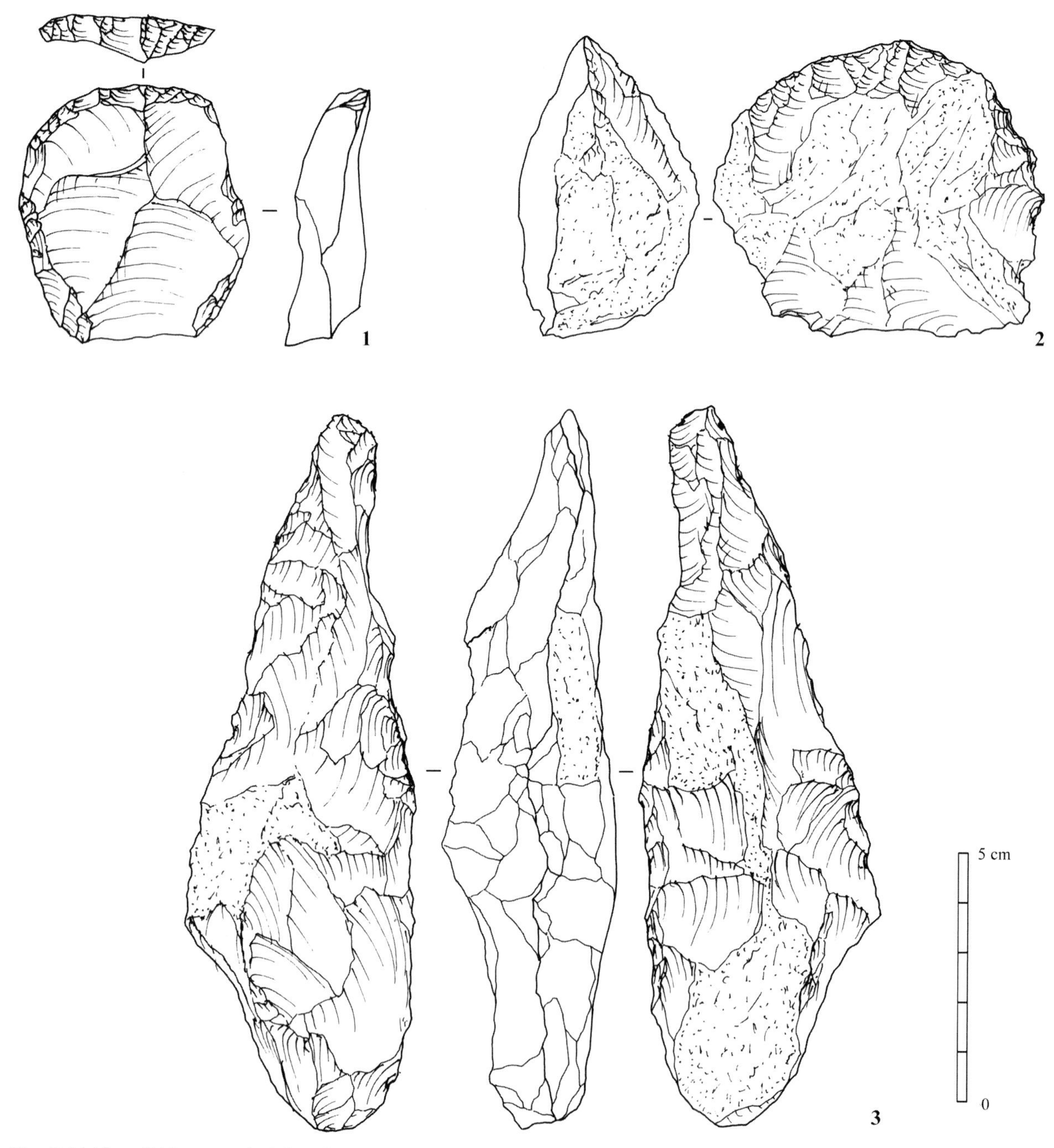

Fig. 8.14 Mesolithic non-obsidian flake and core tools (Unit 5). 1, 2: Endscrapers; 3: Pick.

8.3.5 Mesolithic (Unit 5)

The Mesolithic flaked stone artifacts constitute the second largest lithic assemblage by units at Damjili Cave, consisting of 1040 specimens (Table 8.1). As the Neolithic ones, they are described as a single assemblage here and are analyzed later by stratigraphic subunits. The Mesolithic assemblages show a significantly lower frequency in obsidian use: 30.1% (313/1040) in Unit 5, whereas over consistently more than 70% in Units 4 (Fig. 8.16).

Obsidian

The Mesolithic obsidian assemblage essentially resembles the Neolithic one in that cores are virtually absent (1/313 or 0.3%) and blades are most common in the unretouched blanks (58/114 or 50.9%; Table 8.2). Although the single core in the obsidian assemblage is an exhausted one, which shows blade removals by percussion (Fig. 8.10: 1): many of the accompanying blade blanks exhibit parallel edges and ridges as well as a straight longitudinal profile (Fig. 8.10: 4), suggesting the popular use of pressure debitage. The core management pieces also indicate pressure core reduction technology (Fig. 8.10: 2, 3; see Chapter 9 of this volume).

The most characteristic standardized tools of obsidian are geometric implements (18 specimens; Fig. 8.12). The technology for their manufacturing

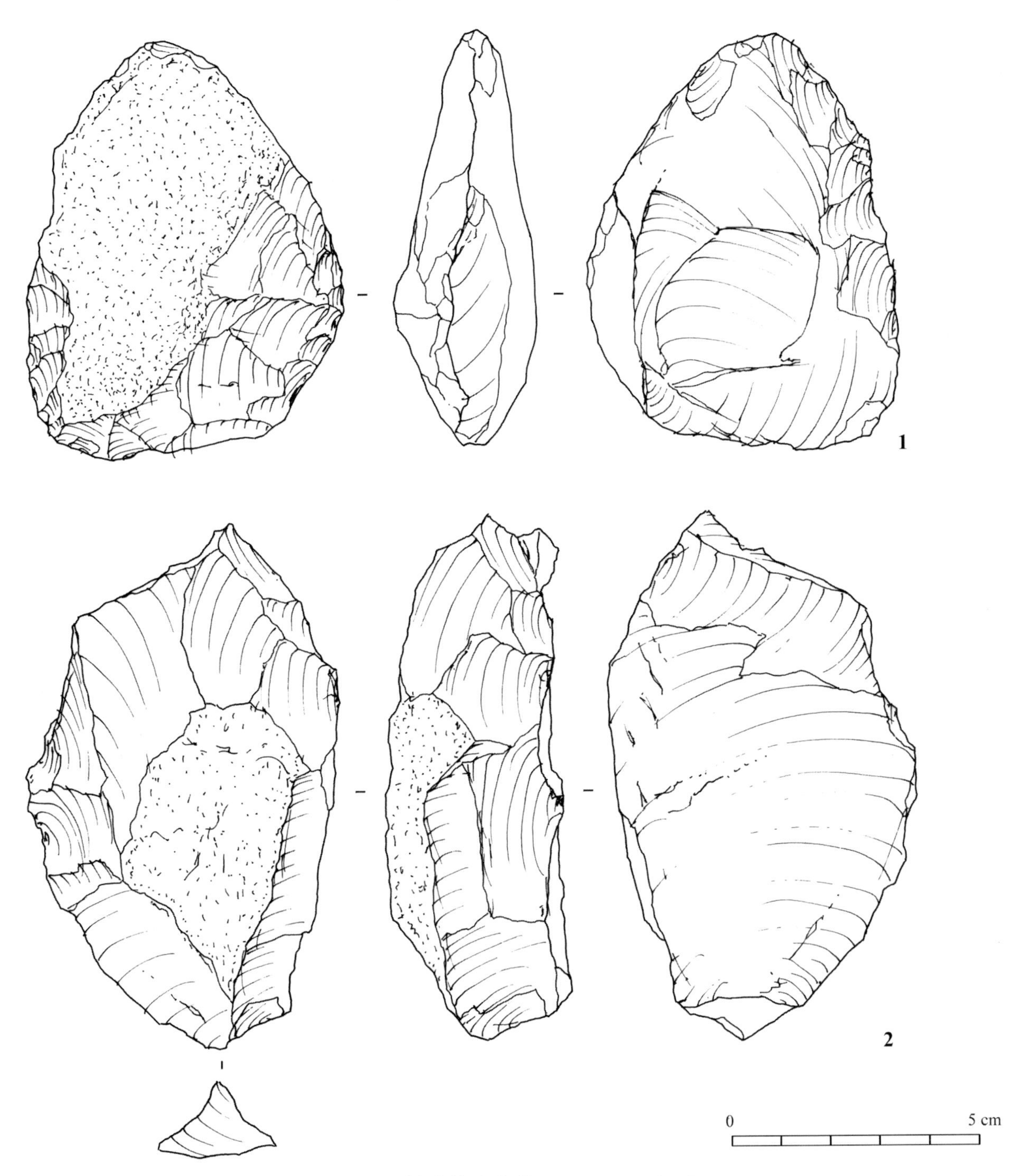

Fig. 8.15 Mesolithic non-obsidian scrapers (Unit 5). 1, 2: Scrapers, massive type.

is quite similar to that of the Neolithic geometrics: obverse retouch was applied to the ends, and inverse or bifacial one to the back.

The obsidian blade tools include a strangulated blade, which may be comparable to the Kmlo tool (Fig. 8.13: 1). Although termed under a separate name, this tool type is virtually indistinguishable from the Çayönü tools common in the Pre-Pottery Neolithic and early Pottery Neolithic of Upper Mesopotamia (Nishiaki et al. 2022). It is characterized by steep obverse pressure retouches applied to the ventral surface often associated with scratches running along the lateral edge. The specimen in question does not display these distinguishing traits. Nevertheless, the steep retouch on this specimen may be related to a similar concept of retouching. However, the irregular retouching on both lateral edges does not allow us to assign it either Kmlo or Çayönü tool.

The other obsidian tools are generally similar to the Neolithic ones, including two borers (Fig. 8.13: 6, 7), nine burins and their byproducts, two endscrapers on blades, three endscrapers on flakes, one massive sidescraper, and two splintered pieces.

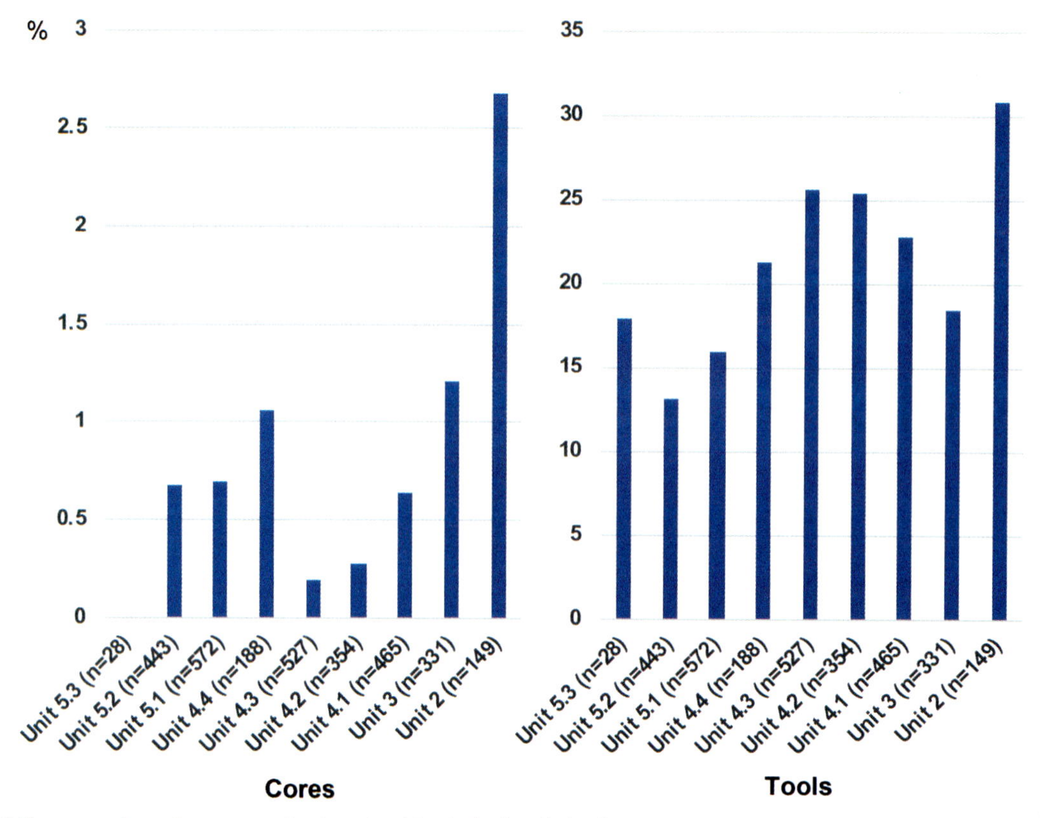

Fig. 8.16 Frequencies of cores and retouched tools in the flaked stone assemblages from Damjili Cave by stratigraphic units. 1: Core frequency; 2: Tool frequency.

Non-Obsidian

The Mesolithic non-obsidian assemblage also includes a small number of cores (6/727 or 0.8%). They predominantly exhibit flake production traces (Fig. 8.11). However, the recovered debitage includes a certain amount of blades (54/321 or 16.8%). Given this, the flake cores with a single-platform (Fig. 8.11: 2, 3) may have been used for blade production in their earlier stage of reduction.

Like the obsidian tool assemblage, the non-obsidian tool assemblage is also characterized by a high frequency of geometric microliths (12 pieces; Fig. 8.12). They include triangles (Fig. 8.12: 10, 11), which were not made on obsidian. Among the other standardized tools, robust tools made on core and thick flake blanks are noticeable. The core tools consist of four choppers and one pick (Fig. 8.14: 3). Six sidescrapers (Fig. 8.15) and four endscrapers on thick flakes (Fig. 8.14: 1, 2) are also diagnostic of the Mesolithic assemblage, which contains only one endscraper on blade.

8.3.6 "Sterile" (Unit 6)

The yellowish gray clay sediments below Unit 5 are archaeologically sterile in principle. However, its top part yielded a small number of obsidian (two pieces) and non-obsidian flaked stone artifacts (16 pieces; Table 8.1), some of which exhibited Middle Paleolithic features. As the lowest part of Unit 5 (Unit 5.3) also yielded comparable artifacts (Chapter 7 of this volume), we consider that this part of the deposits of Damjili Cave to represent secondary materials. There seems to have been a considerable stratigraphic disturbance at the beginning of the deposition of Unit 5.

8.4 Discussion

8.4.1 Occupational intensity

Our field investigations in 2016–2022 revealed stratified lithic assemblages of the Mesolithic to the Bronze Age. It is a unique dataset allowing us to examine the early-mid Holocene cultural evolution from a lithic perspective. At the same time, the dataset provides an opportunity to understand the role of the Damjili Cave in a prehistoric residential system in the South Caucasus.

The ratios of cores to debitage or tools to debitage are often regarded as useful indicators to estimate the occupational intensity. According to the model developed in the anthropological studies of lithic industry (e.g., Odell 2003; Nishiaki et al. 2012), the less cores and more tools in relation to debitage, a stronger mobility is surmised because the core reduction was likely to have been carried out on-site by mobile communities at a minimum scale due to their frequent mobility. They probably took out the usable cores to the next camp.

For comparison, referencing to the lithic data obtained from the early Neolithic levels of Hacı Elamxanlı Tepe, where a comparable excavation method (dry-sieving) was employed for sample recovery, is useful. It shows the cores at 1.5% and tools at 18.5% (Kadowaki 2021: 63). The data from the Mesolithic, Neolithic, and Chalcolithic layers of Damjili Cave shows a lower core frequency (0.5 to 1.0%) and a higher one (15 to 25%) for retouched tools (Fig. 8.17) consistently than at Hacı Elamxanlı Tepe, suggesting a stronger mobility model. However, the patterns seem to have changed through time. They may indicate that the mobility became even more intensive in the later period especially in the Bronze Age. This finding is worthy of further investigation in the future study.

8.4.2 Chronological changes of the lithic industries

Raw material use

The raw material use for lithic artifacts at Damjili Cave experienced a chronological change. Impressive is the changing patterns of the use of obsidian, whose sources should be located to the south or west in the South Caucasus Mountains, at least 100 km away from the Damjili Cave (see Nishiaki et al. 2019b). The exploitation of obsidian was so common from the Neolithic onwards: its proportion remained almost the same until the Bronze Age, about 70 to 80% of the entire assemblages (Fig. 8.16). The high frequency of obsidian artifacts is the phenomenon also noted at Neolithic settlements in the plain (Nishiaki 2020; Kadowaki 2021). It suggests that the use of the imported obsidian was established in the Neolithic as a norm maintained in the later prehistoric periods.

Yet, what intrigues us is that the rapid change of obsidian use occurred during the late Mesolithic period. Its proportion jumped from some 20% to more than 50% within in the latest period of Unit 5 (5.1). It implies that the common use of obsidian in the Neolithic (Unit 4) was not introduced along with the Neolithic socioeconomy. Instead, there was some sort of significant socioeconomic transformation during the Mesolithic at the end of the seventh millennium BC.

Blank production technology

The Mesolithic core assemblages of Damjili Cave exhibit a strong emphasis on blade production for tool blanks. The resultant debitage assemblages show a high proportion of blade blanks of roughly 20 to 35% (Fig. 8.18). The technology for blade production is principally based on pressure debitage. The common use of pressure debitage in the Neolithic has been well known at the Neolithic settlements in the South Caucasus, as demonstrated by a fracture wing analysis of the physio-mechanical traits of the blade removal (Takakura and Nishiaki 2020). The analysis of the blade blanks from Damjili Cave Unit 5 using the same methodology has demonstrated that the pressure debitage was already a common practice in the Mesolithic period (Chapter 9 of this volume). In other words, the pressure debitage was a part of the local core reduction technology at Damjili Cave before the introduction of the Neolithic.

The stratigraphic data reveal significant chronological patterns. One is a more common production of blade blanks in the Neolithic than in the earlier

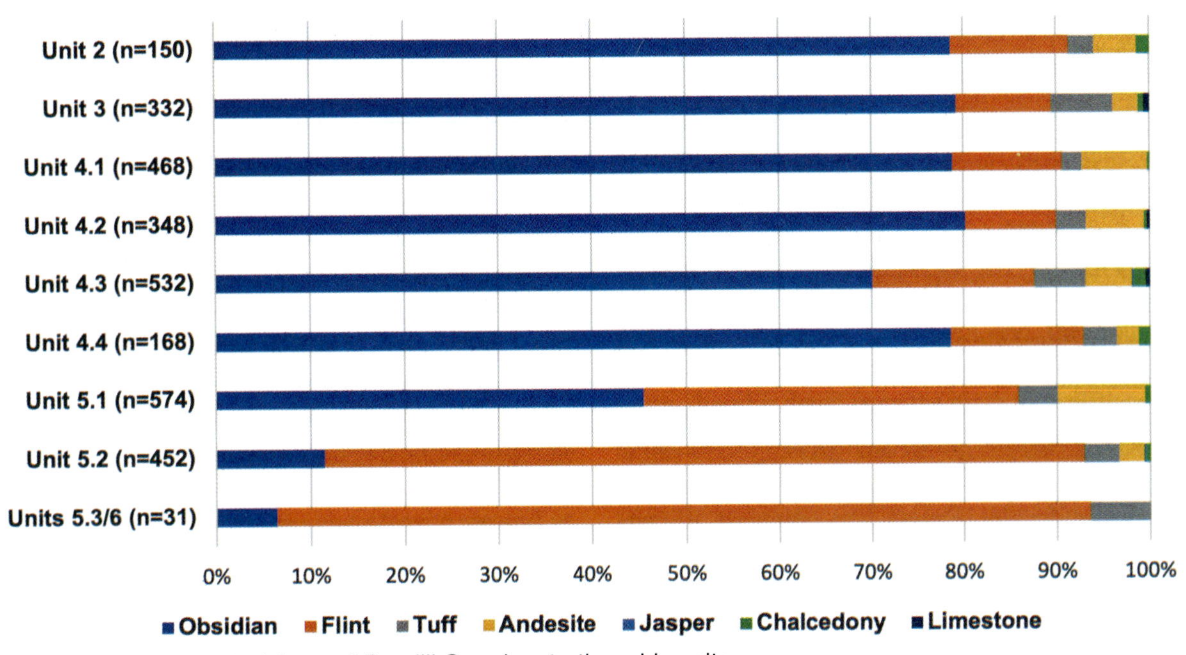

Fig. 8.17 Lithic raw material use at Damjili Cave by stratigraphic units.

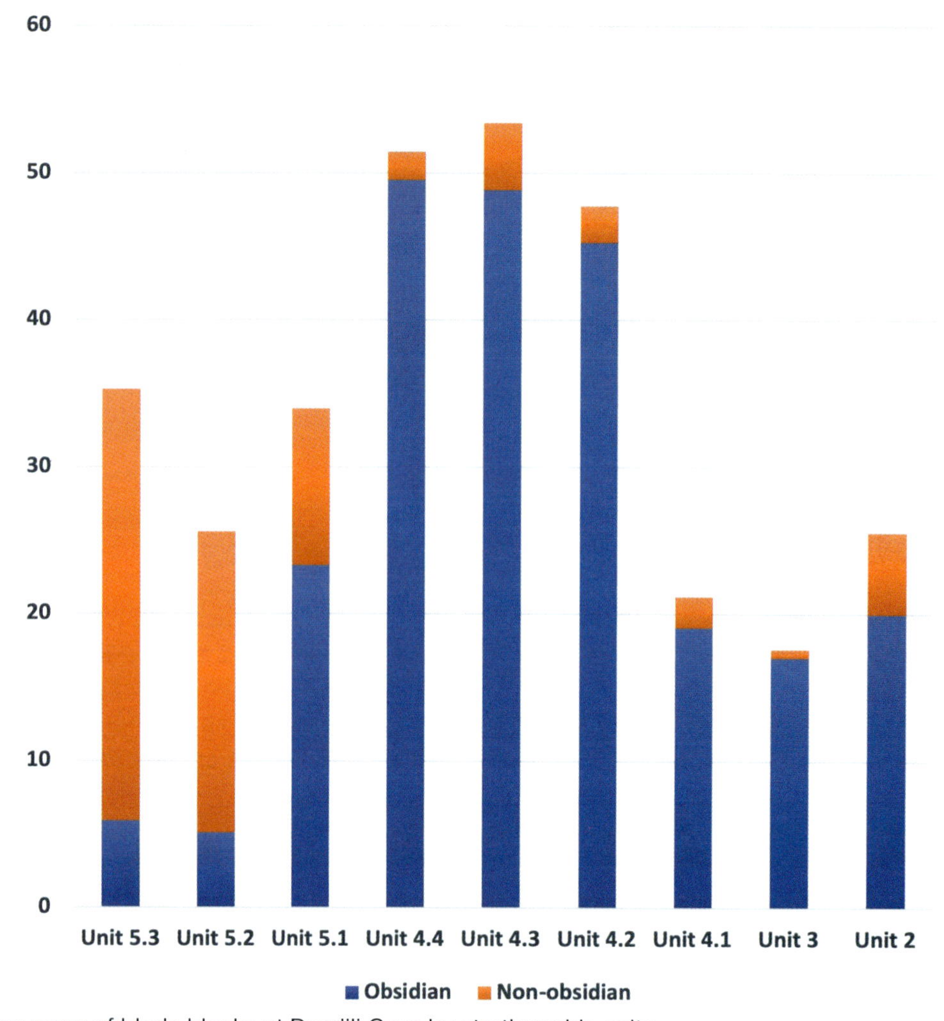

Fig. 8.18 Frequency of blade blanks at Damjili Cave by stratigraphic units.

and later periods. The frequency of blade blanks jumped to over 50% at the beginning of the Neolithic and dropped to approximately 20% at its last phase (Fig. 8.18). A chronological change is also seen in the raw materials for blade production: the more common use of flint in the Mesolithic was replaced by the use of obsidian in the Neolithic. Importantly, these changes in blade production and obsidian use did not occur at the beginning of the Neolithic and Chalcolithic periods. Instead, the changes were already foreshadowed in the late Mesolithic level (Unit 5.1) and the latest Neolithic level (Unit 4.1). These findings strongly suggest a continuity over these periods.

Additionally, we should also pay attention to the chronological change of the obsidian blade widths, which indicate the production of wider blades in the Neolithic period and later (Fig. 8.19). The increase of blades wider than 20 mm is noteworthy. Although the obsidian blade production was carried out mainly with the use of pressure debitage since the Mesolithic (Chapter 9 of this volume), the wide blade production in the Neolithic afterwards is likely to signify an introduction of new core reduction technology, namely, the employment of lever pressure (Ikeyama et al. 2022).

Tool typology

The tool assemblages of the Mesolithic to Bronze Age at Damjili Cave display a couple of specific correspondences between tool and raw material types (Fig. 8.20: 1). One is that burins and splintered pieces were almost exclusively manufactured on obsidian. The tie with obsidian is even stronger for manufacturing splintered pieces. Also distinct is the more common production of scrapers and heavy duty tools on non-obsidian raw materials. These contrasting use of raw materials for specific tool types likely reflects a selective decision to make the best use of the available rocks with distinct physical properties.

However, in general, the typological differences between the raw material groups seem rather small when compared to the contemporaneous seventh to sixth millennium BC lithic assemblages of Upper Mesopotamia. At the latter settlements of North

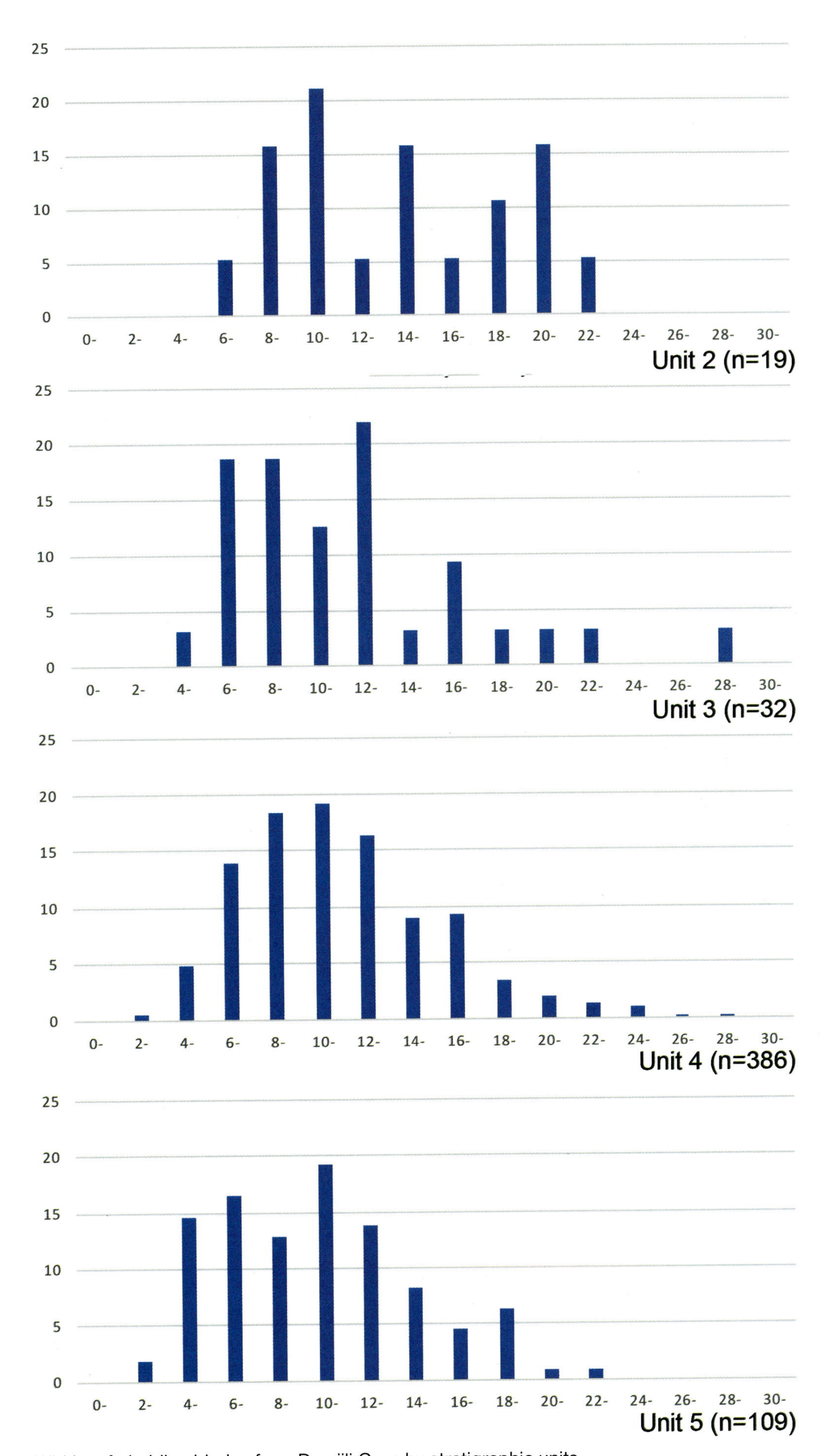

Fig. 8.19 Widths of obsidian blades from Damjili Cave by stratigraphic units.

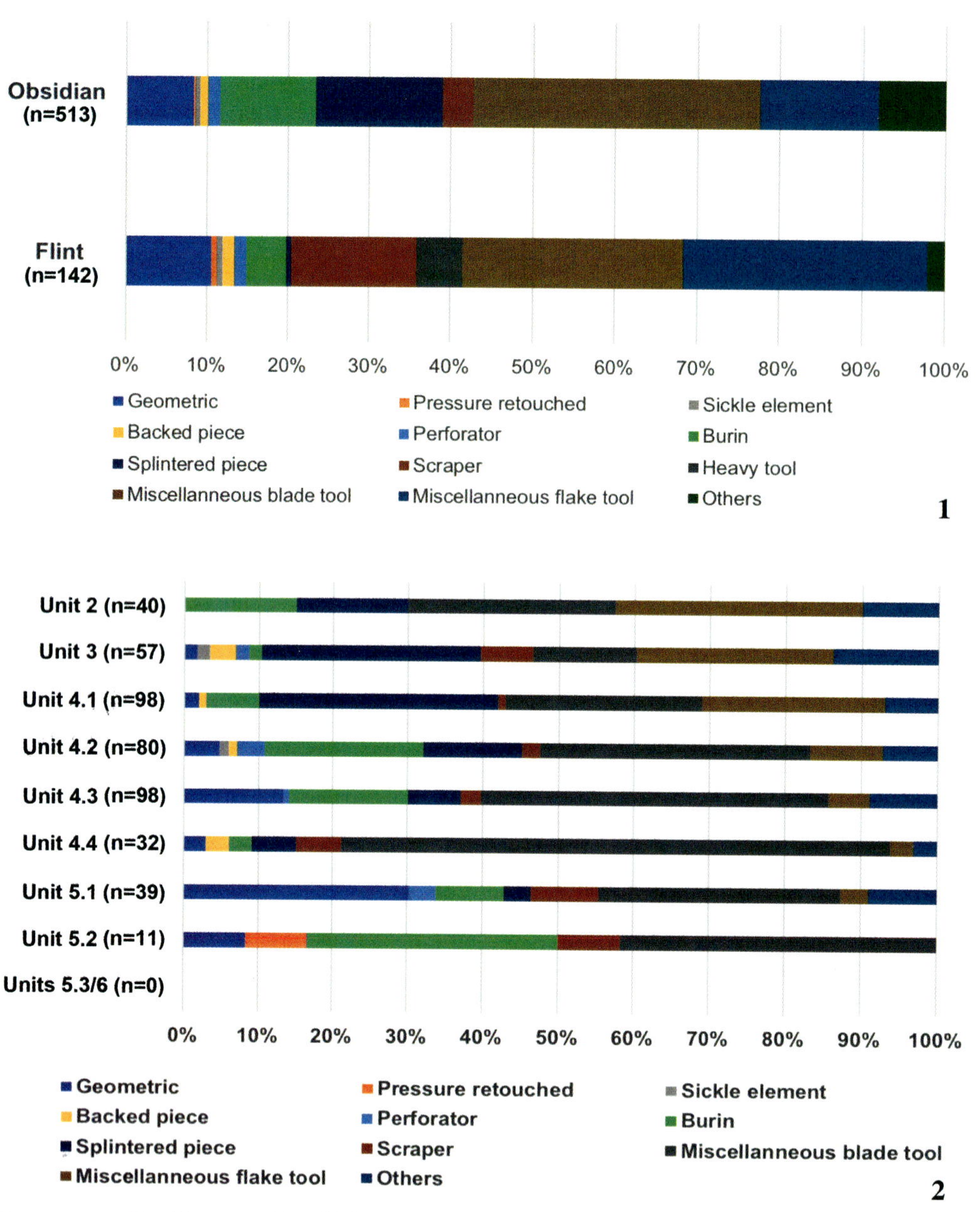

Fig. 8.20 Frequency of tool types at Damjili Cave by raw materials and stratigraphic units. 1: Comparison by raw material types; 2: Comparison by stratigraphic units.

Iraq and Syria, obsidian and local non-obsidian materials were dealt with in a totally different way. For instance, obsidian was firmly associated with particular tool types such as corner-thinned blades, side-blow blade-flakes, and Çayönü tools, which were virtually never manufactured on non-obsidian materials (e.g. Nishiaki 1993, 2000). This way of the obsidian use in Upper Mesopotamia is a probably reflection of the site locations away from the obsidian sources, at least 100 to 200 km to the south. The case at Damjili Cave supports our previous study on the anomaly consumption patterns of obsidian irrelevant to the distance of the original outcrops (Nishiaki et al. 2019b). Obsidian originated from sources more than 300 km away and less than 100 km from the settlement of Göytepe was treated in a similar way (Nishiaki et al. 2019b). Although we have not determined the obsidian sources exploited by the Damjili communities, the approaches to obsidian consumption in the South Caucasus appear to be very different from those known in the Fertile Crescent of Southwest Asia.

In terms of the chronological changes in tool typology, we see at least three important changes at Damjili Cave (Fig. 8.20: 2). Firstly, geometrics, so popular in the Mesolithic, sharply decreased in the Neolithic. This is quite understandable given the significant subsistence changes occurring in the Neolithization processes, especially in relation to abandoning hunting practices (Chapter 15 of this volume). Secondly, the proportion of burin manufacturing also became unpopular at the expense of the increasing splintered pieces in the Neolithic (Fig. 8.20: 2). This pattern has been already noted in the stratified Neolithic assemblages from Göytepe (Nishiaki et al. 2019a, 2020). Our traceological analysis of the Göytepe materials suggests that both tool groups were used mainly for hide-working (Takase 2020). If so, the same tasks appear to have been carried at both plain settlements and Damjili Cave. Thirdly, the use of flake tools significantly became common also in the late Neolithic. The trend continued into the Chalcolithic and Bronze Age.

8.5 Conclusions

The lithic data from stratified Mesolithic-Bronze Age deposits at Damjili Cave provided us with a valuable opportunity to examine the cultural evolution over these periods, which had not been available in the South Caucasus. The strongest point of our data lies in that the lithic assemblages were collected from radiometrically dated stratigraphic contexts. The lithic assemblages reported in this chapter stand as the first dataset unavailable in the South Caucasus.

Among a number of important insights obtained from the present analysis, two findings are worthy to be emphasized. One is that the lithic assemblages indicate the use of this cave as an ephemeral occupation spot throughout the periods of the Mesolithic to the Bronze Age. The rare occurrences of cores and retouched tools are the good indicators for this interpretation. The very small number of cores throughout the sequence suggests that the principal way of the cave use did not change over millennia. The usable cores were probably taken to the next camp or the home base by the communities with a more mobile settlements system. The second important finding is concerned with the chronological changes. Significant changes in lithic industries occurred in the late Mesolithic (Unit 5.1) and the late Neolithic (Unit 4.1). The techno-typological features diagnostic of those periods were fore-shadowed in the previous periods. The data from Damjili Cave indicates a continuity from the Mesolithic to Neolithic and the Neolithic to Chalcolithic. These findings are new to the prehistory of the South Caucasus. The data of the Damjili Cave deserves further analysis to understand the cultural changes in the important periods of the Mesolithic to Bronze Age from a lithic perspective (Chapter 17 of this volume).

References

Arimura, M., K. Martirosyan-Olshansky, A. Petrosyan, and B. Gasparyan (2021) Exploring the changes in lithic industries during the Neolithisation in Armenia (7th to 6th millennium BCE): A comparison of chipped stone tools from Lernagog-1 and Masis Blur. In: *Tracking the Neolithic of the Near East*, edited by Y. Nishiaki, O. Maeda, and M. Arimura, pp. 489–502. Leiden: Sidestone Press.

Ikeyama, F., F. Guliyev, and Y. Nishiaki (2022) Variability in Obsidian Pressure Blade Technology of the Neolithic Southern Caucasus: New Data from Göytepe and Hacı Elamxanlı Tepe, Azerbaijan. *Orient* 57: 125–143.

Kadowaki, S. (2021) Neolithic chipped stone artifacts from Hacı Elamxanlı Tepe. In: *Hacı Elamxanlı Tepe – The Archaeological Investigations of an Early Neolithic Settlement in West Azerbaijan*, edited by Y. Nishiaki, F. Guliyev, and S. Kadowaki, pp. 59–96. Berlin: ex oriente.

Nishiaki, Y. (1993) Anatolian obsidian and the Neolithic obsidian industries of North Syria: a preliminary review. In: *Essays on Anatolian Archaeology*, edited by H.I.H. Prince Takahito Mikasa, pp. 140–160. Harrassowitz Verlag, Wiesbaden.

Nishiaki, Y. (2000) *Lithic Technology of Neolithic Syria*. Oxford: Archaeopress.

Nishiaki, Y. (2020) Neolithic flaked stone industry of Göytepe. In: *Göytepe – The Neolithic Excavations in the Middle Kura Valley, Azerbaijan*, edited by Y. Nishiaki and F. Guliyev, pp. 169–190. Oxford: Archaeopress.

Nishiaki, Y. and F. Guliyev (2019) Neolithic lithic industries of the Southern Caucasus: Göytepe and Hacı Elamxanlı Tepe, West Azerbaijan (Early 6th Millennium BC). In: *Near Eastern Lithics on the Move: Interaction and Contexts in Neolithic Traditions*, edited by L. Astruc, F. Briois, C. McCartney, and L. Kassianidou, pp. 471–483. Nicosia: Astrom Editions.

Nishiaki, Y., Y. Kanjo, S. Muhesen, and T. Akazawa (2012) Temporal variability of Late Levantine Mousterian assemblages from Dederiyeh Cave, Syria. Eurasian Prehistory 9(1/2): 3–27.

Nishiaki, Y., A. Zeynalov, M. Mansrov, C. Akashi, S. Arai, K. Shimogama, and F. Guliyev (2019a) The Mesolithic-Neolithic Interface in the Southern Caucasus: 2016–2017 Excavations at Damjili Cave, West Azerbaijan. *Archaeological Research in Asia* 19: 100140.

Nishiaki, Y. O. Maeda, T. Kannari, M. Nagai, E. Healey, F. Guliyev, and S. Campbell (2019b) Obsidian provenance analyses at Göytepe, Azerbaijan: Implications for understanding Neolithic socioeconomies in the Southern Caucasus. *Archaeometry* 61(4): 765–782.

Nishiaki, Y., A. Zeynalov, M. Munsrov, and F. Guliyev (2022) Radiocarbon chronology of the Mesolithic-Neolithic sequence at Damjili Cave, Azerbaijan, Southern Caucasus. *Radiocarbon* 64(2): 309–322.

Odell, G. (2003) *Lithic Analysis*. New York: Springer.

Takakura, J. and Y. Nishiaki (2020) Fracture wing analysis for identification of obsidian blank production techniques at Göytepe. In: *Göytepe – The Neolithic Excavations in the Middle Kura Valley, Azerbaijan*, edited by Y. Nishiaki and F. Guliyev, pp. 209–221. Oxford: Archaeopress.

Takase, K. (2020) Use-wear analysis of chipped stone artifacts from Göytepe. In: *Göytepe – The Neolithic Excavations in the Middle Kura Valley, Azerbaijan*, edited by Y. Nishiaki and F. Guliyev, pp. 191–208. Oxford: Archaeopress.

Chapter 9

Mesolithic obsidian blade production technology at Damjili Cave

Fumika Ikeyama

9.1 Introduction

Pressure blades/bladelets are widely acknowledged as a fundamental part of obsidian lithic assemblages in the Neolithic South Caucasus. In contrast, studies on the details of Mesolithic blade production technology in the South Caucasus are relatively scarce, partly due to the rarity of securely dated sites of this time period; however, a few sites are now known. For example, recent excavations at Lernagog and Apnagyugh-8 (previously referred to as Kmlo-2) in Ararat Plain reportedly revealed the employment of pressure detachment in obsidian blade production during the Mesolithic Period (Arimura et al. 2018, 2022). Damjili Cave, the only Mesolithic site in the Middle Kura Valley with secure stratigraphic and radiocarbon dates, is a unique addition to the Mesolithic dataset of the South Caucasus. This is because the Mesolithic layers at this cave date back to the late seventh millennium BC, immediately before the advent of the Neolithic Period.

The preliminary report mentioned the application of pressure detachment in obsidian blade production in the Mesolithic assemblage at Damjili Cave based on technological observations of lithic artifacts with the naked eye (Nishiaki et al. 2019). This study aims to evaluate this interpretation with the verifiable method of fracture wing analysis, which has been employed to demonstrate the presence of pressure detachment in Göytepe and Hacı Elamxanlı Tepe assemblages in previous studies (Ikeyama et al. 2021; Takakura and Nishiaki 2020).

9.2 Methods

A fracture wing (FW) is a distinctive V-shaped micro-trace left after detachment and is found in flake scars on vitreous materials such as obsidian. The relationship between different detachment techniques and the crack velocity calculated from the divergence angles between the wings has been examined through experimental studies (e.g., Hutchings 1999; Takakura and Izuho 2004; also see Takakura and Nishiaki 2020). Takakura and Izuho (2004) conducted a series of comprehensive replication studies and established a method for systematically identifying detachment techniques by analyzing FWs. Their results showed that the crack velocity obtained from replicated samples consisted of at least three groups representing different detachment techniques: I) pressure, II) direct percussion with soft hammers and indirect percussion, and III) direct percussion with hard hammers. Based on their replication experiments, Takakura (2007) proposed the following ranges of crack velocity corresponding to each group:

- Group I (pressure): < 620 m/s
- Group II (direct percussion with soft hammers or indirect percussion): 460–930 m/s
- Group III (direct percussion with hard hammers): > 760 m/s

As indicated above, the crack velocities of the different groups overlapped: Groups I and II (460–620 m/s) and Groups II and III (760–930 m/s), and thus multiple FWs should be considered together to identify the applied knapping techniques for each specimen. These overlaps likely indicate a number of factors to determine the crack velocity, including not only knapping technique but also the hardness of the hammer and potentially the strength of knapping, to mention a few. Therefore, the crack velocity calculated by FW analysis cannot be considered an absolute proxy for identifying knapping techniques.

Nevertheless, previous studies have demonstrated a statistically significant correlation between FW velocity and knapping technique. This is particularly important for our study, which is concerned with objectively and verifiably distinguishing blades/bladelets produced by pressure detachment from those produced by other detachment techniques. Furthermore, the FW approach can identify the applied knapping techniques for each specimen or scar on a core, enabling a quantitative evaluation of the selection of detachment techniques within an assemblage. It is also worth mentioning that this analytical method allows the identification of detachment techniques regardless of the preservation status of each sample; conventional techno-morphological approaches are often not applicable to fragmentary samples. However, FW analysis is applicable even to tiny flakes and core fragments as long as the surface of the scars has not weathered too much. We employed this method of analysis to determine whether the Mesolithic knappers at Damjili Cave employed pressure debitage for core reduction.

9.3 Materials

We analyzed all obsidian blades/bladelets and tools on blades/bladelets excavated in the 2019 season at Damjili Cave. The Mesolithic specimens referred to in this chapter constitute 36 samples: 33 from Unit 5.1 and three from Unit 5.2. Owing to the limited time and facilities available in the field laboratory, the sample surfaces were replicated with silicon impressions and observed with a metallographic microscope at our laboratory in Japan. An optimal effort was made to locate two or more measurable FWs on each specimen.

Many samples showed surfaces unsuitable for FW analysis compared to previously studied materials at other sites such as Göytepe and Hacı Elamxanlı Tepe (Ikeyama et al. 2021). The obstacles included heavy surface weathering and trampling in a sample bag after excavations, severely preventing the measurement of FWs. For some of the samples, FWs were difficult to locate on the surface because of the scarcity of those that were clear enough for measurement. Those samples were categorized as "unmeasurable" (Table 9.1). In addition, based on the analysis of multiple FW crack velocities of a specimen, the results were assigned as follows. For example, if FW crack velocities were found only within the range of Group I (< 620 m/s), regardless of overlap with Group II (460–620 m/s), the sample was considered to belong to "Group I." However, if all crack velocities fell within the overlap between Group I and II, the sample could not be attributed to either group and thus classified as "Group I/II." Moreover, if some FW crack velocities were in Group I and others were in Group II, but these were out of the overlap, "unidentifiable" was assigned to the sample.

9.4 Results

These results support our interpretation that pressure debitage to obtain blades/bladelets was conducted during the Mesolithic period at Damjili Cave. Fig. 9.1 presents all the Mesolithic samples whose detachment techniques were identified. Fig. 9.2 shows representative observed FWs. Table 9.1 summarizes the results of this analysis. Despite the small sample size, the lithics assigned to Group I (pressure debitage) accounted for nearly half of the identified specimens (Fig. 9.3). Another noteworthy feature was the high proportion of samples in Group II, suggesting the use of either direct percussion with a soft hammer or indirect percussion. Sixteen samples were assigned to "unmeasurable." However, it should be noted that they contained several regular blades with parallel ridges, which could be easily identified as pressure blades from the conventional techno-morphological perspective.

The most remarkable result is the demonstration of the use of pressure debitage among Mesolithic communities during the late seventh millennium BC by an FW analysis for the first time. This is significant because it demonstrates that pressure blade/bladelet production technology was already practiced before the arrival of Neolithic culture from the East Wing in the Fertile Crescent.

9.5 Conclusion

This study aimed to explore Mesolithic blade production technology based on FW analysis. Despite the small sample size, this study demonstrated the use of pressure debitage for blade/bladelet produc-

Table 9.1 Results of FW analysis of obsidian blades/bladelets from Damjili Cave Unit 5. Unidentifiable: specimens with disparate FW results; Unmeasurable: specimens without identifiable/measurable FWs; Not analyzed: specimens not analyzed for FW.

Type	Count	Description
I	5	Pressure detachment
I or II	2	
II	5	Soft hammer percussion or indirect percussion
II or III	0	
III	1	Hard hammer percussion
Unidentifiable	5	Inconsistent results of FWs
Unmeasurable	16	No identifiable/measurable FWs
Not Analyzed	2	Unavailable for the present study
Total	36	

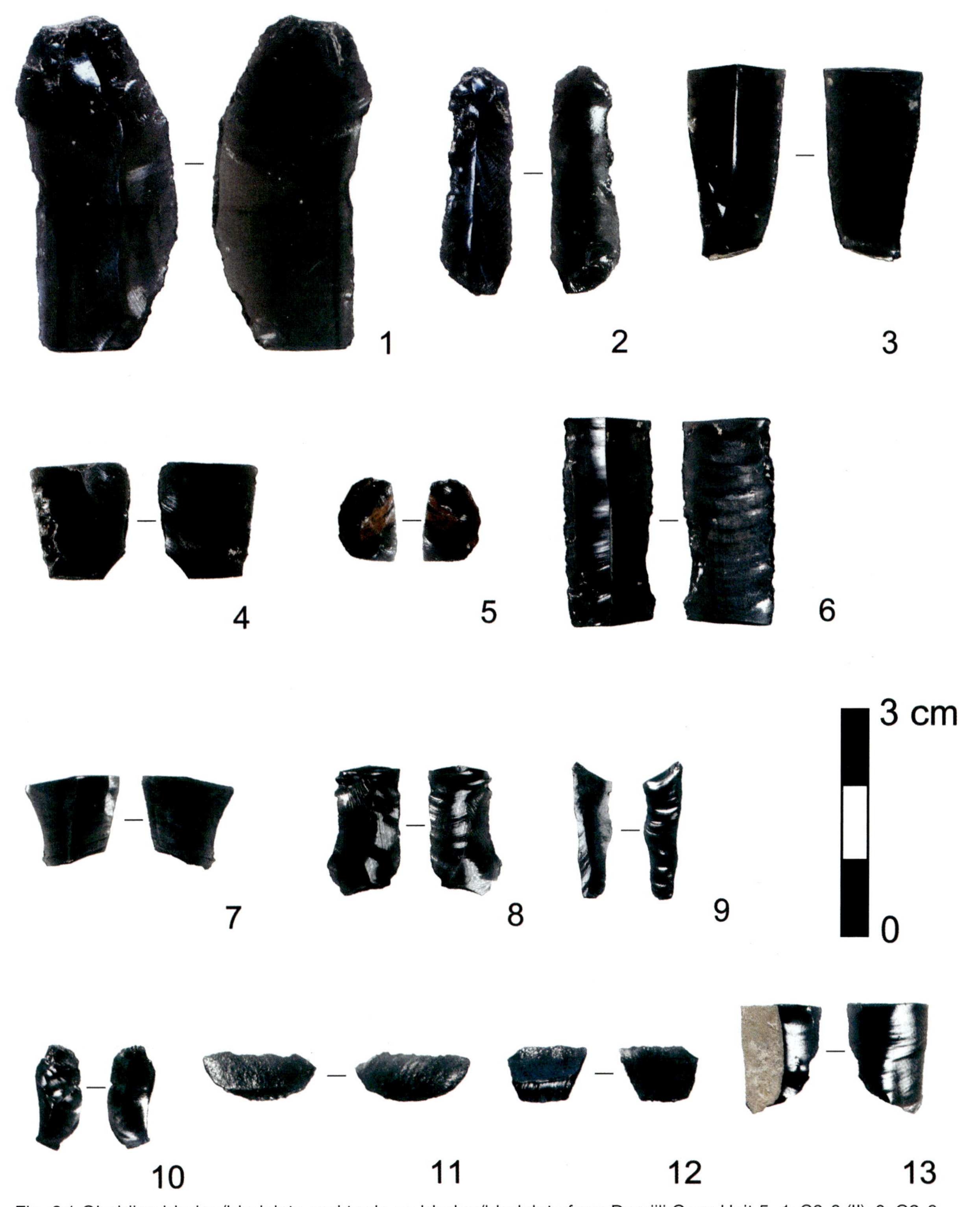

Fig. 9.1 Obsidian blades/bladelets and tools on blades/bladelets from Damjili Cave Unit 5. 1: C2-8 (II); 2: C2-8 (III); 3: C2-8 (I/II); 4: C2-8 (I); 5: C2-8 (II); 6: D2-11 (I); 7: D2-11 (II); 8: D2-11 (II); 9: D2-11 (I); 10: D2-11 (II); 11: D2-11 (I); 12: D2-11 (I/II); 13: C1-7 (I).

tion in Damjili Cave. The results support the hypothesized cultural interaction between the South Caucasus and the East Wing of the Fertile Crescent before the sixth millennium BC, presumably one of the preconditions enabling the abrupt acceptance of the Neolithic lifestyle. Investigation using this verifiable method provides a basis for future comparisons between different periods and regions. A comparable analysis of the related obsidian artifacts, including the assemblages of Hacı Elamxanlı Tepe and Göytepe, would greatly help us monitor the relationship between the changes in lithic technology and the introduction of the Neolithic economy (Takakura and Nishiaki 2020; Ikeyama et al. 2021).

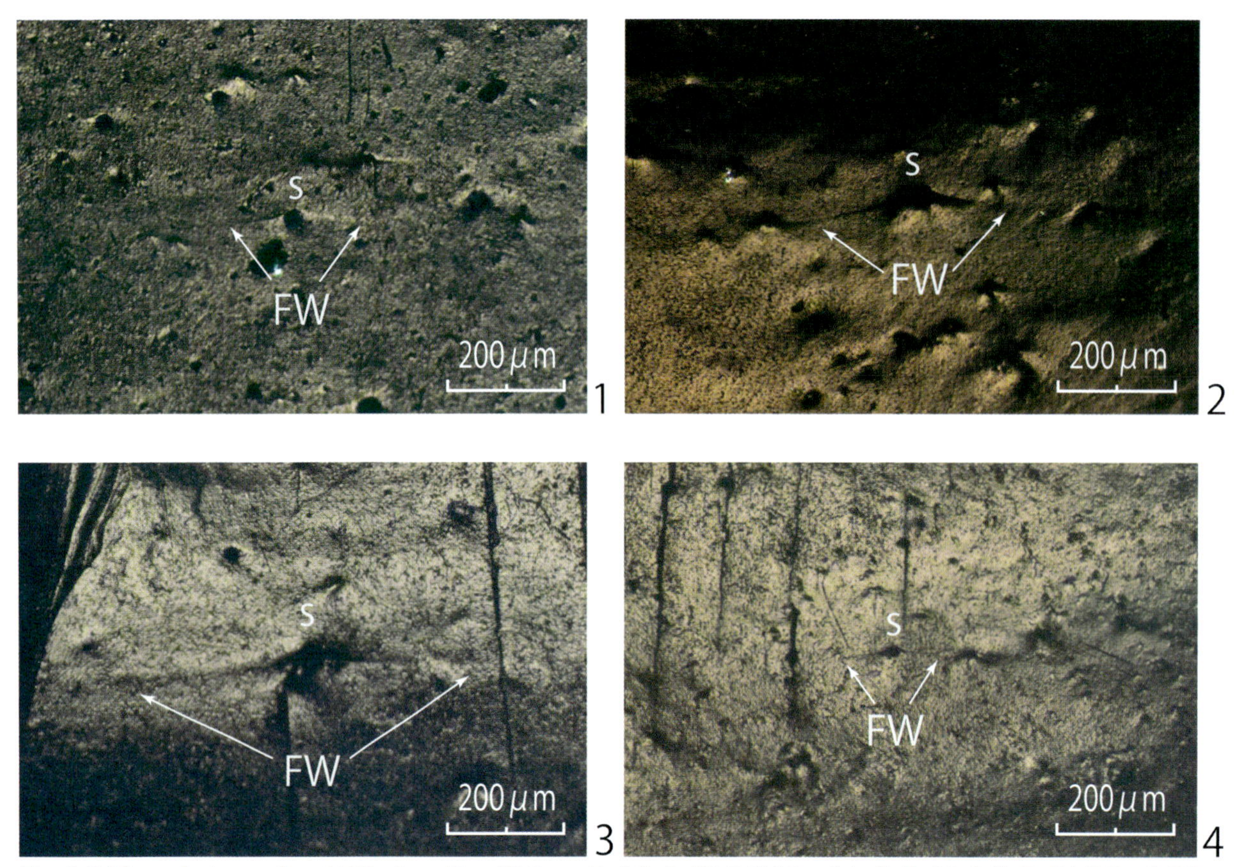

Fig. 9.2 Microphotographs of FWs from Damjili Cave Unit 5. 1: C2-8 (I) [Fig. 9.1: 4]; 2: C2-8 (I); 3: C1-8 (I);. 4: D2-11 (I) [Fig. 9.1: 9]. S: point of origin of the elastic wave; FW: fracture wing. Categories in parentheses represent the results of each FW analysis.

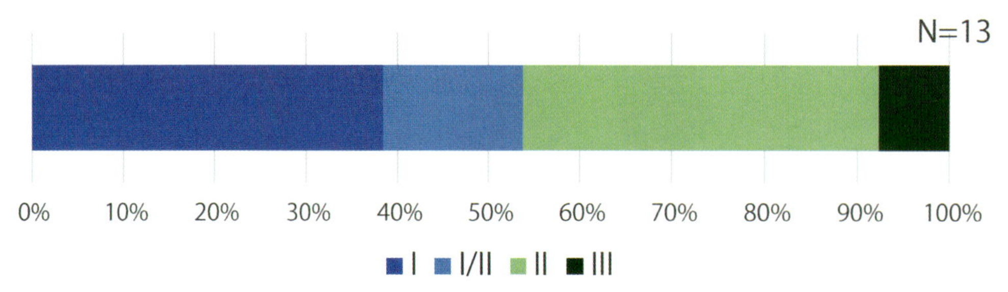

Fig. 9.3 Results of FW analysis.

Acknowledgments

This study was supported by the JSPS KAKENHI (No. JP20J23506), and the Takanashi Foundation for Historical Science, the Mishima Kaiun Memorial Foundation.

References

Arimura, M., A. Petrosyan, D. Arakelyan, S. Nahapetyan, and B. Gasparyan (2018) A preliminary report on the 2015 and 2017 field seasons at the Lernagog-1 site in Armenia. *Aramazd* 12(1): 1–18.

Arimura, M., K. Martirosyan-Olshansky, A. Petrosyan, and B. Gasparyan (2022) Exploring the change in lithic industries during the Neolithisation in Armenia (7th–6th Millennium BCE). In: *Tracking the Neolithic in the Near East: Lithic Perspectives on its Origins, Development and Dispersals*, edited by Y. Nishiaki, O. Maeda, and M. Arimura, pp. 489–502. Leiden: Sidestone Press.

Hutchings, W. K. (1999) Quantification of fracture propagation velocity employing a sample of Clovis channel flakes. *Journal of Archaeological Science* 26(12): 1437–1447.

Ikeyama, F. (2021) Blade production technology in the South Caucasian Neolithic: Examining the application of pressure technology and the potential use of a lever. *Journal of West Asian Archaeology* 22: 17–36 (in Japanese with English abstract).

Nishiaki, Y., A. Zeynalov, M. Mansrov, C. Akashi, S. Arai, K. Shimogama, and F. Guliyev (2019) The Mesolithic-Neolithic interface in the Southern Caucasus: 2016–2017 excavations at Damjili Cave, West Azerbaijan. *Archaeological Research in Asia* 19: 100140.

Takakura, J. (2007) Identification of blade and microblade flaking techniques in the lithic assemblage of the Okushirataki-1 Site, Hokkaido, Japan). *Cultura Antiqua* 59(IV): 98–109 (in Japanese with English abstract).

Takakura, J. and M. Izuho (2004) Identification of flaking techniques: From the analysis of fracture wings. *The Quaternary Research*, 43(1): 37–48 (in Japanese with English summary).

Takakura, J. and Y. Nishiaki (2020) Fracture wing analysis for identification of obsidian blank production techniques at Göytepe. In: *Göytepe: Neolithic Excavations in the Middle Kura Valley, Azerbaijan*, edited by Y. Nishiaki and F. Guliyev, pp. 209–221. Oxford: Archaeopress.

Chapter 10

Ground stone artifacts from Damjili Cave

Yoshihiro Nishiaki

10.1 Introduction

This chapter describes the ground stone artifact assemblages recovered during the 2016–2022 excavations at Damjili Cave. Although collectively referred to as ground stone artifacts, they include grinding, pounding, and even unmodified stones. In other words, the artifacts described in this chapter are non-flaked stone artifacts.

We collected 37 potential ground stone artifacts through fieldwork. However, our laboratory studies did not always confirm the manufacture and/or traces of use. Moreover, many were fragments (25 specimens), including some that were thermally fractured, making artifact identification even more complicated. Consequently, I report here a collection of 23 specimens that are considered worthy of publication: 23 pieces with tangible traces of manufacture and/or use (Table 10.1). They include six natural stones that resemble a type of stone often described as sling stones in Neolithic settlements (e.g., Kadowaki 2020, 2021).

10.2 Ground stone artifacts from Damjili Cave

The raw materials of the 23 ground stone artifacts featured in the present study consisted of rhyolite (10), andesite (5), flint (1), limestone (1), and sand-

Table 10.1 Inventory of the ground stone artifacts from Damjili Cave. * In the column for contexts, "Pit" denotes one of the trenches opened in the 2016 season.

Unit	Context*	Description	Raw material	Size (L×W×T) (mm)	Preservation	Figures
Unit 1	DMJ16.Pit 9.1	Handstone	Sandstone	116×59×49	Broken	
Unit 2	DMJ17-A2	Door socket	Andesite	482×452×136	Complete	Fig. 10.1
Unit 2/3	DMJ19.C0-6	Pebble	Limestone	44×38×22	Complete	Fig. 10.2: 1
Unit 2/3	DMJ19.C0-6	Pebble	Rhyolite	45×29×23	Complete	Fig. 10.2: 1
Unit 2/3	DMJ19.C0-6	Pebble	Rhyolite	50×48×34	Complete	Fig. 10.2: 1
Unit 2/3	DMJ19.C0-6	Pebble	Rhyolite	45×33×22	Complete	Fig. 10.2: 1
Unit 3	DMJ17.B3-4	Handstone	Rhyolite	98×81×23	Broken	
Unit 3	DMJ18.B0-5	Handstone	Sandstone	79×55×35	Broken	
Unit 3/4	DMJ19.C0-9	Pounder	Rhyolite	80×33×19	Broken	
Unit 4.1	DMJ19.C1-1	Pebble	Rhyolite	53×40×28	Complete	
Unit 4.1	DMJ19.C1-1	Pebble	Rhyolite	49×35×22	Complete	
Unit 4.1	DMJ19.D1-8	Slab	Andesite	144×117×43	Broken	
Unit 4.2	DMJ19.D1-10	Handstone	Rhyolite	50×55×55	Broken	
Unit 4.2	DMJ16.Pit 7.14	Handstone	Sandstone	192×78×70	Complete	Figs. 10.2: 2; 10.3: 1
Unit 4.2	DMJ17.A2-6	Pounder	Andesite	72×70×63	Broken	
Unit 4.2	DMJ16.Pit 7.14	Pounder	Flint	53×50×36	Complete	Fig. 10.3: 3
Unit 4.2	DMJ17.A1-6	Pounder	Sandstone	65×58×49	Complete	Fig. 10.3: 5
Unit 4.2	DMJ17.A1-7	Slab	Sandstone	88×125×28	Broken	Fig. 10.4: 1
Unit 4.2	DMJ16.Pit 7.14	Slab	Sandstone	72×112×25	Broken	
Unit 4.3	DMJ19.D0-13	Slab	Rhyolite	141×102×33	Complete	Figs. 10.2: 3; 10.4: 2
Unit 4.4	DMJ18.A0-13	Handstone	Rhyolite	66×56×46	Broken	
Unit 5.1	DMJ16.Pit 9.14	Pounder	Andesite	62×70×47	Broken	Fig. 10.3: 2
Unit 5.1	DMJ16.Pit 9.14	Pounder	Andesite	94×81×26	Complete	Fig. 10.3: 4

stone (6) (Table 10.1). The composition demonstrates a specific choice of raw materials for those heavy tools, following a criterion different from that of flaked stone artifacts, which were heavily dependent on siliceous stones, such as obsidian and flint (Chapter 8 of this volume).

The assemblages are described according to the following stratigraphic units (Table 10.1).

Unit 1 (Medieval period)

This unit yielded only one ground stone artifact–a broken loaf-shaped handstone (Table 10.1). As the objects from this unit contained a variety of intrusive materials from the underlying units, including Paleolithic artifacts, it is uncertain whether the handstone belongs to Unit 1.

Unit 2 (Bronze Age)

The only ground stone artifact reliably assigned to this unit comprised a large limestone slab with a circular hollow on one face (Fig. 10.1: 1), which appeared to be a door socket. The "socket" is approximately 10 cm in diameter and 7 cm in depth (Fig. 10.1: 2).

In addition to this object, other objects assigned to Units 2/3 are worth mentioning. They include a group of round pebbles with diameters of approximately 3–5 cm (Fig. 10.2: 1) that show no signs of artificial modification or use. However, their shape and size are reminiscent of sling balls widely found across prehistoric Southwest Asia (Nishiaki 2003; Kadowaki 2020, 2021).

Unit 3 (Chalcolithic period)

Two specimens were obtained from Unit 3—both broken handstones with a flat oblong shape (Table 10.1). In addition, a broken, elongated pounder with traces of battery at the edges, discovered in the context of Units 3/4, may also be related to this period.

Unit 4 (Neolithic period)

Unit 4 yielded the largest ground stone assemblage uncovered at Damjili Cave, constituting more than half of the total assemblage (12/23 or 52.2%). Interestingly, most of them were recovered from the upper two subunits, Units 4.1 and 4.2 (10 pieces), while the lower two subunits, Units 4.3 and 4.4, yielded only one specimen each (Table 10.1). The significant increase in ground stone artifacts during the later phase of the Neolithic period deserves further discussion.

The Unit 4 assemblage consisted of three handstones, three pounders, four slabs, and two natural pebbles. The handstones included one complete loaf-shaped specimen with ground surfaces on the sides and pounding traces at the ends (Figs. 10.2: 2; 10.3: 1). A circular hollow with signs of battery is recognizable on one side. Two of the three pounders were complete, with both showing a globular shape (Fig. 10.3: 3, 5). These are likely to include cores reused for flake production (Fig. 10.3: 3). Their surfaces exhibited flaking scars and heavy signs of battery. The specimens described as slabs were flat stones with traces of grinding on at least one surface that was almost flat or slightly hollowed (Figs. 10.2:

1

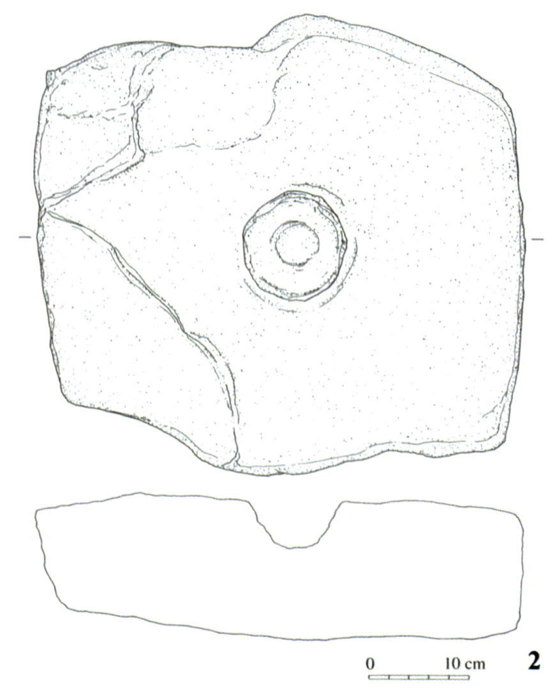

2

Fig. 10.1 Door socket from Unit 2 of Trench 9, Damjili Cave. 1: Photo of the discovery context; 2: Drawing.

Fig. 10.2 Ground stones from Damjili Cave. 1: Pebbles (Unit 2/3); 2: Handstone (Unit 4.2); 3: Slab (Unit 4.3).

3; 10.4). The two natural pebbles were relatively flat rather than ball-like and showed no signs of use.

Unit 5 (Mesolithic period)
Unit 5.1 yielded two pounders. One was a circular disc-shaped stone (Fig. 10.3: 4) made from a half-split pebble, whose surface showed scars from being flaked by percussion. Such stones can also be classified as choppers. The other specimen was stamp-shaped, with two steeply truncated ends (Fig. 10.3: 2).

10.3 Discussion

The small ground stone artifact assemblages of Damjili Cave are dominated by Neolithic specimens (Unit 4), the interpretation of which provides a good starting point for discussion. In archaeological literature, ground stone artifacts such as handstones

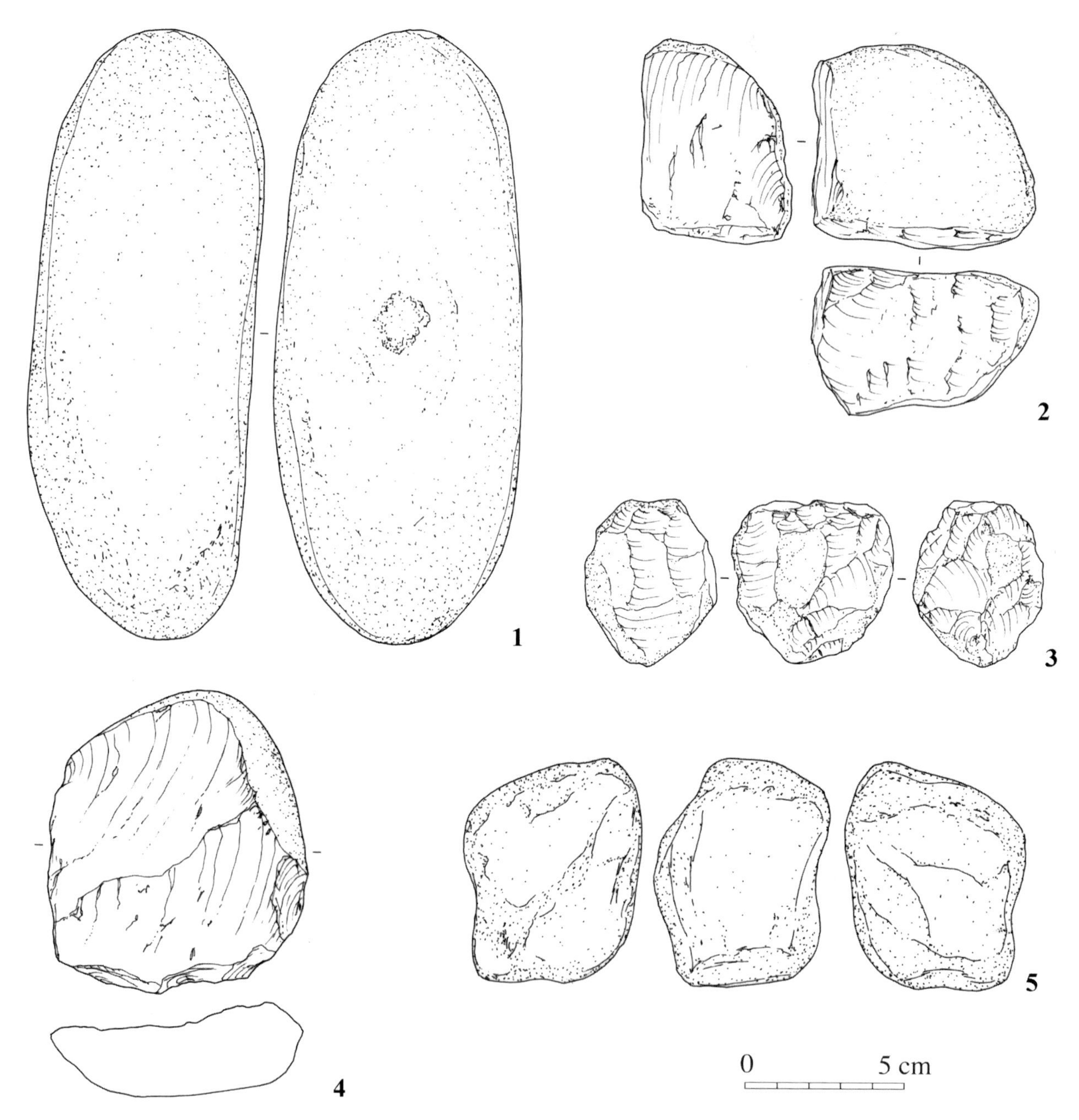

Fig. 10.3 Handstone and pounders from Damjili Cave. 1: Handstone (Unit 4.2; see Fig. 10.2: 2); 2: Pounder (Unit 5.1); 3: Pounder (Unit 4.2); 4: Pounder/Chopper (Unit 5.1); 5: Pounder (Unit 4.2).

and ground slabs are often considered to represent the employment of farming or a sedentary lifeway of exploiting plant resources. Researchers agree that ground stone technology clearly originated from mobile Paleolithic hunter-gatherers (Wright 1994; Piperno et al. 2004; Paixão et al. 2022). However, they have also acknowledged that, at least in Southwest Asia, the significant manifestation of ground stone technology since the final Pleistocene suggests that increasing sedentism likely occurred in response to post-glacial climatic changes (Natufian). Furthermore, ground stone technology flourished after the introduction of farming (Neolithic) in the early Holocene.

If this is the case, was the relatively common occurrence of ground stone tools in Unit 4 at Damjili Cave a result of the introduction and development of farming in the Neolithic?

To evaluate this possibility, we compared the relative frequencies of ground and flaked stone artifacts by stratigraphic unit, which are generally correlated with each other; both stone artifacts were recovered most frequently in Unit 4 (Fig. 10.5). Their occurrence appears to reflect the relative occupational intensity or volume of excavated cultural deposits by period. At the same time, the comparison reveals a clear chronological pattern: ground stone artifacts were less common than flaked stones

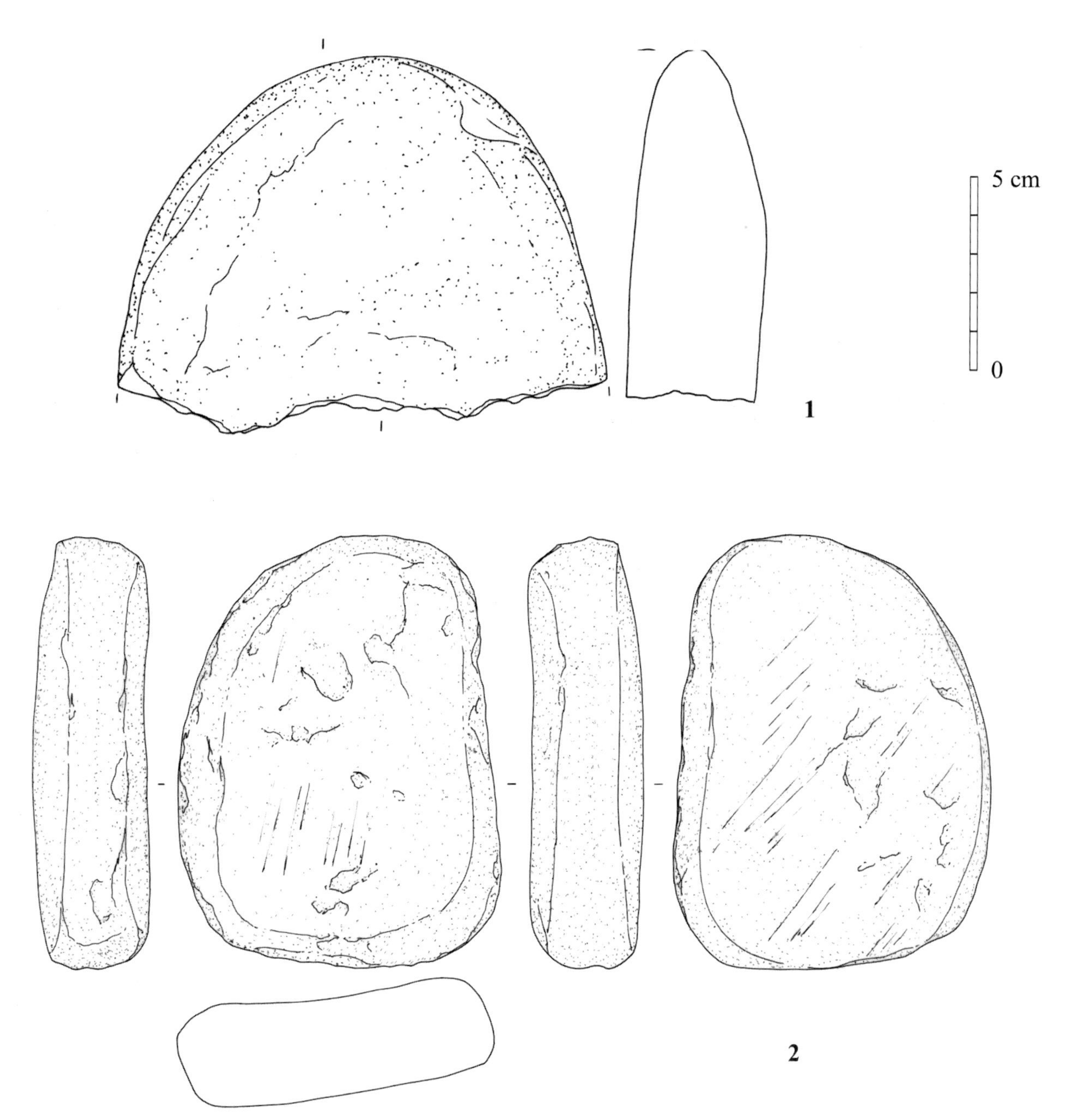

Fig. 10.4 Ground slabs from Damjili Cave. 1: Slab (Unit 4.2); 2: Slab (Unit 4.3; see Fig. 10.2: 3).

in Unit 5 (Mesolithic), equally as common in Units 4 (Neolithic) and 3 (Chalcolithic), and far more common than flaked stone artifacts in Units 2 and 1 (Fig. 10.5). Although it is not surprising that ground stone artifacts were more prevalent than flaked stone artifacts in the post-Stone Age (Units 1 and 2), their relative increase in the Neolithic (Unit 4) can be considered a signifier of the introduction of a new form of subsistence style exploiting different resources (Chapters 14 and 15 of this volume). More specifically, this change seems to have occurred in the later stages of the Neolithic Period (Unit 4.1/2).

The next important finding was that the use of ground stone artifacts was rather uncommon at Damjili Cave, even during the Neolithic period. This trend is enhanced in a comparison with data from the contemporaneous mound settlements of Göytepe (Nishiaki and Guliyev 2020) and Hacı Elamxanlı Tepe (Nishiaki et al. 2021) in the Ganja-Gazakh plain. In addition to the evident differences in sample size, the assemblages at Damjili Cave exhibited a significantly lower percentage of ground stone artifacts than those from the Neolithic settlements (Fig. 10.6). The strong similarity between Göytepe and Hacı Elamxanlı Tepe in the plain, and Damjili Neolithic and Damjili Mesolithic at the mountain foothills, suggests that this type of variability is likely related to each site's function: sedentary settlements versus temporary camps.

An additional observation has been made. A

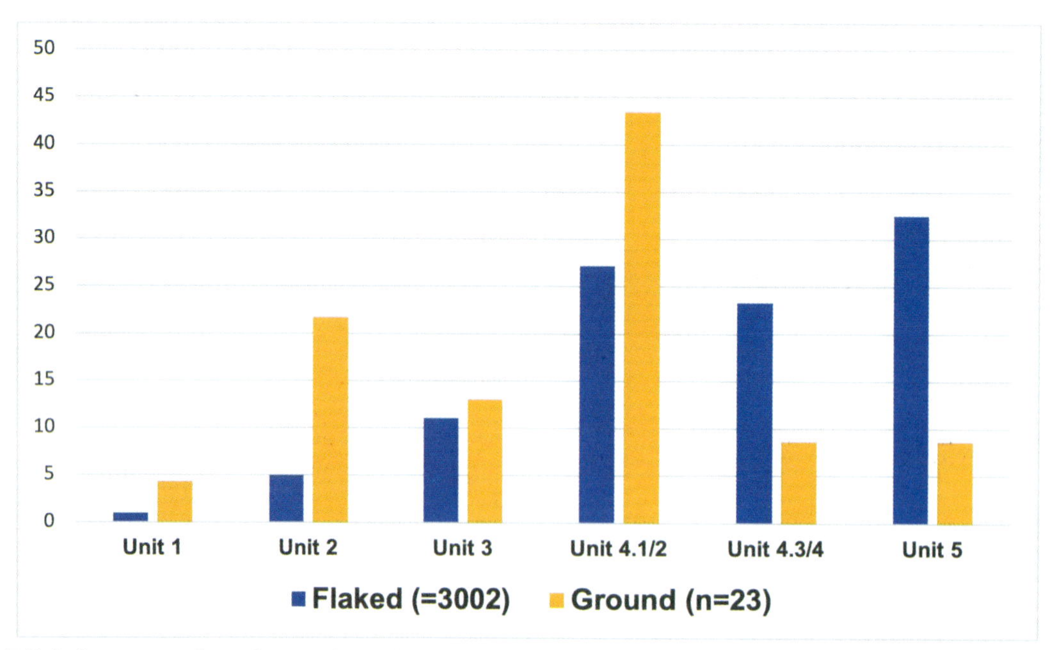

Fig. 10.5 Relative proportion of ground and flaked stone artifacts by stratigraphic units at Damjili Cave.

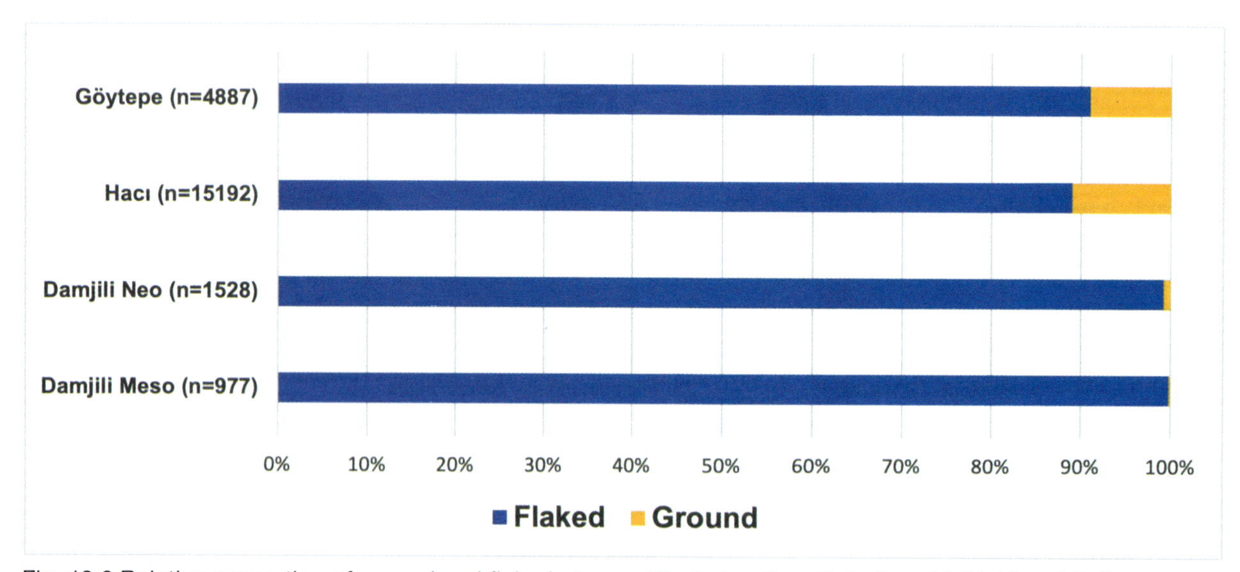

Fig. 10.6 Relative proportion of ground and flaked stone artifacts by sites. Data from Nishiaki and Guliyev (2020), Nishiaki et al. (2021), and the present study.

much smaller typological and technological diversity characterizes the ground stone assemblages of Damjili Cave. Its repertoire is well-represented in the Neolithic ground stone tool typology. Parallels of the handstones, pounders, slabs, and pebbles from Damjili Cave can be easily identified at Göytepe (Kadowaki 2020) and Hacı Elamxanlı Tepe (Kadowaki 2021). However, the Damjili assemblages lacked important elements common in plain settlements. One was a group of large and heavy tools, such as stone mortars and querns up to 40 cm long, at Göytepe and Hacı Elamxanlı Tepe. These tools were completely absent from Damjili; the largest ground stone at Damjili Cave was merely 19.2 cm long. The second important difference was the lack of polished stone tools, such as polished axes and chisels. Uncommon at Hacı Elamxanlı Tepe, they became more common at the later Neolithic settlement of Göytepe (Kadowaki 2020, 2021). However, the Damjili assemblages derived from both the early and late stages of the Neolithic period contained none.

10.4 Conclusion

Ground stone assemblages from Damjili Cave are described in this chapter. They provide useful insights into prehistoric human lifestyles in mountain environments in this area of the South Caucasus, particularly during the Neolithic period (the primary focus of the present research project). First, the analysis reveals an increase in ground stone use during the late Neolithic stage (Unit 4.1/2). The relative frequency of ground stone artifacts in the early stage of the Neolithic (Unit 4.3/4) did not differ greatly from that of the Mesolithic (Unit 5). This view is based on a small sample size. Nevertheless, it is important to recall that, at Damjili Cave, the contrast between the early and late Neolithic phase ground stone assemblages aligns with data from studies of other archaeological records, such as architecture (Chapter 4), pottery (Chapter 11), and botanical remains (Chapter 14), all of which point to a cultural change in the later stage of the Neolithic.

Second, the ground stone assemblages from Damjili Cave expand our view of the Neolithic way of life in the Middle Kura Valley, which has always been discussed using data from mound settlements in the plain. The dataset from Damjili is the first to show how ground stone artifacts were used in a cave in a mountainous environment. The recovered assemblage likely represents a small functional part of the Neolithic system of tool production and use that spread throughout the South Caucasus. Conversely, this study can lead us to generate a fresh view for deciphering the supposedly complicated cultural dynamism during the introduction of the Neolithic socioeconomy in the South Caucasus. Further fieldwork to obtain more reliable datasets from these periods will undoubtedly lead to the continued exploration of the research potential of Damjili Cave.

References

Kadowaki, S. (2020) Neolithic ground stone typology and technology at Göytepe. In: *Göytepe: The Neolithic Excavations in the Middle Kura Valley, Azerbaijan*, edited by Y. Nishiaki and F. Guliyev, pp. 223–259. Oxford: Archaeopress.

Kadowaki, S. (2021) Neolithic ground stone artifacts from Hacı Elamxanlı Tepe. In: *Hacı Elamxanlı Tepe: The Archaeological Investigations of an Early Neolithic Settlement in West Azerbaijan*, edited by Y. Nishiaki, F. Guliyev, and S. Kadowaki, pp. 107–132. Berlin: ex oriente.

Nishiaki, Y. (2003) Functional and morphological observations on the Chalcolithic grinding stones from Tell Kosak Shamali. In: *Tell Kosak Shamali – The Archaeological Investigations on the Upper Euphrates, Syria Vol. 2: Chalcolithic Technology and Subsistence*, edited by Y. Nishiaki and T. Matsutani, pp. 121–183. UMUT Monograph 2. Oxford: Oxbow Books.

Nishiaki, Y. and F. Guliyev (2020) *Göytepe: The Neolithic Excavations in the Middle Kura Valley, Azerbaijan*. Oxford: Archaeopress.

Nishiaki, Y., F. Guliyev, and S. Kadowaki (2021) *Hacı Elamxanlı Tepe: The Archaeological Investigations of an Early Neolithic Settlement in West Azerbaijan*. Berlin: ex oriente.

Paixão, E., J. Marreiros, L. Dubreuil, W. Gneisinger, G. Carver, M. Prévost, and Y. Zaidner (2022) The Middle Paleolithic ground stones tools of Nesher Ramla unit V (Southern Levant): A multi-scale use-wear approach for assessing the assemblage functional variability. *Quaternary International* 624: 94–106.

Piperno, D. R., E. Weiss, I. Holst, and D. Nadel (2004) Processing of wild cereal grains in the Upper Palaeolithic revealed by starch grain analysis. *Nature* 430(7000): 670–673.

Wright, K. I. (1994) Ground-stone tools and hunter-gatherer subsistence in southwest Asia: implications for the transition to farming. *American Antiquity* 59(2): 238–263.

Chapter 11

Neolithic and Chalcolithic pottery from Damjili Cave

Takehiro Miki and Kazuya Shimogama

11.1 Introduction

After adopting the agro-pastoral economy in the Southern Caucasus, the population set out to settle in the alluvial plain near the Kura, resulting in the appearance of mound settlements. One of the oldest traces of such sedentism has been found at Hacı Elamxanlı Tepe (Nishiaki et al. 2021). The amount of pottery was very few in Hacı Elamxanlı Tepe (Miki and Shimogama 2021). However, not all the inhabitants of the Southern Caucasus rapidly shifted their lifestyles in the Neolithic and the subsequent Chalcolithic periods. During these periods, the presence of a more mobile population in a cave site than those in the alluvial plain was demonstrated by the radiocarbon evidence obtained from recent excavations at Damjili Cave (Nishiaki et al. 2019, 2022). In addition, we found as many as 455 possible Neolithic-Chalcolithic potsherds from mainly stratigraphic Units 3 and 4 and other units of Damjili Cave (Table 11.1). Especially from Unit 3 of Square A1, we found 132 potsherds, although the number of potsherds collected was less than 10 in most squares of each unit (Table 11.1). These pieces of pottery were likely to be brought by the Neolithic and Chalcolithic farmers who practiced pottery production. In addition, the evidence of the Chalcolithic occupation in the Southern Caucasus still needs to be discovered. Henceforth, the study of Neolithic and Chalcolithic pottery from Damjili Cave is relevant for discussing 1) the pottery use during the Neolithic period and 2) the Chalcolithic ways of life in the Southern Caucasus.

Below, we will describe representative ware types mainly found in Units 3 and 4 of Damjili Cave,

Table 11.1 The total count of Neolithic and Chalcolithic pottery by units and squares of Damjili Cave.

Unit	Square	Count
unstratified		11
Topsoil	D3	1
	E1	2
Unit 1	E0	3
	D1	26
Unit 2	C2	1
	D1	4
Unit 2/3	D0	1
	C0	14
Unit 3	Trench 7 (2016 season)	8
	B1-C1-B2-C2 (2016 season)	17
	A1	132
	A2	2
	A3	1
	B0	13
	C2	8
	C3	1
	D1	20
	D2	20
	D3	1
Unit 4	Trench 7 (2016 season)	2
	B1-C1-B2-C2 (2016 season)	23
	A0	36
	A2	55
	A3	8
	B0	2
	B3	14
	C1	2
	C2	1
	C3	2
	D0	2
	D1	1
	D2	3
	D3	10
	D99	8
Total		455

then present a general view of diagnostic examples and their description (Section 11.2). We will then approach the comparison with published pottery data from other Neolithic and Chalcolithic sites (Section 11.3) and summarize this chapter (Section 11.4).

11.2 Wares

From Unit 3 (Chalcolithic period) and Unit 4 (Neolithic period), five ware types were frequently observed; that is, mineral tempered ware, plant tempered ware, coarse red brown ware, gray powdery ware, coarse orange ware, and fine orange ware (Fig. 11.1). Below, we will describe the detail of these ware characteristics with diagnostic drawings (Figs. 11.2–11.4; Table 11.2).

Mineral tempered ware (Figs. 11.1: 1; 11.2: 4–7; 11.3: 2, 6; 11.4: 1, 3, 7–8)
The presence of mineral inclusions is the primary criterion of this ware type. Its fabric colors are dark brown, grayish brown to reddish brown. Given the presence of grayish-black to black cores, this ware seems to have been fired at a low firing temperature. The inclusions consist primarily of fine sand and rarely of coarse sand. Although the vessel-shape information is limited, flat bases were present in this ware type (Fig. 11.2: 7; 11.3: 6; 11.4: 7–8). A flared rim (Fig. 11.3: 2) and a simple rounded rim (Fig. 11.4: 1) were confirmed in this ware type. One example with a knob on its body part (Fig. 11.4: 3) also exists. Its forming technique is difficult to discern due to the absence of well-preserved materials. Its vessel surface is usually smoothed or horizontally smoothed. One potsherd shows a trace of burnishing on its exterior surface (Fig. 11.2: 4).

Plant tempered ware (Figs. 11.1: 2; 11.2: 2, 8–10; 11.3: 3, 5)
Plant/vegetal temper (approximately 1–5 mm in length) was intentionally added to the fabric of this ware type. Sometimes medium sand or coarse sand was also added. Its fabric color ranges from dark brown to grayish brown. Plant tempered ware is also poorly fired, showing grayish-black to black cores in sections. An incurved rim (Fig. 11.2: 2), a slightly flared thick rim (Fig. 11.3: 3), and flat bases (Figs. 11.2: 8–10) were confirmed in plant tempered wares. Its vessel surface is smoothed or horizontally smoothed. Some sherds show traces of finger impressions on their exterior surfaces (Fig. 11.2: 10).

Coarse red brown ware (Figs. 11.1: 3–5; 11.2: 3, 11; 11.3: 1, 4; 11.4: 2, 4–6, 9)
Coarse red brown ware is distinguished from other

Fig. 11.1 Pictures of representative ware types. 1: Mineral tempered ware; 2: Plant tempered ware; 3–5: Coarse red brown ware; 6: Gray powdery ware; 7: Coarse orange ware.

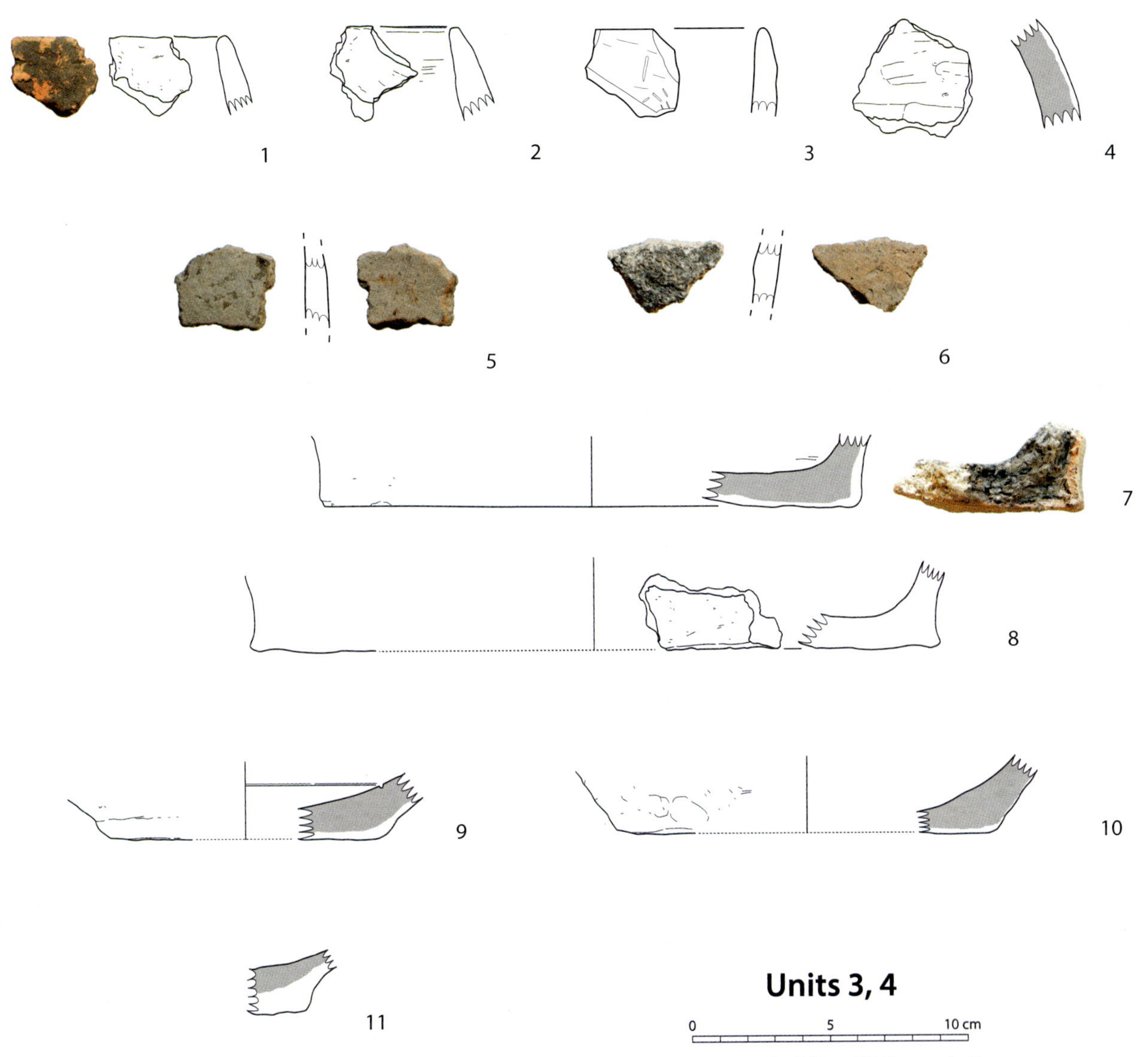

Fig. 11.2 Pottery from Units 3 (Chalcolithic context) and 4 (Neolithic context) of Damjili Cave. 1: Coarse orange ware; 2, 8–10: Plant tempered ware; 3, 11: Coarse red brown ware; 4–7: Mineral tempered ware.

ware types by the distinctive white coarse sand (Fig. 11.1: 5) and other types of coarse sand. This white sand temper is emphasized by its reddish-brown surface color. Black cores were sometimes present (Fig. 11.1: 5). Incurved rims (Figs. 11.2: 3; 11.4: 2) and a spherical body part (Fig. 11.3: 4) suggest a vessel shape of incurved vessels. Flat bases were confirmed (Figs. 11.2: 11; 11.4: 9). Fig. 11.4: 6 shows a pseudo-rim in its upper part, implying coil-making. Traces of smoothing and finger impressions (Fig. 11.3: 4) are present.

Gray powdery ware (Fig. 11.1: 6)
This ware type was confirmed only in body sherds. This ware type has powdery, grayish-white surfaces. No remarkable temper was observed in this ware type.

Coarse orange ware (Figs. 11.1: 7; 11.2: 1)
The color of this ware type is orange to orange-brown, its fabric is very crumbly, containing medium sand, and an incurved rim was confirmed. Orange-colored wares which derived from the upper units of the Bronze Age and the medieval period appear to be similar to this ware in surface color and inclusions, but they can be unequivocally distinguished from the coarse orange ware with its greater friable fabric and different vessel shapes (see Chapter 16 of this volume).

11.3 Discussion

In this discussion, we consider 1) the inter-layer comparison of ware types and 2) the comparison with other Neolithic and Chalcolithic sites. First, Table 11.3 shows the count of potsherds belonging to the five ware types explained above. These five ware types concentrate on Units 3 and 4 although only some potsherds from mixed contexts were confirmed at the units other than Units 3 and 4. We found

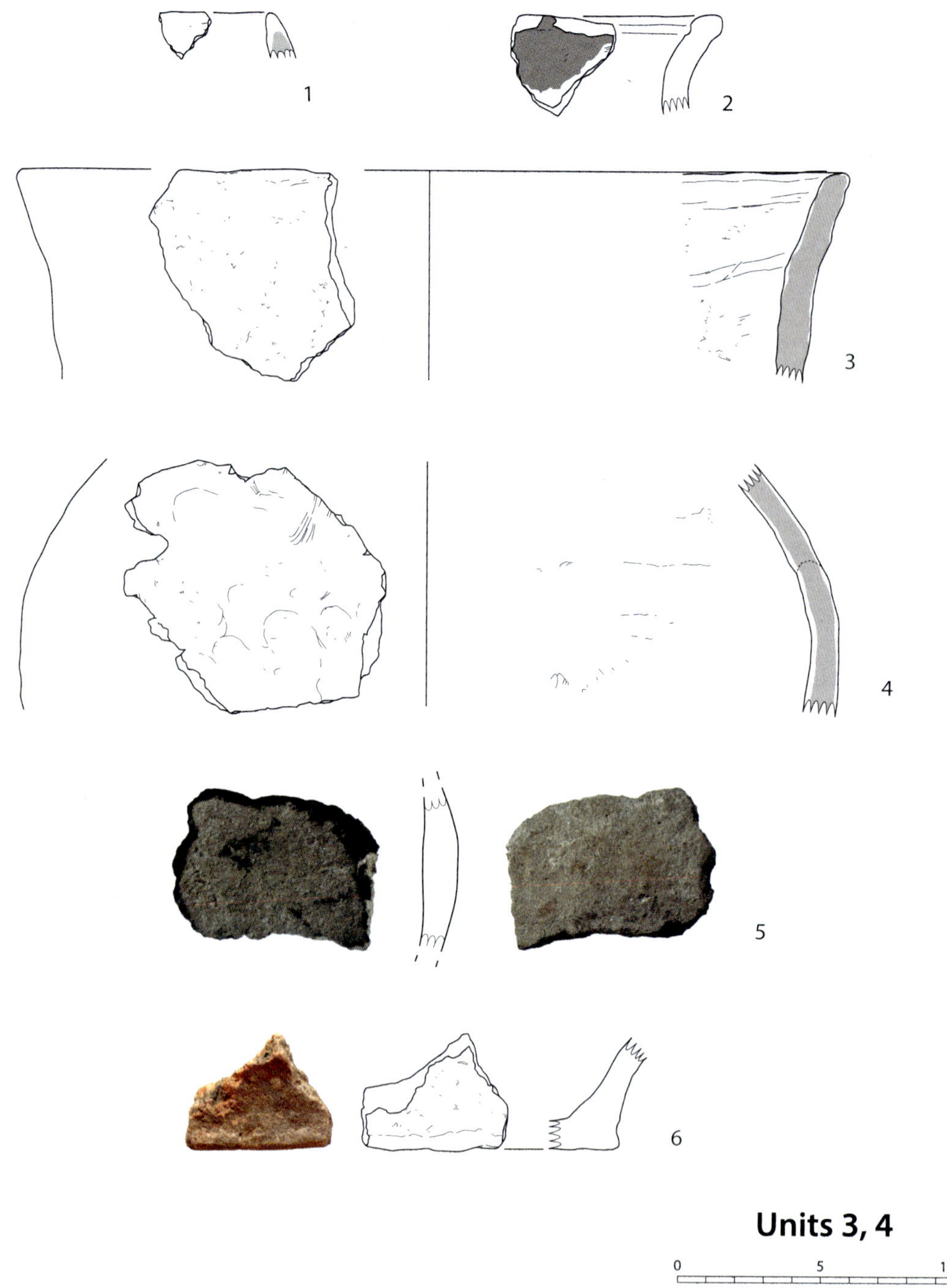

Fig. 11.3 Pottery from Units 3 (Chalcolithic context) and 4 (Neolithic context) of Damjili Cave. 1, 4: Coarse red brown ware; 2, 6: Mineral tempered ware; 3, 5: Plant tempered ware.

the largest number of potsherds from Unit 3 (over 200 potsherds). In terms of pottery count, coarse red brown ware was the most predominant in Units 3 and 4. Mineral-tempered and plant-tempered wares are also common in Units 3 and 4, with mineral tempered ware slightly more than plant-tempered ware. Coarse orange ware and gray powdery ware are the minority in the pottery assemblages of Units 3 and 4. A remarkable difference between the pottery assemblage of Units 3 and 4 is the proportion of coarse red brown ware. However, it should be noted that in Unit 3, most coarse red brown wares were found from Square A1 (especially context A1-5), enlarging the proportion of this ware type in the Chalcolithic layer. Although the related contexts including A1-5 yielded a very few fragments assignable to Neolithic mineral-tempered wares, they derive from pit fills which are considered to be disturbed from Unit 3 (Chalcolithic layer). The pottery counts were extremely small in Units 4.4 and 4.3, but Table 11.3 shows an abrupt increase in the subsequent Units 4.2 and 4.1. The ware assemblage of Unit 4.1 is slightly different from that of Unit 4.2 in the ratio of mineral-tempered wares and plant-tempered ones and the presence of coarse red brown wares.

Second, we conduct an inter-site comparison

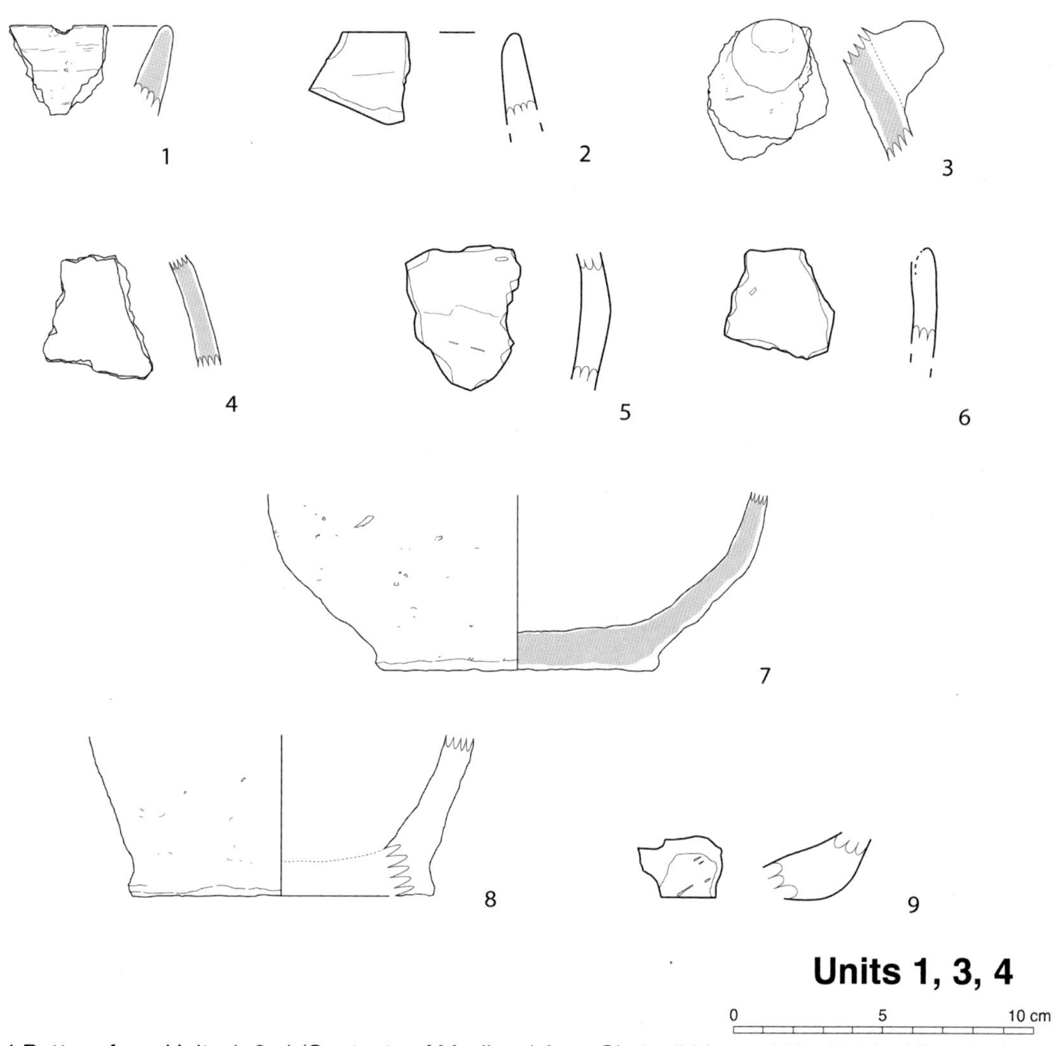

Fig. 11.4 Pottery from Units 1, 3, 4 (Contexts of Medieval Age, Chalcolithic, and Neolithic) of Damjili Cave. 1, 3, 7-8: Mineral tempered ware; 2, 4–6, 9: Coarse red brown ware.

of ceramic assemblages with other Neolithic and Chalcolithic sites in the Southern Caucasus. First, the ceramic assemblage from Hacı Elamxanlı Tepe (c. 5950–5800 BC) shows the coexistence of mineral tempered ware and plant tempered ware, and the presence of imported painted pottery (Miki and Shimogama 2021; Nishiaki et al. 2013; Nishiaki et al. 2015). A similar tendency (the coexistence of mineral-tempered ware and plant-tempered ware) was also confirmed in the earliest level of Göytepe (c. 5650–5460 BC; Arimatsu 2020). The coexistence of mineral tempered ware and plant tempered ware in Hacı Elamxanlı Tepe and Göytepe is similar to that in Unit 4 of Damjili, except for the absence of painted ware. It should be also noted that there is an outstanding typological analogy between the ceramic assemblage from the farming settlement of Göytepe and that of Damjili Cave; flat-base closed pots with inverted rims and everted rim bowls (Arimatsu 2020). However, the frequent occurrence of relief decorations and slip applications in the later sequence of Göytepe are not attested among the Damjili sherds. Considering radiocarbon dates from Units 4.4 (c. 6000–5800 BC; Chapter 5 of this volume) and 4.2/1 (c. 5600–5400 BC; Chapter 5 of this volume), they overlap with the time range of Hacı Elamxanlı Tepe (c. 5950–5800 BC) and Göytepe (c. 5650–5460 BC), and the apparent similarity in ware assemblages and typological characteristics implies the import from lowland settlements. The difference in pottery counts between Unit 4.4/3 (4 sherds) and 4.2/1 (168 sherds) also fits well with that between Hacı Elamxanlı Tepe and Göytepe.

In Unit 3 (Chalcolithic) of Damjili Cave, we confirmed its radiocarbon time range between 4600 and 3500 BC. The ceramic assemblage of Unit 4 is characterized by the presence of coarse red brown ware, which was never confirmed in Neolithic sites such as Hacı Elamxanlı Tepe and Göytepe. We did not find any parallels to the coarse red brown ware yet. In Sondages C and D (Chalcolithic) of Aknashen in the Ararat Plain, chaff and grit-tempered pottery groups having rims with notches along their edges were reported (Harutyunyan 2022: 87–88).

Table 11.2 Inventory of the Neolithic and Chalcolithic pottery from Damjili Cave. * In the column for contexts, "Pit" denotes one of the trenches opened in the 2016 season.

Fig.	Context*	Unit	Ware type	Part	Color (exterior)	Color (interior)	Color (core)	Temper
11.2: 1	Pit 9.9 sieved	4.1	coarse orange ware	rim	light orange	light orange	light orange	plant; grit
11.2: 2	A2.6	4.2	plant-tempered ware	rim	gray-brown	gray-brown	dark brown	plant; sand; grit
11.2: 3	C1.1	4.1	coarse red brown ware	rim	orange	orange	orange	white coarse sand (ø 0.2–0.5 mm)
11.2: 4	A3.6 sieved	4.2	mineral-tempered ware	body	dark gray - brown	dark gray - brown	dark gray - black	fine sand; grit
11.2: 5	D99.1	4.1	mineral-tempered ware	body	light yellow orange	light yellow orange	yellow orange	white, brown coarse sand (ø 0.5–1 mm)
11.2: 6	D99.2	4.1	mineral-tempered ware	body	pale orange	brownish gray	brownish gray	dark, gray coarse sand (ø 0.25–0.5 mm)
11.2: 7	Pit 9.10 sieved	4.1	mineral-tempered ware	base	orange brown	brown	dark gray - black	sand; grit (ø 1–2 mm)
11.2: 8	A2.6 sieved	4.2	plant-tempered ware	base	gray-brown	dark brown	gray - black	plant; grit
11.2: 9	A2.6	4.2	plant-tempered ware	base	gray-brown	gray-brown	dark brown	plant; sand; grit
11.2: 10	Pit 9.9 sieved	4.1	plant-tempered ware	base	dark brown	red-brown	brown-black	plant; fine sand
11.2: 11	A1.6 sieved	3	coarse red brown ware	base	red-brown	red-brown	dark brown	sand; white grit (ø 1–3 mm)
11.3: 1	A1.5	3	coarse red brown ware	rim	red-brown	red-brown	dark brown	sand; white grit (ø 1–3 mm)
11.3: 2	B3.5	4.1	mineral-tempered ware	rim	gray-brown	red-brown	dark gray	fine sand (< ø 1 mm)
11.3: 3	B3.5	4.1	plant-tempered ware	rim	light buff	light buff	gray-brown	fine sand (< ø 1 mm); plant
11.3: 4	A1.5	3	coarse red brown ware	body	red-brown	red-brown	dark brown	sand; white grit (ø 1–3 mm)
11.3: 5	D3.8	4.1	plant-tempered ware	body	light brown	light brown	dark brown	coarse sand (ø 1 mm)
11.3: 6	Pit 9.8	3	mineral-tempered ware	base	dark brown	dark brown	gray-brown	fine sand (< ø 1 mm)
11.4: 1	A3.4	3	mineral-tempered ware	rim	orange brown	orange brown	gray	fine sand (< ø 1 mm); white particle
11.4: 2	D1.7	3	coarse red brown ware	rim	dull reddish brown	dull reddish brown	dark reddish brown	very coarse sand (ø 1 mm), dark brown sand, vegetal temper 3–4 mm
11.4: 3	A.04	4.1	mineral-tempered ware	body	light orange	brown	brown	fine sand (< ø 1 mm)
11.4: 4	B0.4 sieved	3	coarse red brown ware	body	orange brown	dark brown	dark gray	plant; grit; mica
11.4: 5	D1.4	1	coarse red brown ware	body	reddish brown	reddish brown	dull orange	white very coarse sand (ø 1–2 mm)
11.4: 6	D1.4	1	coarse red brown ware	body	reddish brown	reddish brown	dull orange	white coarse sand (ø 0.2–0.5 mm)
11.4: 7	A0.4	4.1	mineral-tempered ware	base	orange brown	brown	dark gray	fine sand (< ø 1 mm)
11.4: 8	B0.4	3	mineral-tempered ware	base	light brown	orange	orange	sand (< ø 1 mm)
11.4: 9	D1.7	3	coarse red brown ware	base	bright reddish brown	dull orange	brownish black	very coarse light brown sand (ø 1–2 mm), vegetal temper 3 mm

These groups show similarities with those found in level 0 of Aratashen (Palumbi 2007: 68, fig. 3: 5, 6). In the middle and late Chalcolithic of Ovçular Tepesi, Nakhchivan (c. 4300–4000 BC), buff, chaff-tempered and chaff-faced pottery with comb impressions were prevalent (Marro et al. 2009). On the other hand, at Sioni, the eponymous site of the Chalcolithic Sinoni culture, the predominant pottery

Table 11.3 Rims, body sherds, and bases of ware types in each unit.

Part	Rim				Body					Base			Undistinguishable					
Ware	Mineral-tempered ware	Plant-tempered ware	Coarse orange ware	Coarse red brown ware	Mineral-tempered ware	Plant-tempered ware	Coarse orange ware	Coarse red brown ware	Gray powdery ware	Mineral-tempered ware	Plant-tempered ware	Coarse red brown ware	Mineral-tempered ware	Plant-tempered ware	Coarse red brown ware	Burnt clay	Clay object?	Total
Topsoil	-	-	-	-	-	2	-	1	-	-	-	-	-	-	-	-	-	3
Unit 1	-	-	-	2	-	3	-	24	-	-	-	-	-	-	-	-	-	29
Unit 2	-	-	-	-	1	-	2	-	1	-	-	1	-	-	-	-	-	5
Unit 2/3	-	-	-	1	3	2	1	-	2	2	2	-	-	2	-	-	-	15
Unit 3	2	-	2	5	24	11	5	151	2	3	-	4	-	5	4	2	-	220
Unit 4.1	1	2	1	1	40	11	16	33	1	3	1	-	1	-	-	-	1	112
Unit 4.2	-	1	-	-	14	32	2	-	4	-	3	-	-	-	-	-	-	56
Unit 4.3	-	-	-	-	1	2	-	-	-	-	-	-	-	-	-	-	-	3
Unit 4.4	-	-	-	-	-	1	-	-	-	-	-	-	-	-	-	-	-	1
Unstratified	-	-	-	-	5	2	-	-	-	-	-	-	2	2	-	-	-	11
Total	3	3	3	9	88	66	26	209	10	8	6	5	3	9	4	2	1	455

is burnished and grit-tempered pottery with wavy rims (Kiguradze 2000: 322; Kiguradze and Sagona 2004: 48). The comparison of the Chalcolithic pottery from Unit 4 with other contemporaneous sites preliminarily suggests the inter-regional or settlement/cave difference in pottery assemblages among the Southern Caucasus during the Chalcolithic period.

11.4 Conclusions

In this chapter, we reported the details of Neolithic and Chalcolithic pottery from Damjili Cave that illustrates the ordinary practices of the cave dwellers in the Southern Caucasus. Five main findings are summarized as follows:

(1) The ceramics found in Units 3 and 4 were subdivided into five ware types: mineral tempered ware, plant tempered ware, coarse red brown ware, gray powdery ware, and coarse orange ware.

(2) Coarse red brown ware was the most predominant in the ware assemblages of Units 3 and 4.

(3) In Unit 4 (Neolithic), mineral-tempered ware and plant-tempered ware were co-present.

(4) The co-presence of mineral-tempered ware and plant-tempered ware in the Neolithic period of Damjili Cave shows a similarity with other Neolithic settlement sites in the Southern Caucasus.

(5) The pottery from Unit 3 (Chalcolithic), especially coarse red brown ware, preliminarily indicates the difference with other contemporaneous sites in the Southern Caucasus.

References

Arimatsu, Y. (2020) Neolithic pottery from Göytepe. In: *Göytepe: Neolithic Excavations in the Middle Kura Valley, Azerbaijan*, edited by Y. Nishiaki and F. Guliyev, pp. 261–285. Oxford: Archaeopress.

Harutyunyan, A. (2022) The pottery of Aknashen. In: *The Neolithic Settlement of Aknashen (Ararat valley, Armenia). Excavation Seasons 2004–2015*, edited by R. Badalyan, C. Chataigner, and A. Harutyunyan, pp. 82–105. Oxford: Archaeopress.

Kiguradze, T. (2000) The Chalcolithic-Early Bronze transition in the Eastern Caucasus. In: *Chronologies des pays du Caucase et de l'Euphrate aux IVème-IIIème millénaires*, edited by C. Marro and H. Hauptmann, pp. 321–328. Istanbul: Institut Français d'Études Anatoliennes.

Kiguradze, T. and A. Sagona (2004) On the origins of the Kura-Araxes cultural complex. In: *Archaeology in the Borderlands. Investigations in Caucasia and Beyond*, edited by A. Smith and K. Rubinson, pp. 38–94. The Cotsen Institute of Archaeology, Monograph 47. Los Angeles: University of California.

Marro, C., V. Bakhshaliyev, and S. Ashurov (2009) Excavations at Ovçular Tepesi (Nakhichevan,

Azerbaijan). First preliminary report: the 2006–2008 seasons. *Anatolia Antiqua* XVII: 31–87.

Miki, T. and K. Shimogama (2021) Pottery from Hacı Elamxanlı Tepe. In: *Hacı Elamxanlı Tepe: The Archaeological Investigations of an Early Neolithic Settlement in West Azerbaijan*, edited by Y. Nishiaki, F. Guliyev, and S. Kadowaki, pp. 133–152. Berlin: ex oriente.

Nishiaki, Y., F. Guliyev, S. Kadowaki, Y. Arimatsu, Y. Hayakawa, K. Shimogama, T. Miki, C. Akashi, S. Arai, and S. Salimbeyov (2013) Hacı Elamxanlı Tepe: Excavations of the earliest Pottery Neolithic occupations on the Middle Kura, Azerbaijan, 2012. *Archäologische Mitteilungen aus Iran und Turan* 45: 1–25.

Nishiaki, Y., F. Guliyev, S. Kadowaki, V. Alakbarov, T. Miki, S. Salimbeyov, C. Akashi, and S. Arai (2015) Investigating cultural and socioeconomic change at the beginning of the Pottery Neolithic in the Southern Caucasus: The 2013 excavations at Hacı Elamxanlı Tepe, Azerbaijan. *Bulletin of the American School of Oriental Research* 374: 1–28.

Nishiaki, Y., F. Guliyev, and S. Kadowaki eds. (2021) *Hacı Elamxanlı Tepe: The Archaeological Investigations of an Early Neolithic Settlement in West Azerbaijan*. Berlin: ex oriente.

Nishiaki, Y., A. Zeynalov, M. Mansrov, C. Akashi, S. Arai, K. Shimogama, and F. Guliyev (2019) The Mesolithic-Neolithic interface in the Southern Caucasus: 2016–2017 excavations at Damjili Cave, West Azerbaijan. *Archaeological Research in Asia* 19: 2252–2267.

Nishiaki, Y., A. Zeynalov, M. Munsrov, and F. Guliyev (2022) Radiocarbon chronology of the Mesolithic-Neolithic sequence at Damjili Cave, Azerbaijan, Southern Caucasus. *Radiocarbon* 64(2): 309–322.

Palumbi, G. (2007) A preliminary analysis on the prehistoric pottery from Aratashen (Armenia). In: *Les cultures du Caucase (VIe – IIIe millénaires avant notre ère). Les relations avec le Proche-Orient*, edited by B. Lyonnet, pp. 63–76. Paris: CNRS Éditions.

Chapter 12

Osseous objects from Damjili Cave

Saiji Arai

12.1 Introduction

The production and use of various bone objects is one of the most important characteristics of early farming societies in the Southern Caucasus. Accordingly, the continuous occupation of Damjili Cave from the early Holocene to the Neolithic period has significant potential for tracing the origin and development of the bone industry in this region. However, renewed excavations at the site lack evidence of bone work; only four modified osseous objects from the Mesolithic period were discovered. In addition, three specimens are represented as small fragments with uncertain modification traces. The last is a unique ornament made of boar tusk. The details of each specimen are described below.

12.2 Finds

12.2.1 Worked carapace (Fig. 12.1)

Two small fragments of burnt tortoise (*Testudo graeca*) carapace having crisscross parallel striations, probably belonging to an identical object, were found in a Mesolithic layer of Trench 9 (Fig. 12.1). These striations ran randomly on the medial surfaces of both specimens. As reported in our preliminary report on Damjili (Nishiaki et al. 2019), similar finds were found in Early Neolithic sites in the Ararat Plain (Balasescu and Radu 2022: 230). As we have not conducted such an analysis on our materials to date, the possibility that filleting activity on the carapace results in traces cannot be excluded. However, the crisscross directions of the striations indicate intensive scraping for cleaning the materials rather than meat acquisition.

12.2.2 Worked antler fragment (Fig. 12.2)

A small piece of modified antler was found in the Mesolithic layer in Trench 7 (Fig. 12.2). Owing to its incomplete shape and worn surface, this speci-

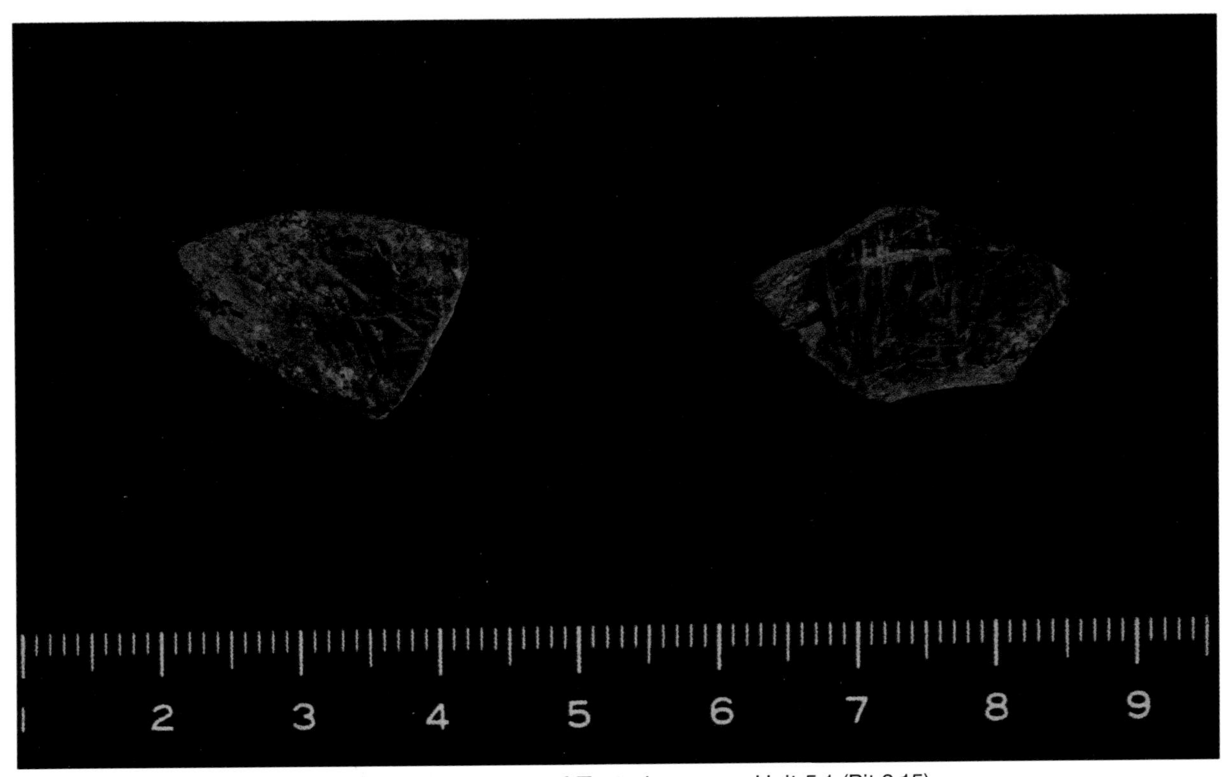

Fig. 12.1 Mesolithic worked and burnt carapaces of *Testudo graeca*. Unit 5.1 (Pit 9.15).

men could not be assigned to a specific typological category.

12.2.3 Boar tusk ornament (Fig. 12.3)

The most notable osseous find is a tusk ornament from the Mesolithic layer in Trench 9 (Fig. 12.3). This half-moon-shaped ornament is processed from a split mandibular canine of *Sus* sp. and perforated at the proximal extremity. Successive pecking marks on the enamel side form a dashed line that represents either decoration or a trace of an attempt to split the raw material. A preliminary microscopic study of the enamel surface could not identify the precise sequence of processing, which is necessary to determine the purpose of the dashed line (Fig. 12.4). Conversely, the dentin side indicates no trace of further decorations but abrading marks on the entire surface resulting from primary shaping. The perforation at the proximal end appears to be caused by unidirectional drilling, as a shallow depression was observed around the hole only on the dentin side.

12.3 Boar tusk ornaments in Late Mesolithic and Early Neolithic: a continuity of identity

Boar tusk ornaments are found at several Neolithic sites in this region. These sites include Shulaveris Gora (Kiguradze 1986: abb.16), Aknashen (Badalyan et al. 2022: fig. 4), Mentesh Tepe (Taha and Le Doeeur 2017: pl. 8), Hacı Elamxanlı Tepe (Arai 2021: fig. 8.9), whereas other sites such as Shomutepe (Ahundov 2012) and Göytepe (Arai 2020) lack evidence for production or use of tusk ornament. As sites with tusk ornaments are distributed throughout the entire Shomutepe/Shulaveri horizon, they appear to be chronological rather than geographical. According to the chronology of Neolithic

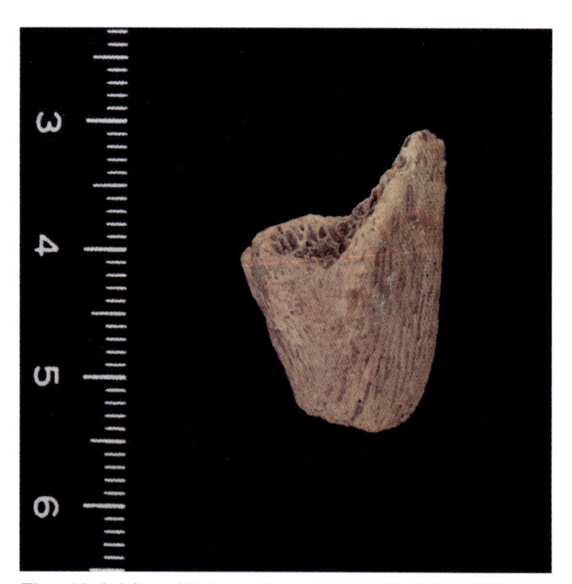

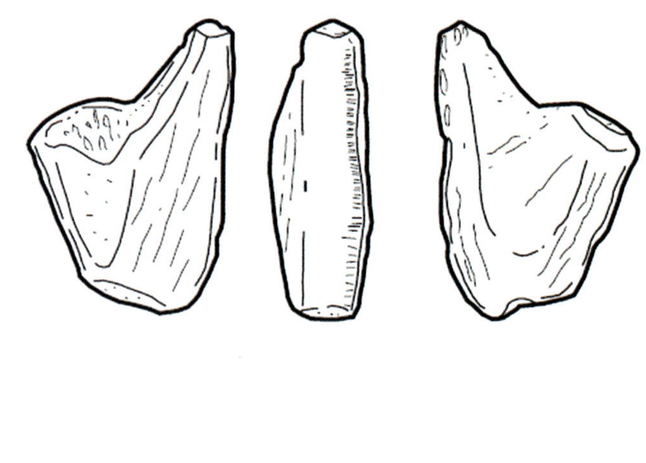

Fig. 12.2 Mesolithic antler object. Unit 5.1 (Pit 7.14).

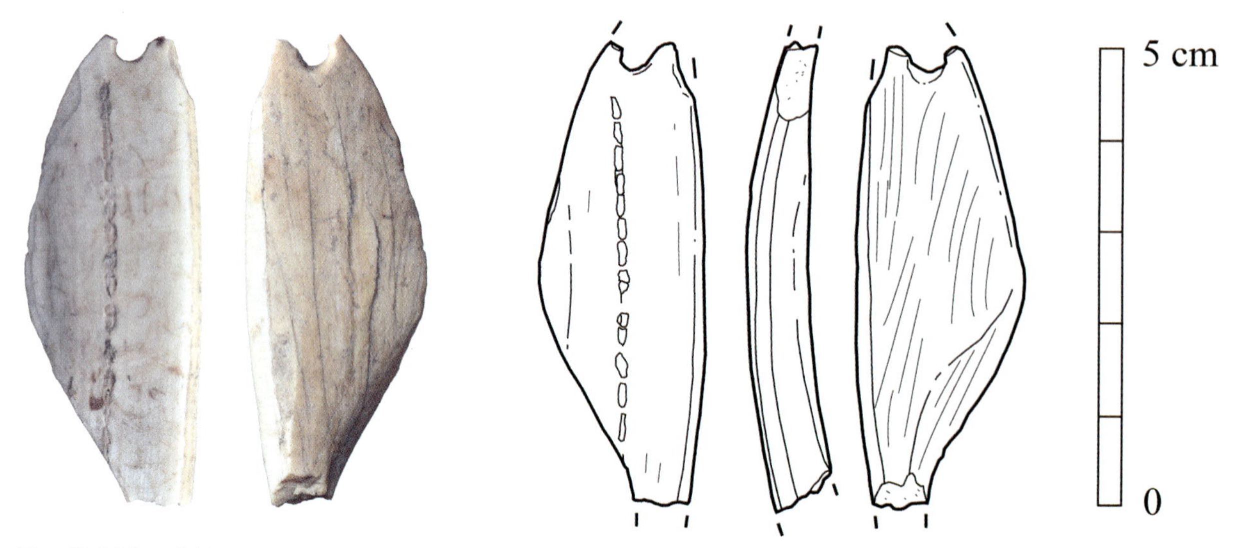

Fig. 12.3 Mesolithic boar tusk ornament. Unit 5.1 (B3.18).

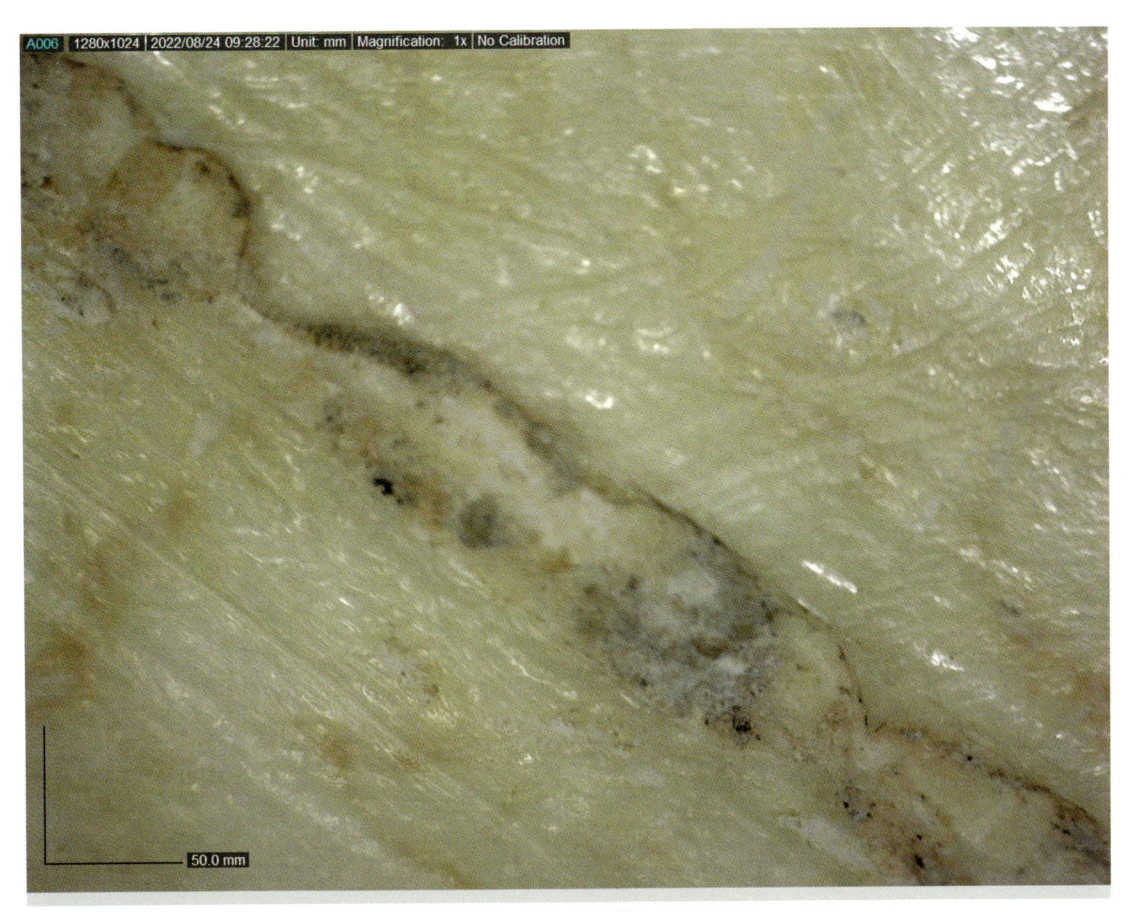

Fig. 12.4 Magnified surface of the boar tusk ornament (Fig. 12.3).

sites in the Kvemo-Kartli region of eastern Georgia established by Kiguradze (1986), boar tusk ornaments appeared during Phase II (Shulaveris Gora III-I and Imiris Gora VII-IV), suggesting that these ornaments represent an early stage of the Neolithic period. Hence, although their final forms differ, the use of boar tusks as a raw material for ornamental purposes appears to have been common in both the Late Mesolithic and Early phases of the Neolithic period.

This is particularly important for understanding the Neolithization process in the Southern Caucasus because the last hunter-gatherers and earliest farmers of the region shared the same belief system associated with boar tusks. In general, personal ornaments function as expressions of a person's (group's) identity. Identity is formed from various backgrounds. Giddens (2006) defines identity as the distinctive characteristics of a person's (group's) character related to who they are and what is meaningful to them. He listed the primary sources of identity, including gender, sexual orientation, nationality, ethnicity, and social class (Giddens 2006: 1020). Regardless of the meaning, the fact that the Late Mesolithic and Early Neolithic people used similar objects made of the same raw material is significant.

12.4 Concluding remarks

The five seasons at Damjili provided only three osseous objects dating to the Mesolithic period. This was surprising, because Neolithic settlements in the alluvial plains offer plenty of bone objects. This fact suggests that both the production and use (e.g., hide working, tilling) of bone tools were conducted only at the settlements.

The discovery of the Late Mesolithic boar tusk ornament was the most important result, which contributes to understanding the material culture of the period and proves the continuity of identity between Mesolithic hunter-gatherers and Neolithic farmers in the Southern Caucasus. This type of ornament disappeared in the later phase of the Neolithic era, implying a transformation of identity from the local Mesolithic tradition to mature Neolithic farmers' mind.

References

Arai, S. (2020) The Neolithic bone and antler industry from Göytepe. In: *Göytepe: The Neolithic Excavations in the Middle Kura Valley, Azerbaijan*, edited by Y. Nishiaki and F. Guliyev, pp. 293–322. Oxford: Archaeopress.

Ahundov, T. (2012) *Shomutepe*. Baku: Institute of Archaeology and Ethnography.

Arai, S. (2021) Neolithic bone tools and ornaments from Hacı Elamxanlı Tepe. In: *Hacı Elamxanlı Tepe: The Archaeological Investigations of an Early Neolithic Settlement in West Azerbaijan*, edited by Y. Nishiaki, F. Guliyev, and S. Kadowaki, pp. 159–176. Berlin: ex oriente.

Badalyan, R., A. Hartyunyan, K. Meliksetian, E. Pernicka, and R. Christidou (2022) Miscellaneous objects from Aknashen. In: *The Neolithic Settlement of Aknashen (Ararat Valley, Armenia). Excavation Seasons 2004–2015*, edited by R. Badalyan, C. Chataigner, and A. Hartyunyan, pp. 212–224. Oxford: Archaeopress.

Balasescu, A. and V. Radu (2022) Animal subsistence economy at the Neolithic site of Aknashen. In: *The Neolithic Settlement of Aknashen (Ararat Valley, Armenia). Excavation Seasons 2004–2015*, edited by R. Badalyan, C. Chataigner, and A. Hartyunyan, pp. 225–248. Oxford: Archaeopress.

Giddens, A. (2006) *Sociology (5th Edition)*. Cambridge: Polity Press.

Kiguradze, T. (1986) *Neolithische Siedlungen von Kvemo-Kartli, Georien*. Materialien zur Allgemeinen und Vergleichenden Archäologie Band 29. Verlag C. H. Beck, München.

Nishiaki, Y., A. Zeynalov, M. Mansurov, C. Akashi, S. Arai, K. Shimogama, and F. Guliyev (2019) The Mesolithic-Neolithic interface in the Southern Caucasus: 2016–2017 excavations at Damjili Cave, West Azerbaijan. *Archaeological Research in Asia* 19. DOI: 10.1016/j.ara.2019.100140

Taha, B. and G. Le Dosseur (2017) Bone tools as records of regional differences during the Neolithic. A preliminary comparative study between the bone industries at Mentesh Tepe and Kamiltepe sites. In: *The Kura Projects: New Research on the Later Prehistory of the Southern Caucasus*, edited by B. Helwing, T. Aliyev, B. Lyonnet, F. Guliyev, S. Hansen, and G. Mirtskhulava, pp. 399–424. Archäologie in Iran und Turan 16. Berlin: Dietrich Reimer Verlag.

Chapter 13

The Mesolithic stone figurine from Damjili Cave

Yoshihiro Nishiaki, Ulviya Safarova, Fumika Ikeyama, and Yagub Mammadov

13.1 Introduction

The excavations at Damjili Cave revealed a stratified cultural sequence covering most of the Holocene period in a primary context. Significantly, it includes Mesolithic and Neolithic cultural layers, which are essential for understanding Neolithization in the South Caucasus. As illustrated in the present volume, a valuable collection of archaeological remains was recovered from these layers to document the transition between these two periods. Simultaneously, this research demonstrated a unique aspect of the Damjili Cave material records. It is the rare occurrence of "small finds" such as clay, stone, and bone objects that are abundantly found at Neolithic settlements in the South Caucasus (Nishiaki and Guliyev 2020; Nishiaki et al. 2021). This characteristic plausibly reflects the functional aspects of human habitation in caves. A restricted repertoire of the remaining materials was expected because of the supposedly limited range of activities in the cave environment. This interpretation is also supported by the fact that Damjili Neolithic records do not indicate the use of mudbricks for construction, which was popular in mound settlements from the earliest stage of the Neolithic period (Nishiaki et al. 2021).

In this regard, the discovery of a stone human figurine from a Mesolithic layer at Damjili Cave, which did not yield any figurines from Neolithic layers, was one of the most impressive findings. Moreover, this figurine is the first example of recovery in the Middle Kura Valley from the late Mesolithic period. Here, we describe the excavation context and morphological features of this valuable specimen.

13.2 Excavation context

The Mesolithic deposits at Damjili Cave, approximately 40–120-cm thick, are divided into Units 5.1–5.3. Unit 5.3, the earliest, was heavily disturbed and contained both Mesolithic and Middle Paleolithic artifacts. Conversely, the occupational traces of Units 5.1 and 5.2 are well preserved. Semicircular rows of limestone blocks with diameters of 3–4 m were found in both strata (see Chapter 4 of the present volume). They appear to represent the remains of campsites—reminiscent of the Mesolithic architecture of the Chokh in the Greater Caucasus (Kushnareva 1997). Fireplaces were identified.

The figurine was discovered in Unit 5.2 outside a stone-walled structure (Square A0). Unit 5.2 was dated to approximately 6400–6100 cal BC, with seven radiocarbon dates (see Chapter 5 of the present volume). The primary architectural features of this layer are the stone alignments and simple fireplaces mentioned above (Figs. 4.26; 4.27, respectively). The deposits represented a brown soil layer (DMJ18-A0-20), approximately 40-cm thick, accumulated in an open space (Square A0) approximately 1 m from these structures. The figurine was identified by dry sieving with a mesh size of 3 mm.

13.3 Morphological features

The figurine is 51-mm long, 15-mm wide, and 9.5-mm thick. Its raw material is sandstone with a smooth surface. Damjili Cave is situated within a Cretaceous limestone and flint formation (see Chapter 3 of the present volume). Nevertheless, within a 20-km radius, particularly in the Agstafa Valley and its tributaries, there are primary sources of tuff volcanic rocks and secondary sources providing various rocks, such as andesite and sandstone (Nishiaki et al. 2021). The figurine was probably made from river pebbles obtained from secondary sources.

No signs of intentional modification are present on the figurine except for engraving, the details of which are visible only through microscopic analysis (Figs. 13.1; 13.2). The overall pattern suggests that the artifact represents a human figure. One of the wider surfaces with the most complicated engravings appeared on the front side (Figs. 13.1: 1b, 2b; 13.2: 5–7). The supposed head is decorated with several vertical lines from the top, probably representing hair (Figs. 13.1: 2; 13.2: 1–3, 5, 8, 10). The vertical lines are short in the middle depicting short-hair bangs. The ends of these lines are explicitly delin-

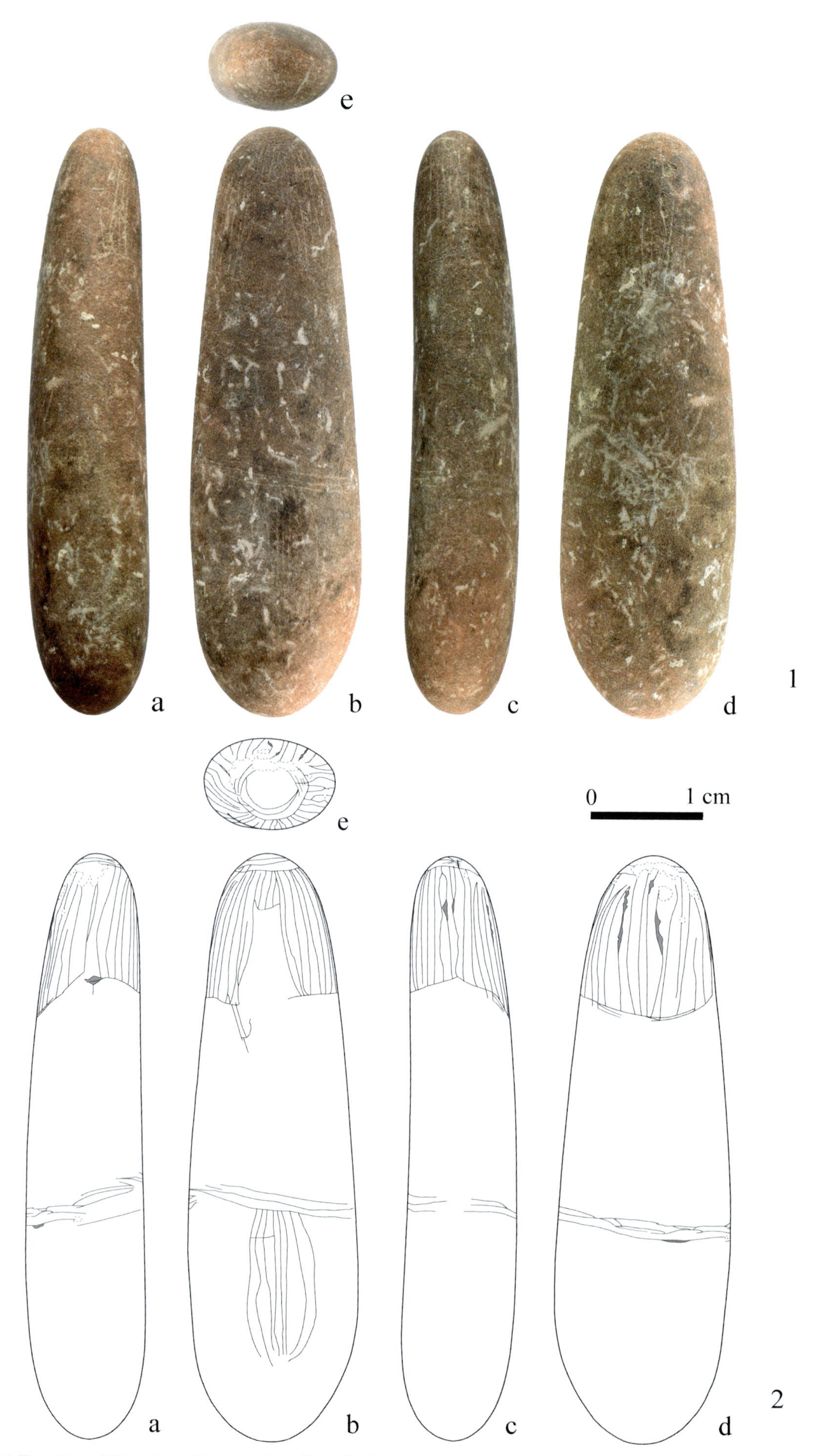

Fig. 13.1 The Mesolithic stone figurine from Damjili Cave. 1: Photograph; 2: Distribution of the curved lines.

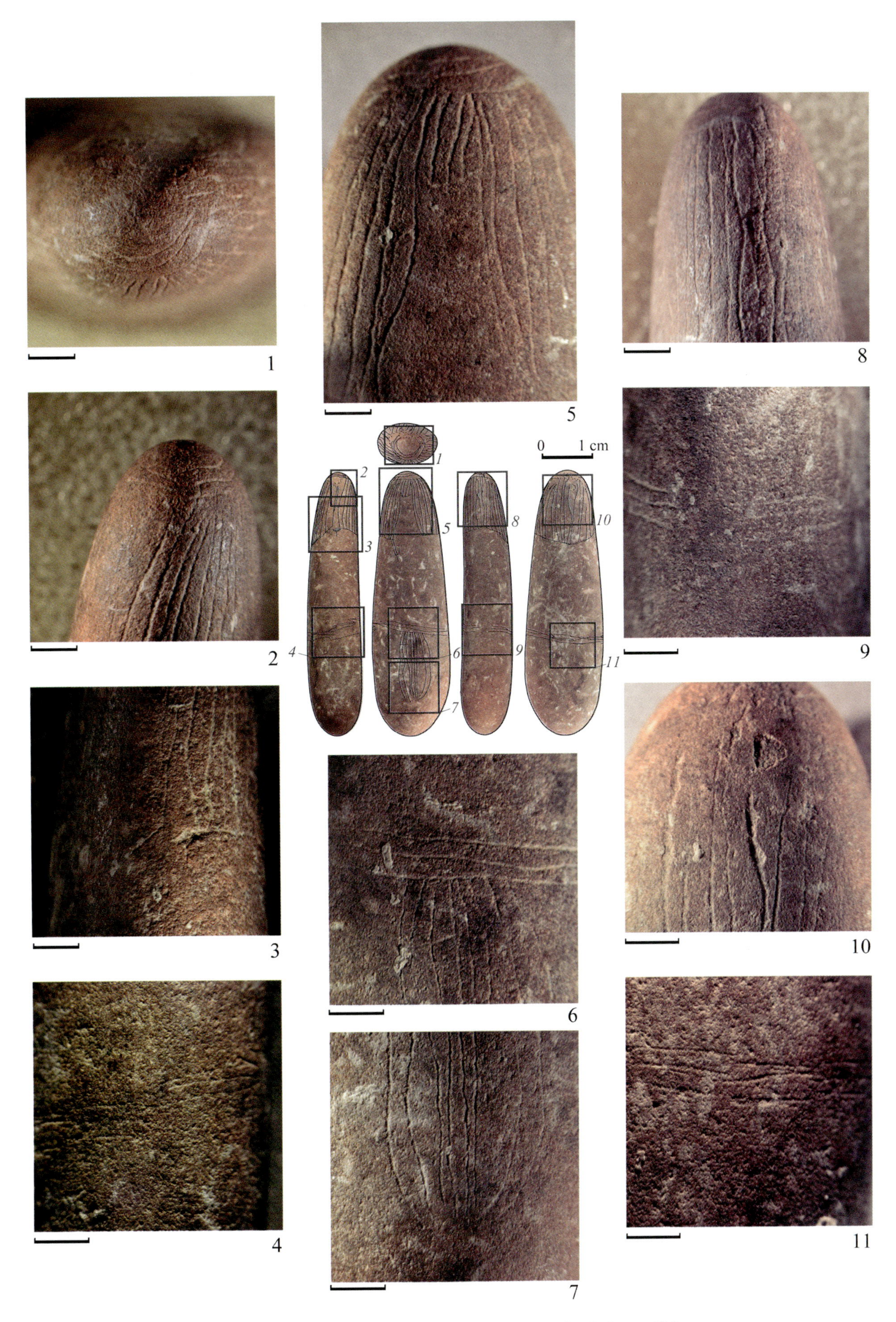

Fig. 13.2 Details of the carving of the Damjili Mesolithic stone figurine. Scale bar = 200μm.

eated using horizontal lines. Furthermore, the face did not show any depiction of facial features, such as the eyes and nose (Figs. 13.1; 13.2). Moreover, the top of the head was unprocessed in an area approximately 4.7 mm in diameter (Figs. 13.1: 2e; 13.2: 1). This area is encircled by 2–3 circular lines that may represent a hairband at the edge of the cap. The top of the head was not decorated.

Contrastingly, the lower body is decorated less intensely than the head. There are three horizontal lines at approximately two-fifths from the bottom, representing a "belt" 1.3-mm wide (Fig. 13.1: 2). In the middle of the belt on the front, a group of vertical lines 6.3-mm wide run downward, ending approximately 6.7 mm upward from the bottom of the figurine (Figs. 13.1: 2b; 13.2: 6, 7). This plausibly represents a loincloth or apron used for hiding the genitals. The lower ends of the head hair are delineated using horizontal lines, whereas the vertical lines on the lower body fade at the ends.

13.4 Discussion

The above observations indicate that the object is a human figurine created from riverstone. There is no representation of the genitals, breasts, or buttocks. Therefore, the figurine's sex is indeterminate. Engraving was most plausibly conducted using stone tools, resulting in varying depths and widths. Nonetheless, the most intense traces of the engraving on the front of the head suggest the manufacturer's emphasis on the face (Figs. 13.1: 2b; 13.2: 5), although the details of the figurine's face are not depicted.

To date, this object is the first Mesolithic human figurine discovered in the Middle Kura Valley—the heartland of Shomutepe Neolithic culture. The known Mesolithic sites in the valley and its neighboring regions have not yielded any recognizable human figurines. A recent study on the Kmlo and Lernagog Caves in Armenia, dating from the ninth to eighth millennium BC, did not refer to portable art (Gasparian and Arimura 2014). Similarly, the literature on the seventh millennium BC site of Bavra Ablari, Georgia (Varoutsikos et al. 2017), and Chokh in the North Caucasus, southernmost Russia (Kushnareva 1997), includes no references to human figurines. Manufacturing of human figurines was rare in the Mesolithic South Caucasus.

At the present research state, examples reportedly from the Mesolithic contexts of Gobustan, on the Caspian Sea coast (Farajova 2011), are the sole specimens for comparison. Gobustan figurines are often called "Venus." They are generally made of flat stones instead of Damjili type stick-shaped gravel (Sigari et al. 2020). Furthermore, they do not exhibit sophisticated engravings such as those on the Damjili figurine. Gobustan figurines are larger (up to 10-cm long). The abundance of petroglyphs similar to figurines in Gobustan suggests the traditional development of regionally different figurines in the South Caucasus during the Mesolithic period. However, the Damjili figurine suggests a different Mesolithic tradition of manufacturing portable art in the inland South Caucasus.

13.5 Conclusion

During the Mesolithic occupations at Damjili—late seventh millennium BC—Neolithic cultures were widely established in the Fertile Crescent of Southwest Asia, equipped with small figurines mostly made of clay as an essential cultural element (Kozlowski and Aurenche 2005). The manufacture of clay figurines in the Neolithic South Caucasus was probably derived from the Fertile Crescent along with the agro-pastoralist socio-economy. The Damjili Mesolithic figurine differs significantly from the Neolithic clay figurines popular in Middle Kura as well as the Neolithic figurines of the Fertile Crescent, which emphasize the buttocks and breasts (Narimanov 1987). Therefore, this specimen evokes a series of intriguing discussions on the similarities and dissimilarities between the symbolic aspects of Mesolithic and incoming farming societies and their meanings.

References

Farajova, M. (2011) Gobustan: rock art cultural landscape. *Adoranten* 2011: 41–66.

Gasparian, B. and M. Arimura (2014) *The Stone Age of Armenia*. Kanazawa: Kanazawa University.

Kozlowski, S. and O. Aurenche (2005) *Territories, boundaries and cultures in the Neolithic Near East*. Oxford: Archaeopress.

Kushnareva, K. K. (1997) *The Southern Caucasus in Prehistory: Stages of Cultural and Socioeconomic Development from the Eighth to the Second Millennium BC*. Philadelphia: University of Pennsylvania Museum of Archaeology.

Narimanov, I. (1987) *The Earliest Agricultural Settlements in the Territory of Azerbaijan*. Baku: Academy of Sciences.

Nishiaki, Y. and F. Guliyev (2020) *Göytepe – The Neolithic Excavations in the Middle Kura Valley, Azerbaijan*. Oxford: Archaeopress.

Nishiaki, Y., F. Guliyev, and S. Kadowaki (2021) *Hacı Elamxanlı Tepe – The Archaeological*

Investigations of an Early Neolithic Settlement in West Azerbaijan. Berlin: ex oriente.

Sigari, D., S. Shirini, and R. Abudllayev (2020) Gobustan rock art cultural landscape (Azerbaijan). *Encyclopedia Global Archaeology*, pp. 4618–4625. Springer, https://doi.org/10.1007/978-3-030-30018-0_2827.

Varoutsikos, B., A. Mgeladze, J. Chahoud, M. Gabunia, T. Agapishvili, L. Martin, and C. Chataigner (2017) From the Mesolithic to the Chalcolithic in the South Caucasus: new data from the Bavra Ablari rock shelter. In: *Context and Connection: Essays on the Archaeology of the Ancient Near East in Honour of Antonio Sagona*, edited by A. Batmaz, G. Bedianashvili, A. Michalewicz and A. Robinson, pp. 233–255. Leuven: Peeters.

Chapter 14

Macro-botanical remains from Damjili Cave

Chie Akashi

14.1 Introduction

This study focuses on the carpological remains of plants excavated from Damjili Cave, Gazakh Province, Western Azerbaijan. The preservation state of the remains is rather poor, but they have provided valuable insights for understanding Neolithization in the South Caucasus region and the relationship between indigenous hunter-gatherers and immigrant farmers.

Damjili Cave resembles a rockshelter, attracting many local people for the cold freshwater from its spring and socializing around it. Its importance began in prehistoric times, and it has provided shelter and water for hunters and livestock herders for a fairly long time.

Damjili has been excavated since 2016 as part of the Azerbaijani-Japanese joint expedition to reconstruct the Neolithization of the South Caucasus and prehistoric human movements in the region. This mission has already produced significant results regarding the establishment of the earliest food-producing economy in Western Azerbaijan by excavating two Neolithic (Shomutepe culture) sites: Hacı Elamxanlı Tepe and Göytepe (Nishiaki et al. 2013; Nishiaki et al. 2015) (Nishiaki and Guliyev 2020; Nishiaki et al. 2021). However, the oldest Shomutepe culture site, Hacı Elamxanlı Tepe, was already a full-fledged farming society (Akashi et al. 2018), dating as far back as 5950 cal BC. We still lacked a transitional site from the Mesolithic to the Neolithic period, which is crucial for discussing the Neolithization of the region.

Damjili Cave has occupations of the very period that we needed the most, which is the seventh and the sixth millennium BC. To date, this is the only site that enables us to trace the changes from the Mesolithic to the Neolithic period in the South Caucasus. The faunal remains of this site indicate that animal husbandry emerged for the first time in Unit 4 in the sixth millennium BC (see Chapter 15 of this volume). This chapter investigates how botanical remains can shed light on the transition from hunting-gathering to food production and the relationship between indigenous and immigrant populations.

14.2 Sampling, flotation, and sorting

More than 70 sediment samples were collected from the main excavation area (Trenches 7 and 9) during the four seasons from 2016 to 2019 for the macro-botanical study. The main targets were Units 4 (Neolithic) and 5 (Mesolithic), with several samples collected from Units 2 (Bronze Age), 3 (Chalcolithic), and 6 (archaeologically sterile).

This paper presents the results of 29 samples, including 1, 2, 18, and 8 from Units 2, 3, 4, and 5, respectively (see Table 14.1). Their contexts are mostly cultural fills unrelated to specific structures, but three samples from Unit 5 come from hearths, and one from Unit 4 comes from charcoal concentration. The total amount of sediment was 343 liters, and all samples were processed by the author using two large water flotation tanks and a 0.3 mm mesh sieve, the same device used for processing the samples from Hacı Elamxanlı Tepe and Göytepe. The light fractions obtained were dried, packed, and shipped

Table 14.1 Chronology of Damjili Cave and number of sediment samples for macro-botanical study (Nishiaki et al. 2022).

Units	Radiocarbon dates*	Periods	Number of samples
Unit 2	4800–4200 cal BP	Bronze Age	1
Unit 3	6500–5700 cal BP	Chalcolithic	2
Unit 4.1/2	7600–7300 cal BP	Neolithic	9
Unit 4.3/4	8000–7600 cal BP	Neolithic	9
Unit 5	8400–8000 cal BP	Mesolithic	8

to Japan for further analyses with permission from the Azerbaijan National Academy of Sciences.

Light fractions were examined under a binocular microscope, and seeds and plant parts were identified. The identification was performed with reference to modern specimens as well as published guides and reports. About 1,200 items were recovered, excluding unidentifiable fragments and uncharred remains (Fig. 14.1). Most plant remains are charred, with mineralized *Celtis* stones. Table 14.2 shows the plant assemblages found in each unit. A detailed list is provided in Table 14.3.

14.3 Results

14.3.1. Identified species

Tree fruits and nuts

Numerous mineralized *Celtis* stones were present in the flotation samples and handpicked seeds during the excavation (not included in this study). They were mostly fragmented; however, 33 stones were intact. *Celtis* was common in both Units 5 and 4, but its number dropped in the upper layers of Unit 4. One *Sambucus* seed was recovered from Layer 9 (Unit 4.4). Pieces of fresh juicy fruit were present in Units 4 and 5 and may belong to either *Celtis* or *Sambucus*. Twenty nutshells were found in all units except Unit 2, but were rather fragmented to be identified.

Possible crops

There was little evidence of agricultural activity in Damjili, except 10 large-seeded Poaceae grains, grouped as "possible cereal grain." They were all heavily charred and broken, and only one, probably barley, could be identified at the genus level. No chaffs were recovered. "Charred lumps" of starchy remains were also picked and counted because they could be overcharred grains, food remains, or tuberous parts of a plant. A detailed examination may enable us to clarify their positions in plant assemblages.

Other wild taxa

Other wild plant assemblages were not diverse. The two major wild taxa identified in Damjili were *Chenopodium album*-type and *Artemisia.* Among them, *Chenopodium album*-type was the predominant species. This species is often intrusive because modern chenopod seeds can easily fall to lower levels. However, the seeds from Damjili seemed charred, and their density per one liter of sediment was higher in Unit 4 than in the upper layers; therefore, it was included in the assemblage.

Artemisia was the second most predominant species. It grows throughout the temperate zone of Eurasia and is abundant in pollen diagrams; however, its seeds are generally scarce among the macro-remains. Damjili produced 221 charred seeds of *Artemisia* from Units 2, 4, and 5. The abundance of *Artemisia* seeds seems characteristic of only prehistoric West Azerbaijan. In terms of other Neolithic

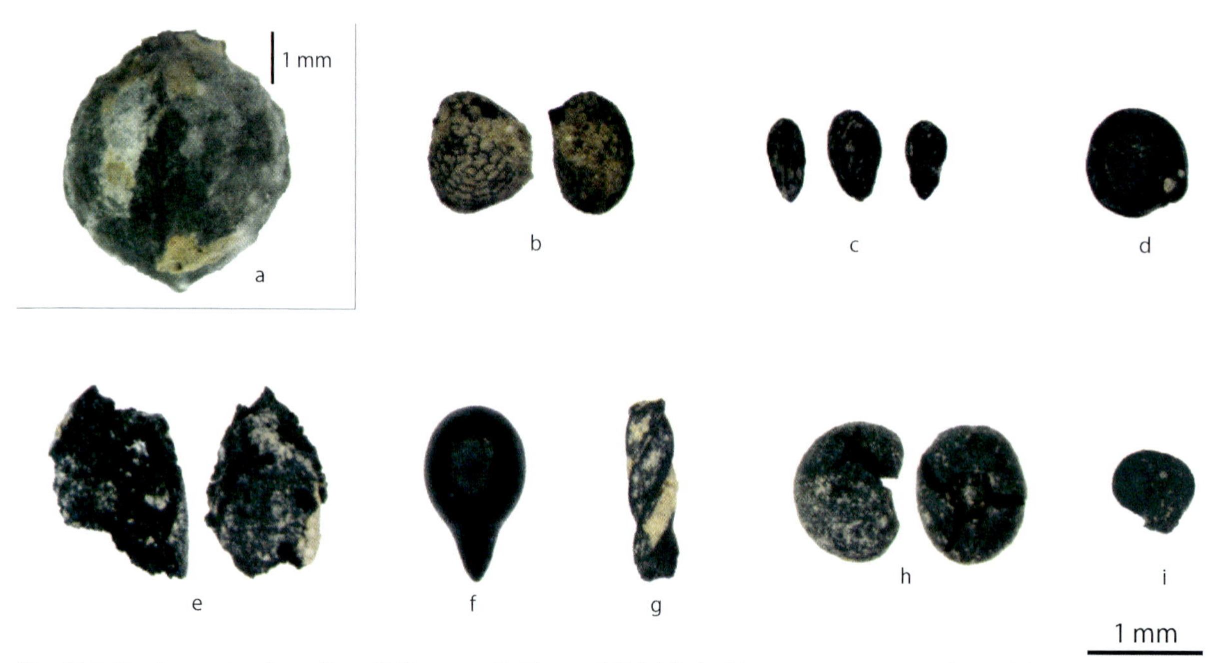

Fig. 14.1 Plant remains from Damjili Cave. a: *Celtis* sp. (Pit 9.15), b: *Hyoscyamus* sp., c: *Artemisia* sp., d: *Chenopodium album*-type (Pit 9.11), e: "possible cereal grain" (B3.6), f: *Thymelaea* sp. (Pit 9.14), g: *Stipa/ Erodium* awn, h: *Galium* sp. (Pit 9.21), i: fungal sclerotia.

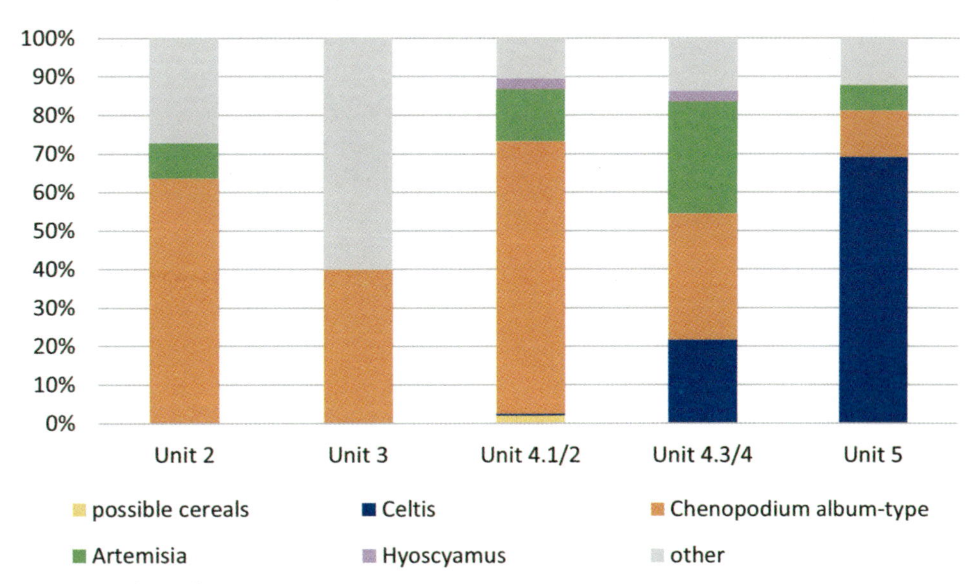

Fig. 14.2 Proportions of dominant species.

sites, Hacı Elamxanlı Tepe and Göytepe yielded hundreds of seeds (Akashi et al. 2018) while no more than a few seeds have been reported from the majority of West Asian sites.

Other (herbaceous) wild plants included *Hyoscyamus*, *Verbascum*, *Thymelaea*, *Malva*, Trifoliae, *Silene*, and *Galium*. Twenty-six *Hyoscyamus* seeds were present in Unit 4 itself. The remaining wild taxa occurred in much smaller quantities, with fewer than 10 items. In addition to the seeds and fruits, awns of *Stipa/Erodium* and unidentified plant parts (stem and rachis) were recovered.

The fungal sclerotia must be mentioned. They are globular shaped, about 0.5 mm in diameter, with a peduncle-like structure (Fig. 14.1: i). It is unclear whether they are charred or not. Considering the characteristic of Damjili as a natural spring, providing a wet environment for fungi to grow, these fungal sclerotia were treated as bioturbation in the present study. Moreover, their occurrence was 20 times higher in the upper units and decreased as the layers descended, as is expected in the case of modern intrusion datasets. Conversely, this may prove that the lower Units 4 and 5 received little bioturbation.

14.3.2 Plant assemblages in each unit

Unit 5

The only edible plant was *Celtis*, and 147 of its mineralized stones, including fragments, were found. *Chenopodium album*-type and *Artemisia* sp. were the only other taxa discovered in substantial numbers with 26 and 14 items, respectively. No large-seeded Poaceae remains were recovered. Three of the Unit 5 samples were hearths, but no plant species were particularly related to the hearth context.

Unit 4.3/4 (Layers 8, 9)

Seed density was the highest in these layers. *Celtis* stones continued to be numerous (108), and other tree fruit remains were common. Two large-seeded Poaceae grains were found, one of which may have been cereal. *Chenopodium album*-type and *Artemisia* existed in equally large quantities, with 166 and 146 items, respectively. *Hyoscyamus* appeared for the first time and was the only other species identified with more than 10 items.

Unit 4.1/2 (Layers 6, 7)

Celtis stones abruptly decreased in these layers to only two fragments. Instead, the cereal grains/charred lumps were more numerous in these layers than in any other layer. *Chenopodium album*-type was predominant, with more than 300 remains, followed by *Artemisia* (60) and *Hyoscyamus* seeds (12).

Unit 3

Two samples yielded 10 nutshell fragments, eight *Chenopodium album*-type seeds, and two charred lumps. *Celtis* stones disappeared from the unit.

Unit 2

To date, one sample has been sorted, and it has produced only wild herbaceous plants: *Chenopodium album*-type, *Artemisia*, *Caryophyllaceae*, and Trifoliae.

14.4 Discussion

The earliest evidence of agriculture in the Middle Kura Valley is the domesticated barley rachises discovered at the lowest level in Hacı Elamxanlı Tepe. Thus, the origin of farming in this region dates back to at least 6000 cal BC. A few pertinent questions

Table 14.2 Plant assemblages in each unit of Damjili Cave.

Unit	Unit 2	Unit 3	Unit 4.1/2	Unit 4.3/4	Unit 5	TOTAL
Layer	-	Layer 5	Layer 6, 7	Layer 8, 9	Layer 10, 11	
Sediment volume (l)	7	16	124	100	96	343
Light fraction volume (ml)	14.8	29.7	185.6	142.4	130.2	502.7
Celtis	-	-	2	108	147	257
Celtis (whole)				(19)	(14)	(33)
Sambucus	-	-	-	1	-	1
possible cereal grain	-	-	9	1	-	10
charred lump	-	2	16	-	-	18
fruit fresh	-	-	1	12	2	15
nut shell	-	10	5	4	1	20
Artemisia	1	-	60	146	14	221
Asteraceae	-	-	-	1	-	1
Asteraceae/Lamiaceae	-	-	-	14	-	14
Brassicaceae/Fabaceae	-	-	-	-	1	1
Silene	-	-	1	-	-	1
Caryophillaceae	2	-	4	-	-	6
Chenopodium album-type	7	8	316	166	26	523
Chenopodiaceae	-	-	3	-	-	3
Coronilla	-	-	1	-	-	1
Trifoliae	1	-	-	-	-	1
Fabaceae	-	-	3	1	2	6
Malva	-	-	-	1	1	2
Papaveraceae	-	-	-	6	-	6
Stipa/Erodium awn	-	-	3	3	2	8
Poaceae, large-seeded	-	-	-	1	-	1
Poaceae, small-seeded	-	-	1	-	2	3
Galium	-	-	-	-	1	1
Verbascum	-	-	-	3	-	3
Hyoscyamus	-	-	12	14	-	26
Thymelaea	-	-	-	2	-	2
plant part	-	-	-	1	2	3
fungal sclerotia	11	17	101	13	3	145
indetermine	17	9	132	365	139	662
total	39	46	670	863	343	1961
total (ex. Indet)	11	20	437	503	213	1184
total wild taxa (ex. tree fruits)	11	8	404	358	49	830
fungi (%)	28.21	36.96	18.69	1.51	0.87	7.39

are whether agriculture was practiced before that, and who were the first farmers in this region; were they indigenous people or immigrants from West Asia? The plant remains of Damjili offer some hints regarding these questions.

The only indication of crop is badly-charred, fragmented large-seeded Poaceae seeds grouped as "possible cereal grains." One of the grains may be barley, although its preservation state does not allow for further identification. Nevertheless, it is suggestive that these "possible cereal grains" appeared in Unit 4 and not Unit 5. Furthermore,

Table 14.3 List of plant remains from Damjili Cave. * In the row for sample number, “Pit” denotes one of the trenches opened in the 2016 season.

Unit	Unit 2	Unit 3	Unit 3	Unit 4	Unit 4	Unit 4	Unit 4	Unit 4	Unit 4	Unit 4	Unit 4	Unit 4	Unit 4	Unit 4	Unit 4	Unit 4	Unit 4	Unit 4	Unit 4	Unit 4	Unit 4	Unit 5	Unit 5	Unit 5	Unit 5	Unit 5	Unit 5	Unit 5	Unit 5
Layer	-	layer 5	layer 5	layer 6	layer 6	layer 6	layer 6	layer 7	layer 7	layer 7	layer 7	layer 7	layer 8	layer 8	layer 8	layer 9	layer 9	layer 9	layer 9	layer 9	layer 9	layer 10	layer 10	layer 10	layer 10	layer 11	layer 11	-	-
Sample no.	Pit7.8	A3.4	B0.4	A1.5	A3.5	Pit9.9	Pit9.10	A2.6	A1.6	A3.6	B3.6	Pit9.11	A3.7	B3.7	Pit9.12	A3.8	B3.8	B3.9	Pit9.13	Pit9.14	Pit9.15	A1.9	A3.12	B3.11	Pit9.17	Pit9.20	Pit9.21	Pit7.14	Pit7.15
																			charcoal concentration				stone structure	hearth ?	hearth		hearth		
sediment volume (l)	7	8	8	7	10	22	24	12	9	11	9	20	10	10	20	5	6	23	5	12	9	11	34	3.5	2.5	9	8	8	20
light fraction volume (ml)	14.8	17.5	12.2	5	7.5	44.5	39.5	8.5	5.6	7.5	4.5	63	5.4	9.4	43	2.8	4	18	10.8	32	17	13	37	3	6.6	10.6	15	15	30
Celtis all	-	-	-	-	-	-	-	-	-	-	-	2	5	6	10	17	10	31	-	15	14	30	64	4	31	-	9	-	9
(whole)														(1)		(2)	(2)	(6)		(5)	(3)		(11)	(1)					(2)
Sambucus	-	-	-	-	-	-	-	-	-	-	-	-	-	-	-	-	1	-	-	-	-	-	-	-	-	-	-	-	-
possible cereal grain	-	-	-	-	-	8	-	-	-	-	1	-	-	-	-	-	1	-	-	-	-	-	-	-	-	-	-	-	-
charred lump	-	2	-	-	-	11	-	1	1	3	-	-	-	-	-	-	-	-	-	-	-	-	-	-	-	-	-	-	-
fruit fresh	-	-	-	1	-	-	-	-	-	-	-	-	-	-	3	-	-	-	-	-	9	-	2	-	-	-	-	-	-
nut shell	-	-	10	-	2	3	-	-	-	-	-	-	-	2	1	-	-	1	-	-	-	-	-	-	-	1	-	-	-
Artemisia	1	-	-	-	-	1	9	-	-	-	1	49	1	1	101	-	4	2	9	14	14	-	3	-	-	1	2	2	6
Asteraceae	-	-	-	-	-	-	-	-	-	-	-	-	-	-	-	-	-	-	1	-	-	-	-	-	-	-	-	-	-
Asteraceae/Lamiaceae	-	-	-	-	-	-	-	-	-	-	-	-	-	-	-	-	-	-	-	14	-	-	-	-	-	-	-	-	-
Brassicaceae/Fabaceae	-	-	-	-	-	-	-	-	-	-	-	-	-	-	-	-	-	-	-	-	-	-	-	-	1	-	-	-	-
Silene	-	-	-	-	-	1	-	-	-	-	-	-	-	-	-	-	-	-	-	-	-	-	-	-	-	-	-	-	-
Caryophillaceae	2	-	-	-	-	1	-	-	-	-	-	3	-	-	-	-	-	-	-	-	-	-	-	-	-	-	-	-	-
Chenopodium album-type	7	8	-	8	-	50	22	4	1	2	-	229	2	2	42	1	9	2	6	54	48	8	16	-	2	-	-	-	-
Chenopodiaceae	-	-	-	-	-	-	-	-	-	1	-	2	-	-	-	-	-	-	-	-	-	-	-	-	-	-	-	-	-
Coronilla	-	-	-	1	-	-	-	-	-	-	-	-	-	-	-	-	-	-	-	-	-	-	-	-	-	-	-	-	-
Trifoliae	1	-	-	-	-	-	-	-	-	-	-	-	-	-	-	-	-	-	-	-	-	-	-	-	-	-	-	-	-
Fabaceae	-	-	-	-	-	-	1	-	-	2	-	-	-	-	-	-	-	1	-	-	-	-	2	-	-	-	-	-	-
Malva	-	-	-	-	-	-	-	-	-	-	-	-	1	-	-	-	-	-	-	-	-	-	1	-	-	-	-	-	-
Papaveraceae	-	-	-	-	-	-	-	-	-	-	-	-	-	-	-	-	-	-	-	5	1	-	-	-	-	-	-	-	-
Stipa/Erodium awn	-	-	-	-	-	-	-	-	-	-	-	3	1	-	-	-	-	-	-	-	2	-	-	-	-	-	2	-	-
Poaceae, large-seeded	-	-	-	-	-	-	-	-	-	-	-	-	-	-	-	-	1	-	-	-	-	-	-	-	-	-	-	-	-
Poaceae, small-seeded	-	-	-	-	-	1	-	-	-	-	-	-	-	-	-	-	-	-	-	-	-	-	2	-	-	-	-	-	-
Galium	-	-	-	-	-	-	-	-	-	-	-	-	-	-	-	-	-	-	-	-	-	-	1	-	-	-	-	-	-
Verbascum	-	-	-	-	-	-	-	-	-	-	-	-	-	1	-	-	-	1	-	1	-	-	-	-	-	-	-	-	-
Hyoscyamus	-	-	-	2	-	1	-	-	-	1	-	8	3	-	8	-	-	2	1	-	-	-	-	-	-	-	-	-	-
Thymelaea	-	-	-	-	-	-	-	-	-	-	-	-	-	-	1	-	-	-	-	1	-	-	-	-	-	-	-	-	-
plant part	-	-	-	-	-	-	-	-	-	-	-	-	-	-	-	-	-	-	-	-	1	-	-	-	-	2	-	-	-
fungal sclerotia	11	14	3	1	5	26	21	3	2	29	2	12	-	-	6	-	-	7	-	-	-	-	-	-	-	-	-	-	3
indetermine	17	5	4	3	2	49	33	10	6	15	1	13	8	3	151	1	1	11	4	76	110	2	46	35	37	-	6	8	5
total	39	29	17	16	9	152	86	18	10	53	5	321	21	15	323	19	27	58	21	180	199	40	137	39	71	4	19	10	23

Unit 5 did not produce any large-seeded Poaceae grains or "charred lumps." The complete absence of large-seeded Poaceae from Unit 5 implies that hunter-gatherers inhabited Damjili before 6000 cal BC. In the next period, that is, Unit 4, the farming society was already well-established in the Middle Kura Valley; therefore, it is only natural to consider that people in Damjili were familiar with agricultural products at this stage.

The lack of cereal chaff in all the units is not contradictory, as agricultural fields from the Neolithic period onwards could not be located near Damjili, and people must have brought cereals to this rockshelter after dehusking or cooking them.

Tree fruit exploitation was active in Damjili until the lower subunits of Unit 4, but dropped at upper Unit 4, subsequently disappeared despite the fact that *Celtis* stones can be preserved without charring. Decrease of tree fruits was also observed at mound settlements in the Middle Kura Valley; *Celtis* and nuts were common in the early sixth millennium BC (Hacı Elamxanlı Tepe) but scarce in the middle sixth millennium (Göytepe).

Among the wild taxa, the dominance of *Artemisia* has drawn attention. *Artemisia* grows throughout Eurasia at present as it did in the past because it appears abundantly in pollen diagrams. However, its seeds are scarce in the macro-botanical remains. The three prehistoric sites of West Azerbaijan, namely Damjili, Hacı Elamxanlı Tepe, and Göytepe, produced a large number of charred *Artemisia* seeds. In Göytepe, more than 600 *Artemisia* seeds were concentrated in a bin storing cereal chaff and were considered to have been used as fungicide or insecticide (Kadowaki et al. 2015; Akashi et al. 2018). If we assume that its existence in Unit 5 of Damjili means that the heavy use of *Artemisia* began as early as the Mesolithic age in Western Azerbaijan, it suggests that traditional medicinal plant use was passed down to a subsequent agricultural society. This implies that the Neolithization of this region involved not only immigrants but also indigenous peoples.

The importance of the other two dominant wild species, that is, *Chenopodium album*-type and *Hyoscyamus* sp., is less clear. The former was predominant in all the units whereas the latter appeared only in Unit 4. *Chenopodium* sp. is edible as a vegetable and has anthelmintic effects on livestock (Jabbar et al. 2007; Bussmann et al. 2016). *Hyoscyamus* contains scopolamine and atropine and has medical efficacy; smoke from burned seeds is often used as a painkiller (Maleki and Akhani 2018). If collected, they might have been used for medicinal or veterinary purposes.

14.5 Conclusion

Highlights of the macro-botanical remains from Damjili can be summarized as follows:

- Possible Evidence of cereals appears in Unit 4.1/2 of Damjili, contemporary to Göytepe
- There is no evidence of farming before 6000 cal BC in the Middle Kura Valley
- Intensive exploitation of *Celtis* sp. declines after Unit 4.3/4
- The variety of wild herbaceous plants is limited; *Chenopodium album*-type, *Artemisia* sp., and *Hyoscyamus* sp. might have been used
- Use of *Artemisia* started as early as Unit 5, Mesolithic period

Based on the archaeobotanical evidence of Damjili, the origin of crop cultivation in the Middle Kura Valley does not date back to earlier than 6000 cal BC. As observed in the Middle Kura Valley, agriculture appears rather abruptly, suggesting that it was introduced by immigrants from Southwest Asia.

However, this does not necessarily mean that immigrants from West Asia played a major role as first-time farmers in the region. The local tradition of utilizing *Artemisia* continued from the Mesolithic to the Neolithic period, which indicates that the Neolithization of the Middle Kura Valley region was accomplished through the interaction between immigrants and local hunter-gatherers.

References

Akashi, C., K. Tanno, F. Guliyev, and Y. Nishiaki (2018) Neolithisation processes of the South Caucasus: As viewed from macro-botanical analyses at Hacı Elamxanlı Tepe, West Azerbaijan. *Paléorient* 44(2): 75–89.

Bussmann, R. W., N. Y. Paniagua Zambrana, S. Sikharulidze, Z. Kikvidze, D. Kikodze, D. Tchelidze, M. Khutsishvili, K. Batsatsashvili, and R. E. Hart (2016) A comparative ethnobotany of Khevsureti, Samtskhe-Javakheti, Tusheti, Svaneti, and Racha-Lechkhumi, Republic of Georgia (Sakartvelo), Caucasus. *Journal of Ethnobiology and Ethnomedicine* 12(43): 18 pages.

Jabbar, A., M. Arfan Zamana, Z. Iqbal, M. Yaseen, and A. Shamima (2007) Anthelmintic activity of *Chenopodium album* (L.) and Caesalpinia crista (L.) against trichostrongylid nematodes of sheep. *Journal of Ethnopharmacology* 114: 86–91.

Kadowaki, S., L. A. Maher, M. Portillo, R. Maria Albert, C. Akashi, F. Guliyev, and Y. Nishiaki

(2015) Geoarchaeological and palaeobotanical evidence for prehistoric cereal storage in the southern Caucasus: The Neolithic settlement of Goytepe (mid 8th millennium BP). *Journal of Archaeological Science* 53: 408–425.

Maleki, T. and H. Akhani (2018) Ethnobotanical and ethnomedicinal studies in Baluchi tribes: A case study in Mt. Taftan, southeastern Iran. *Journal of Ethnopharmacology* 217: 163–177.

Nishiaki, Y., F. Guliyev, S. Kadowaki, Y. Arimatsu, Y. Hayakawa, K. Shimogama, T. Miki, C. Akashi, S. Arai, and S. Salimbeyev (2013) Hacı Elamxanlı Tepe: Excavations of the earliest Pottery Neolithic occupations on the Middle Kura, Azerbaijan, 2012. *Archäologische Mitteilungen aus Iran und Turan* 45: 1–25.

Nishiaki, Y., F. Guliyev, S. Kadowaki, V. Alakbarov, T. Miki, S. Salimbeyev, C. Akashi, and S. Arai (2015) Investigating cultural and socioeconomic change at the beginning of the Pottery Neolithic in the Southern Caucasus: The 2013 excavations at Hacı Elamxanlı Tepe, Azerbaijan. *Bulletin of the American School of Oriental Research* 374: 1–28.

Nishiaki, Y. and F. Guliyev (2020) *Goytepe: Neolithic Excavations in the Middle Kura Valley, Azerbaijan*. Oxford: Archaeopress.

Nishiaki, Y., F. Guliyev, and S. Kadowaki (2021) *Hacı Elamxanlı Tepe: The Archaeological Investigations of an Early Neolithic Settlement in West Azerbaijan*. Berlin: ex oriente.

Chapter 15

Faunal remains from Damjili Cave

Saiji Arai

15.1 Introduction

Excavations at the Damjili Cave offer a critical dataset for the history of animal exploitation in the northern foothills of the Lesser Caucasus during the Holocene. This chapter presents the results of a study of faunal materials obtained during five seasons (2016 to 2019 and 2022) at the site. Thereafter, the trend of animal economy during each period with special reference to the introduction of domestic ungulates in this region, is discussed.

15.2 Materials and methods

The materials analyzed here are confined to specimens from Trenches 7 and 9, as excavations in other trenches did not reach prehistoric periods (Chapter 4 of this volume). Only Trenches 7 and 9 comprised well-stratified layers from the Mesolithic to medieval periods, with a substantial number of animal bones. Most materials were obtained from Trench 9; however, notably, different recovery methods were used for each unit of cultural layer. The materials from Units 4 (Neolithic) and 5 (Mesolithic) were collected by dry sieving (3 mm mesh), whereas those from Units 1 (medieval) and 2 (Bronze Age) were handpicked during the excavation. Materials from Unit 3 were partially gathered using a more flexible method; half of the sediments were dry sieved. Different recovery methods between the units resulted in large assemblages of animal bones from the Mesolithic and Neolithic periods, whereas those from the later periods were much smaller. Hence, we report the results of the analysis focusing on materials from Units 5 to 4. Collected materials were analyzed at an excavation house in Gazakh or the National Academy of Sciences of Azerbaijan in Baku using atlases (e.g., Boessneck et al. 1964) and photographs of animal bones, as necessary. A total of 12,498 fragments of faunal materials weighing approximately 10 kg were recorded from all the trenches (Tables 15.1 and 15.2). Of these, 11,529 fragments (c. 9.1 kg) were obtained from Trench 9. Owing to severe fragmentation, less than 10% of the assemblage was identified at genus or subfamily levels. For the caprine remains from Unit 4, the culling profiles were established based on their tooth wear status.

15.3 Species identified

Caprine

The total number of caprine, sheep (*Ovis orientalis* and *O. aries*), and goat (*Capra aegagrus* and *C. hircus*) bones predominated the faunal assemblage from each cultural unit of Damjili, except for Unit 6. Based on the morphological identification of cranial and postcranial bones, sheep bones overwhelmed those of goats in each unit. Metrical analysis of postcranial materials demonstrated that sheep bones from Unit 5 (Fig. 15.1) were much larger than those from Unit 4 (Fig. 15.2) and other Neolithic settlements in the lowlands (see below).

Bovine

Bos taurus and *B. primigenius* (or perhaps *Bison bonasus*) were rarely found in the faunal assemblage of Damjili, except in Units 3 and 1. The scarcity of cattle bones in Unit 4 does not match the frequency of cattle in the Neolithic South Caucasus (Kushnareva 1997: table 7). This would reflect either the site's function as a short-term camp or the chronological position of Damjili, since cattle rarely appeared in the early phase of the Neolithic (Vila et al. 2017; Arai 2021), or both. In addition, bovines are virtually absent in Unit 5, except for two fragments of postcranial elements, which do not allow us to conclude the exploitation of bovines during the Mesolithic period. The measured value of the distal radius in Unit 4 demonstrates the presence of domestic cattle in the Neolithic layer.

Suid

Several boar bones (*Sus scrofa*) were identified in Units 5, 4, and 1. None of these belonged to domestic species, at least those from Units 5 and 4, as indicated by the large astragalus from Unit 4.

Table 15.1 Number of identified specimens from Damjili (2016–2022).

	Unit 6		Unit 5		Unit 4		Unit 3		Unit 2		Unit 1	
	NISP	%NISP	NISP	%NISP	NISP	%NISP	NISP	%NISP	NISP	%NISP	NISP	%NISP
Ovis sp.	1	5%	28	9%	80	11%	5	11%	9	18%	0	0%
Capra sp.	0	0%	4	1%	19	3%	2	4%	0	0%	0	0%
Ovis / Capra	0	0%	72	23%	382	54%	22	49%	26	52%	13	52%
Ovis / Capra / Gazella	0	0%	8	3%	5	1%	0	0%	0	0%	0	0%
Bos sp.	0	0%	2	1%	11	2%	8	18%	4	8%	5	20%
Sus scrofa	0	0%	2	1%	7	1%	0	0%	0	0%	1	4%
Cervus elaphus	0	0%	19	6%	14	2%	0	0%	0	0%	0	0%
Bos / Cervus	0	0%	0	0%	5	1%	0	0%	0	0%	0	0%
Gazella subgutturosa	9	43%	10	3%	0	0%	0	0%	2	4%	0	0%
Ursus arctos	0	0%	1	0%	0	0%	0	0%	0	0%	0	0%
Panthera pardus	0	0%	4	1%	0	0%	0	0%	0	0%	0	0%
Felis sp.	0	0%	3	1%	3	0%	0	0%	0	0%	0	0%
Vulpes vulpes	0	0%	6	2%	12	2%	0	0%	0	0%	0	0%
Lepus sp.	0	0%	2	1%	8	1%	0	0%	0	0%	0	0%
Vulpes / Lepus	0	0%	0	0%	7	1%	0	0%	0	0%	0	0%
Meles meles	0	0%	0	0%	1	0%	0	0%	0	0%	0	0%
Small carnivore	2	10%	5	2%	14	2%	0	0%	0	0%	0	0%
Small rodent	0	0%	5	2%	5	1%	1	2%	2	4%	0	0%
Aves	0	0%	12	4%	10	1%	0	0%	1	2%	0	0%
Fish	1	5%	8	3%	10	1%	0	0%	0	0%	0	0%
Snake	0	0%	23	7%	4	1%	0	0%	0	0%	0	0%
Tortoise	8	38%	82	26%	81	11%	1	2%	0	0%	2	8%
Frog	0	0%	1	0%	2	0%	0	0%	0	0%	0	0%
Mollusc	0	0%	15	5%	32	4%	6	13%	6	12%	4	16%
Total	21	100%	312	100%	712	100%	45	100%	50	100%	25	100%

Table 15.2 Weights of identified specimens from Damjili (2016–2022).

	Unit 6		Unit 5		Unit 4		Unit 3		Unit 2		Unit 1	
	Weight(g)	%Weight	Weight(g)	%Weight	Weight(g)	%Weight	Weight(g)	%Weight	Weight(g)	%Weight	Weight(g)	%Weight
Ovis sp.	1.2	4%	144.8	16%	332.5	19%	16	4%	32.9	13%	0	0%
Capra sp.	0	0%	10.6	1%	63.3	4%	5.4	1%	0	0%	0	0%
Ovis / Capra	0	0%	125.7	14%	723.1	41%	134.6	35%	65.1	25%	23.1	18%
Ovis / Capra / Gazella	0	0%	6.5	1%	5.3	0%	0	0%	0	0%	0	0%
Bos sp.	0	0%	18.5	2%	52.3	3%	221.5	57%	83.3	32%	83.3	67%
Sus scrofa	0	0%	7.7	1%	13.2	1%	0	0%	0	0%	6.1	5%
Cervus elaphus	0	0%	158.3	18%	129.2	7%	0	0%	0	0%	0	0%
Bos / Cervus	0	0%	0	0%	86.1	5%	0	0%	0	0%	0	0%
Gazella subgutturosa	5.8	20%	16.6	2%	0	0%	0	0%	20.3	8%	0	0%
Ursus arctos	0	0%	7.8	1%	0	0%	0	0%	0	0%	0	0%
Panthera pardus	0	0%	19.6	2%	0	0%	0	0%	0	0%	0	0%
Felis sp.	0	0%	1.2	0%	5.7	0%	0	0%	0	0%	0	0%
Vulpes vulpes	0	0%	1.4	0%	7	0%	0	0%	0	0%	0	0%
Lepus sp.	0	0%	0.7	0%	7.8	0%	0	0%	0	0%	0	0%
Vulpes / Lepus	0	0%	0	0%	6.1	0%	0	0%	0	0%	0	0%
Meles meles	0	0%	0	0%	2.6	0%	0	0%	0	0%	0	0%
Small carnivore	1.6	6%	1.8	0%	5.4	0%	0	0%	0	0%	0	0%
Small rodent	0	0%	1.8	0%	0.1	0%	<0.1	0%	0.1	0%	0	0%
Aves	0	0%	2.7	0%	7.3	0%	0	0%	<0.1	0%	0	0%
Fish	0.3	1%	3.2	0%	2.6	0%	0	0%	0	0%	0	0%
Snake	0	0%	2.7	0%	0.4	0%	0	0%	0	0%	0	0%
Tortoise	19.6	69%	348.4	40%	266.8	15%	0.9	0%	0	0%	4.8	4%
Frog	0	0%	<0.1	0%	0.4	0%	0	0%	0	0%	0	0%
Mollusc	0	0%	1.7	0%	35.2	2%	8.6	2%	56.4	22%	7.9	6%
Total	28.5	100%	881.7	100%	1752.4	100%	387	100%	258.1	100%	125.2	100%

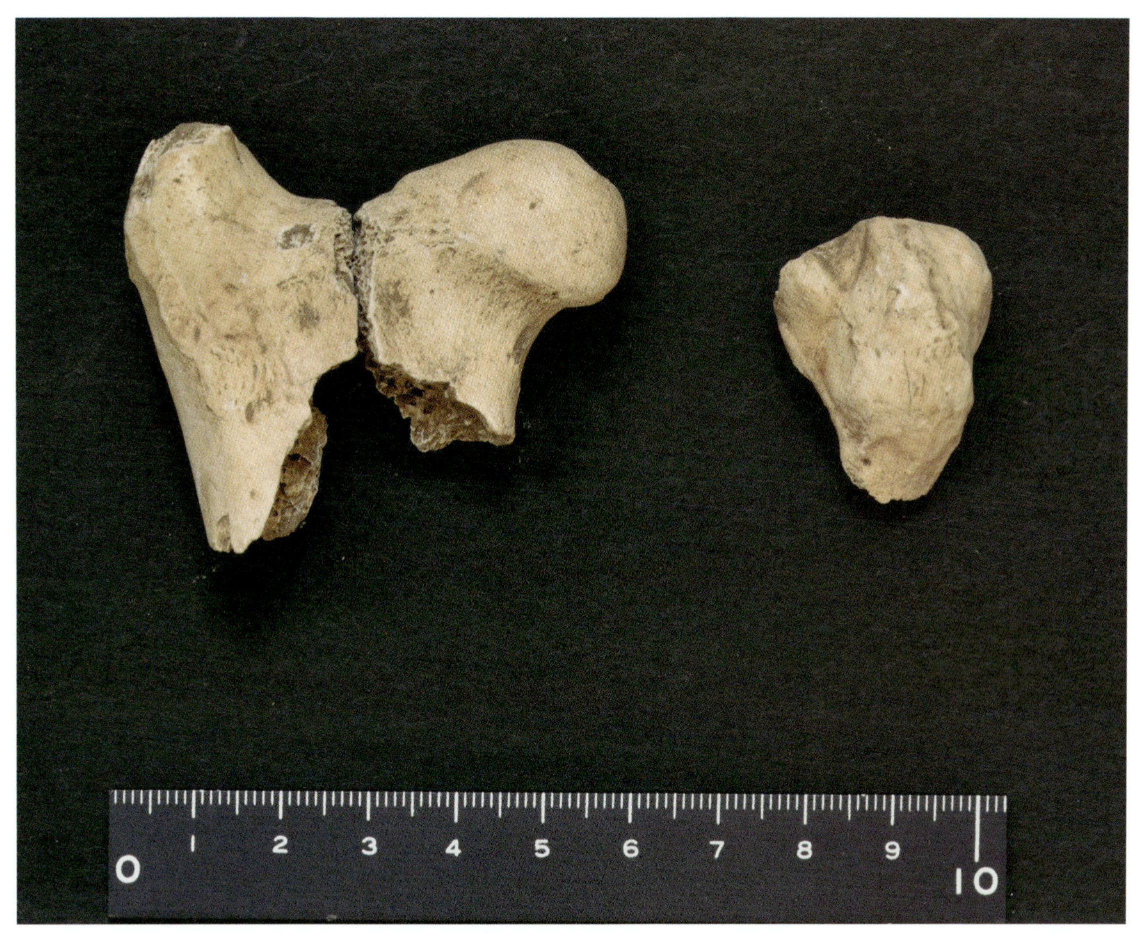

Fig. 15.1 Proximal femur and patera of *Ovis* sp. from Unit 5.

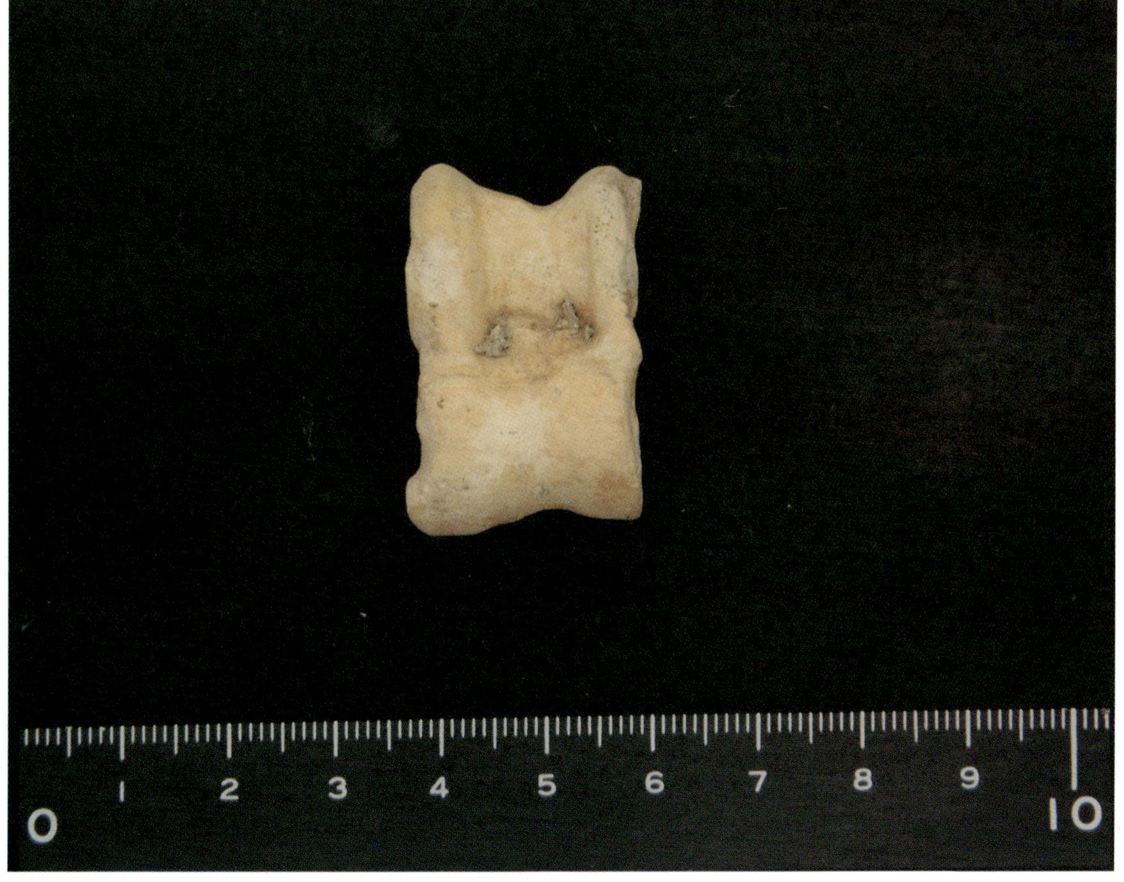

Fig. 15.2 Astragalus of *Ovis* sp. from Unit 4.

Deer

Red deer (*Cervus elaphus*) appeared in the assemblages from Units 5 and 4. While both assemblages primarily comprised lower limb bones (metapodia and phalanges) (Fig. 15.3), the Unit 4 assemblage included more cranial and antler fragments (Fig. 15.4).

Fig. 15.3 Second and third phalanges of *Cervus elaphus* from Unit 5.

Fig. 15.4 Mandibular dP4 of *Cervus elaphus* from Unit 4.

Gazelle

Most goitered gazelle (*Gazella subgutturosa*) remains were found in Units 6 and 5 (Fig. 15.5). Particularly, gazelle bones accounted for more than 40% of the assemblages in Unit 6. However, they virtually disappear from the later assemblage, except for two fragments from Unit 2, which would reflect a change in the hunting strategy of the prehistoric visitors.

Other mammalian taxa

The faunal assemblage from Damjili comprises various large, medium, and small mammalian remains, which include bear (*Ursus* cf. *arctos*), leopard (*Panthera pardus*) (Figs. 15.6; 15.7), wild cat (*Felis* cf. *sylvestris*) (Fig. 15.8), fox (*Vulpes vulpes*) (Fig. 15.9), hare (*Lepus* cf. *europaeus*) (Fig. 15.10), badger (*Meles meles*), and unidentified small rodents. They were only found in Units 5 and 4, indicating a decline in hunting activity during the later periods. Unit 5 assemblage includes large games, such as bears and leopards, whereas Unit 4 assemblage includes only small games, such as foxes and hares.

Aves

Although 23 fragments of bird bones were included in the assemblage, none have been identified to date.

Fish

Nineteen fragments of fish (possibly Cyprinidae) bones were recovered; however, no vertebrae were observed. This biased result may reflect cooking activities and the pattern of spatial use at the site.

Reptiles

Both the carapace and limb bones of tortoise (*Testudo graeca*) were regularly recovered from Units 6 to 4. Different proportions of tortoise remains between periods may reflect different durations of stay at the site (Stiner et al. 2000). Mandibles of snakes (or lizards) also frequently appear in Unit 5, and to a lesser extent, in Unit 4 (Fig. 15.11). Although none were identified to genus level, many snakes inhabit the cave and use the excavation trench as a nest every year.

Amphibians and mollusks

In addition to the aforementioned animals, frog

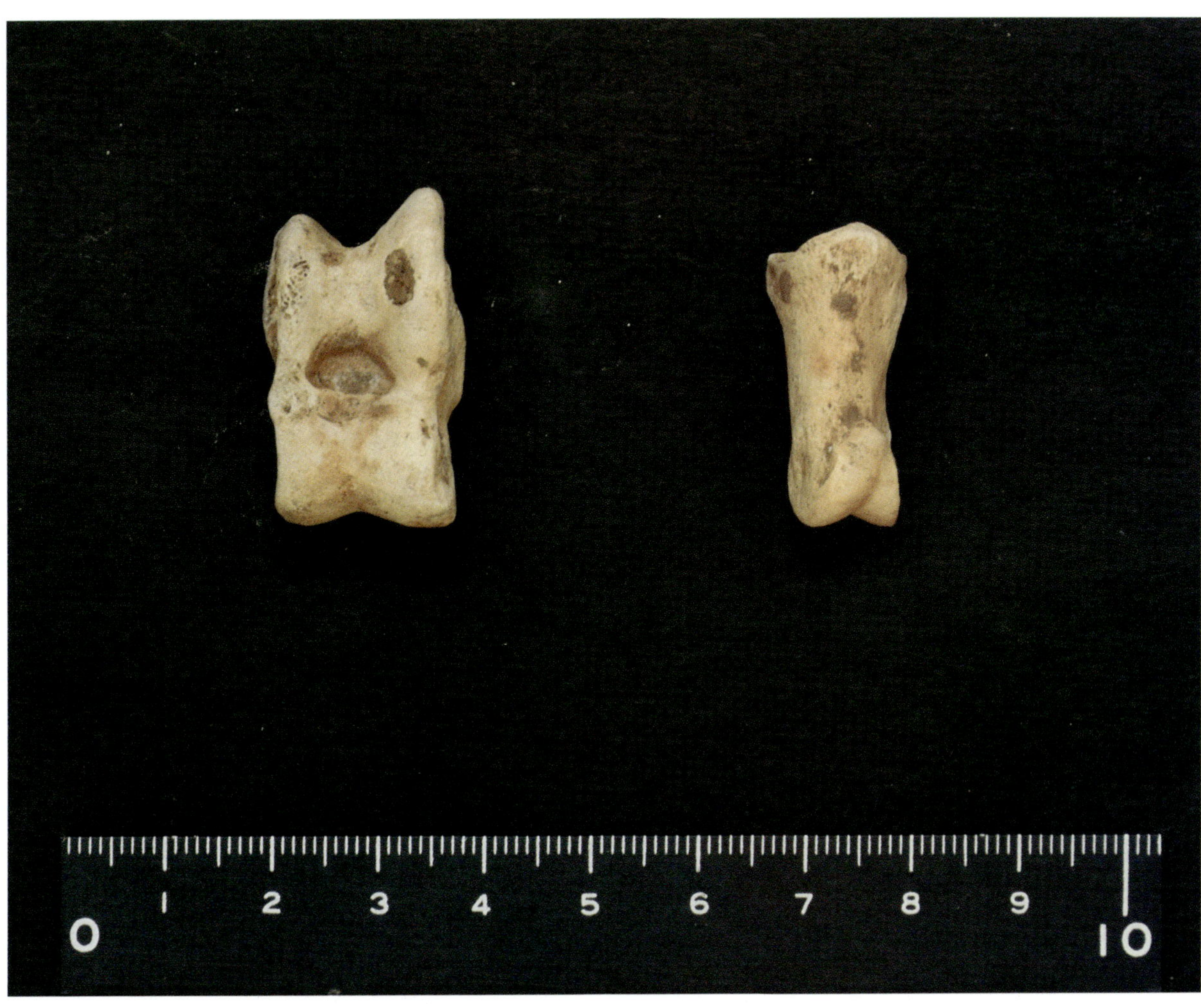

Fig. 15.5 Astragalus and proximal phalanx of *Gazella subgutturosa* from Unit 6.

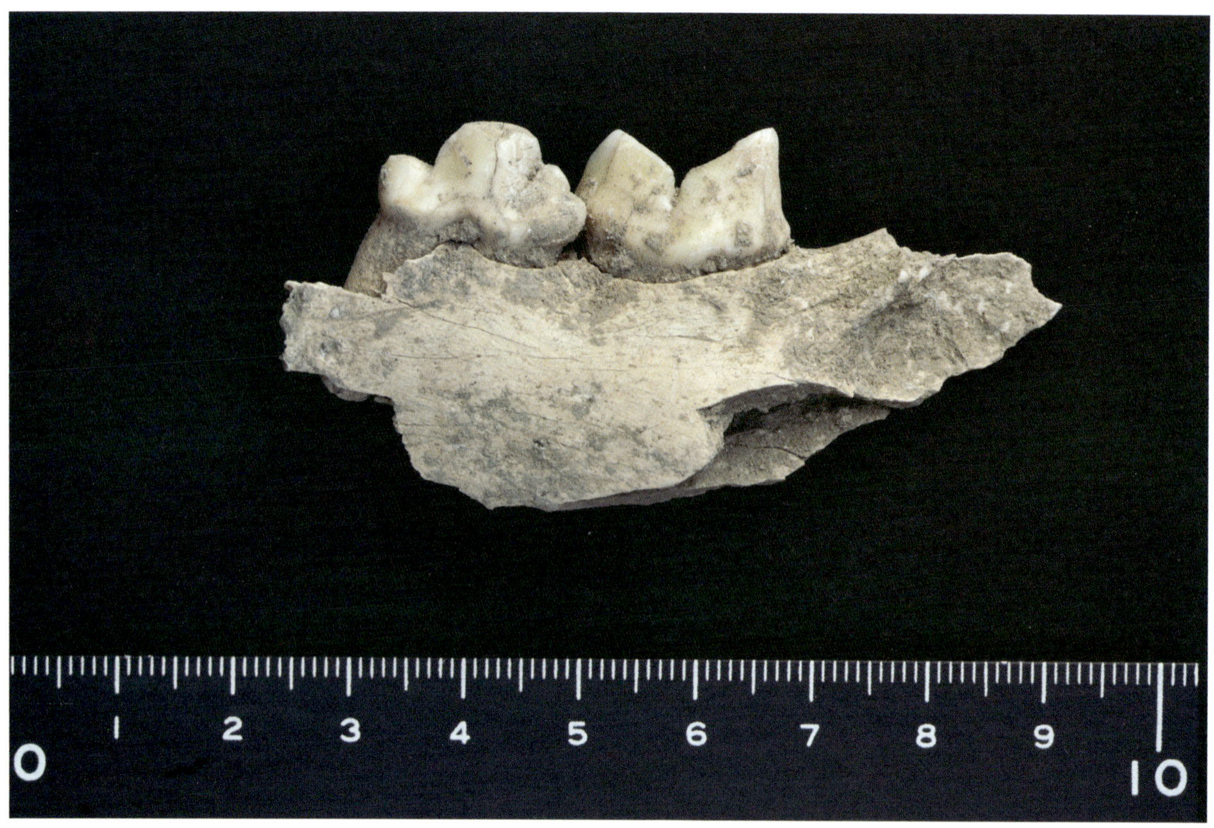

Fig. 15.6 Mandible of *Panthera pardus* from Unit 5.

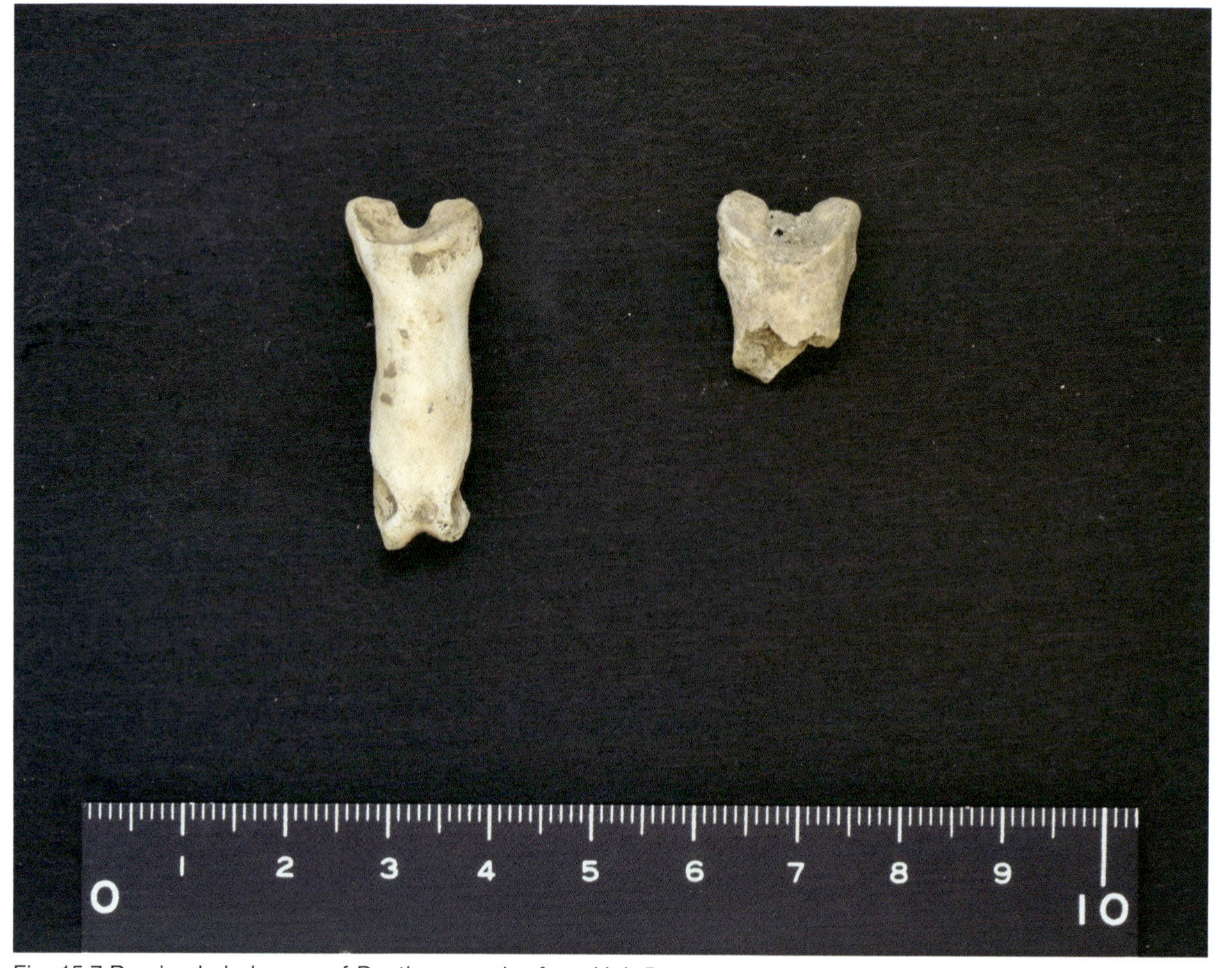

Fig. 15.7 Proximal phalanges of *Panthera pardus* from Unit 5.

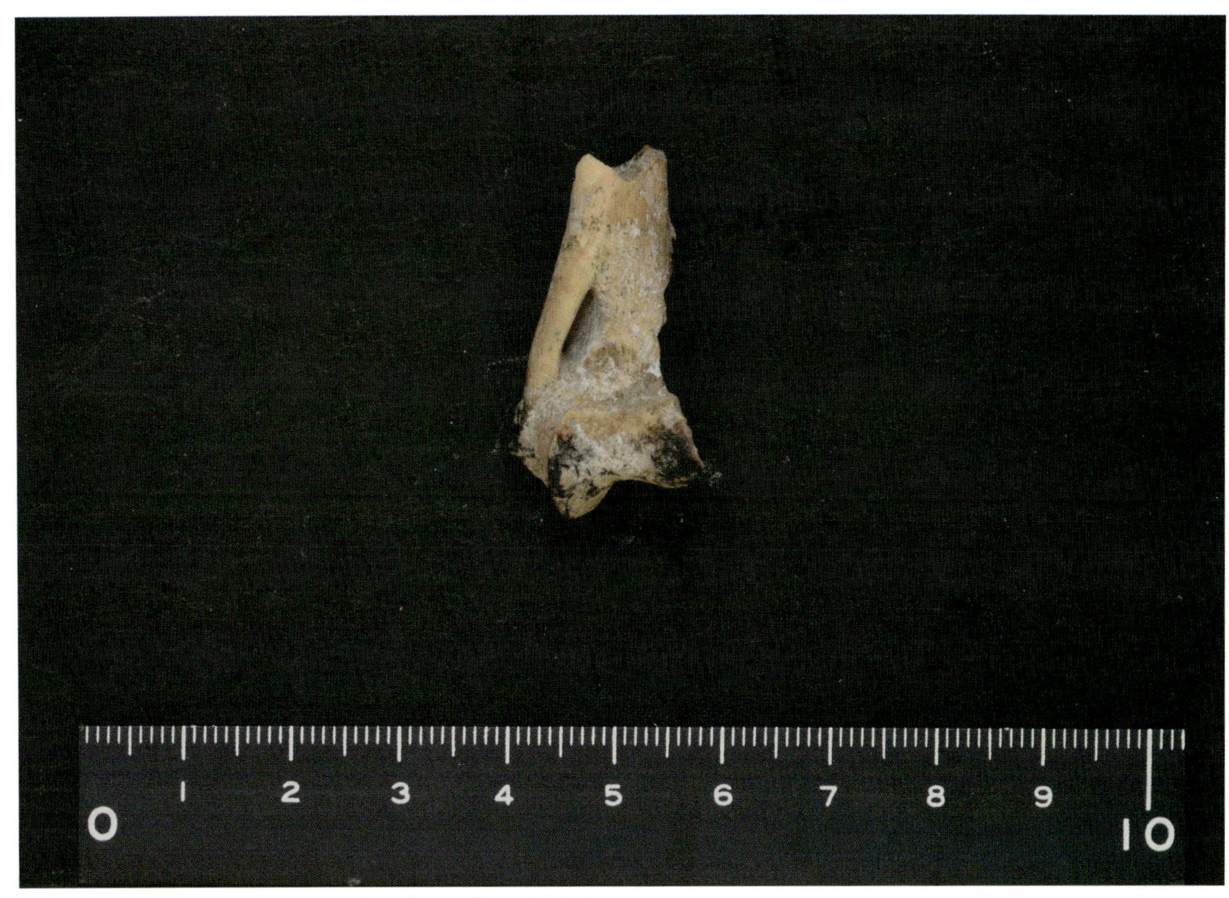

Fig. 15.8 Burnt distal humerus of *Felis* sp. from Unit 4.

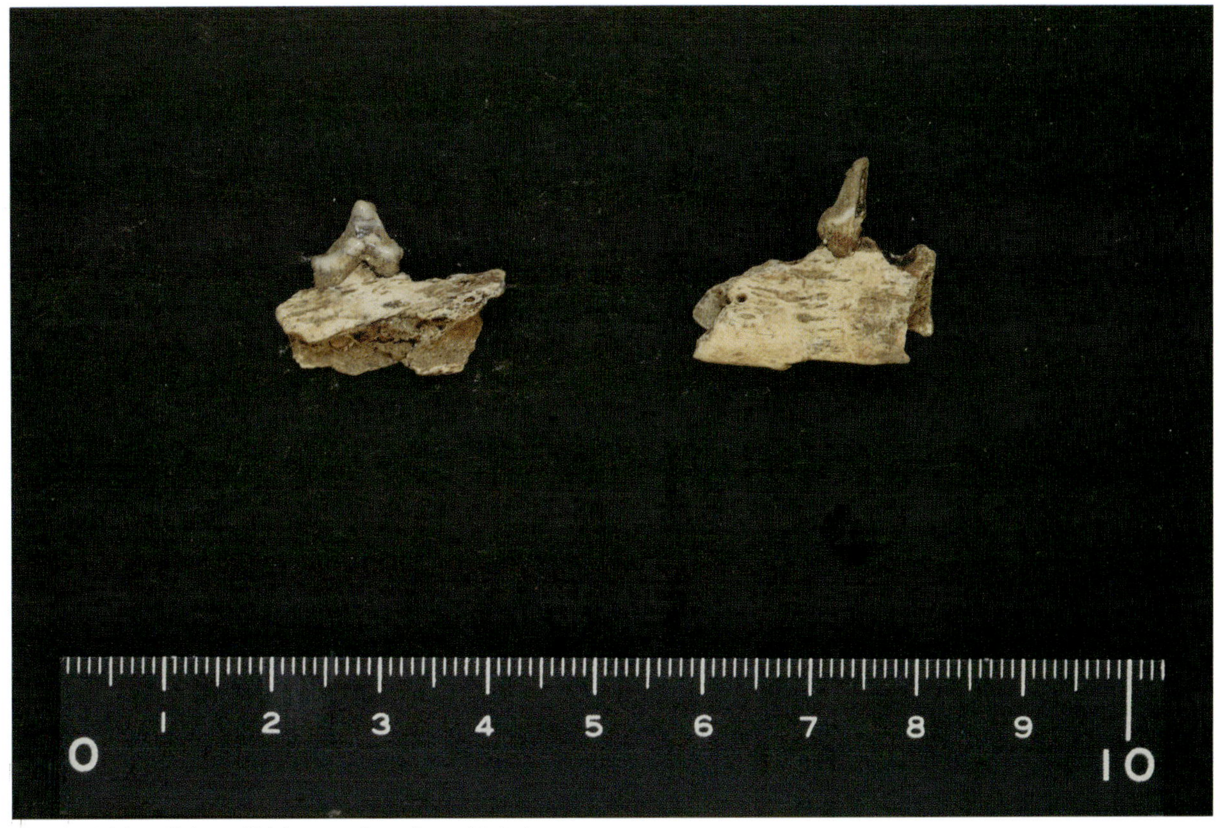

Fig. 15.9 Mandibles of *Vulpes vulpes* from Unit 4.

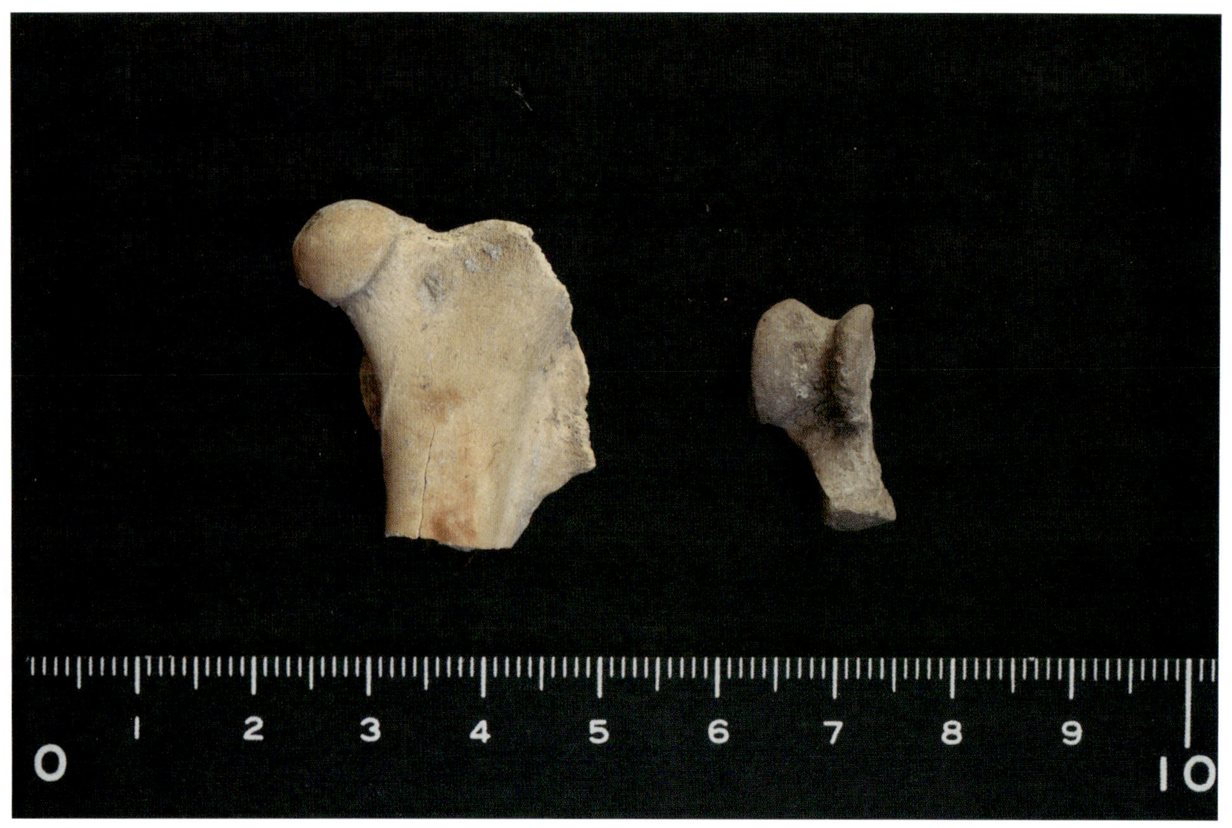

Fig. 15.10 Proximal femur and astragalus of *Lepus europaeus* from Unit 4.

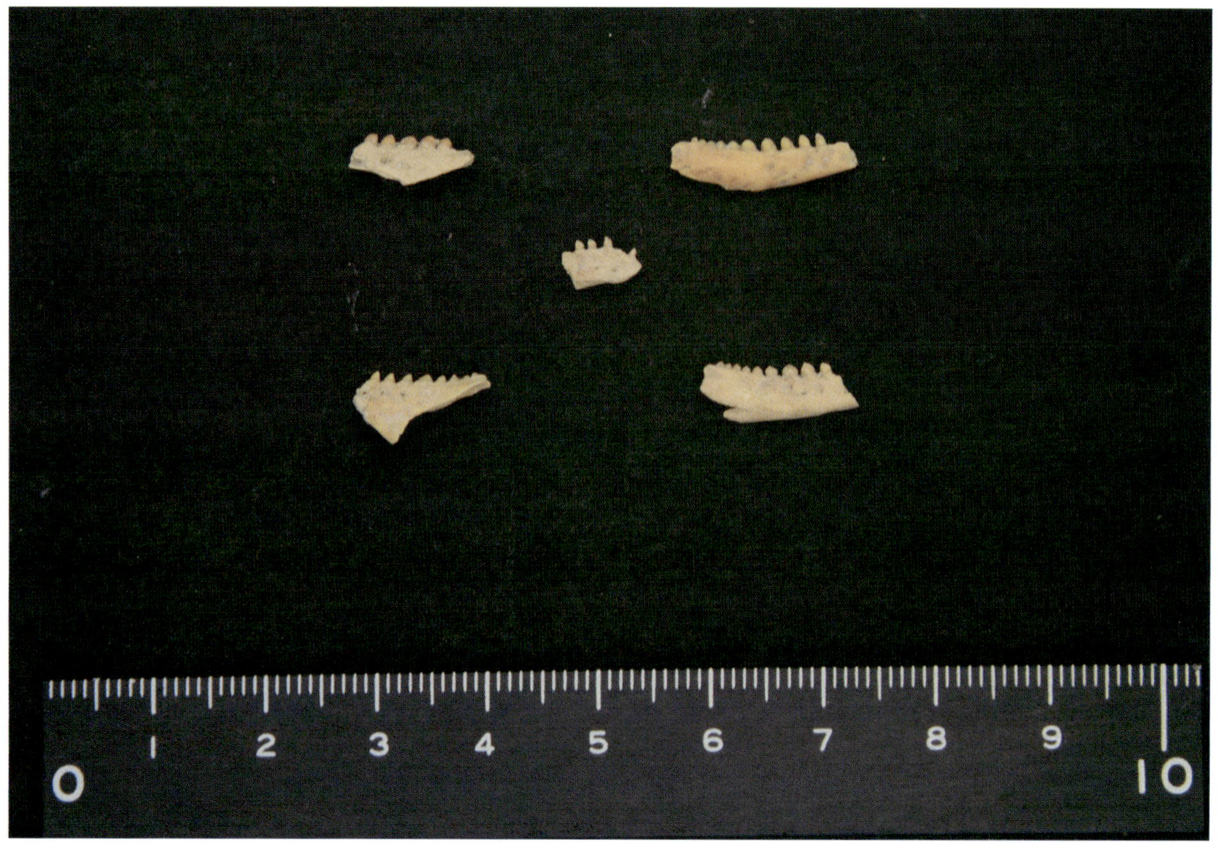

Fig. 15.11 Mandibles of snakes from Units 5 and 4.

bones and land snail shells were included in the assemblage.

15.4 Change of animal exploitation between Units 5 and 4

A critical change in the frequency of species was observed between Units 5 and 4: a dramatic increase in caprines and a decline in large game hunting. Caprines from Unit 5 comprised approximately 30% of the assemblage, whereas those from Unit 4 comprised approximately 70% (Fig. 15.12). In addition, comparing the metrical data of *Ovis* sp. between the units, those from Unit 4 were much smaller than those from Unit 5 and were in accordance with the size range of sheep from Neolithic settlements in the lowland (Fig. 15.13). This result can be interpreted as the introduction of domestic sheep to Damjili during the Neolithic period. Alternatively, the change from large to small game in Unit 4 reflects the different roles of hunting during the Mesolithic and Neolithic periods. Large games from Unit 5 possibly offered both meat and skin, whereas small games from Unit 4 do not appear to contribute to the meat supply, however, contribute to the skin. Considering the presence of large and dangerous species such as bears and leopards in the Unit 5 assemblage, it appears plausible that hunting activity was important in both the economic and social aspects of the Mesolithic people (Bar-Oz et al. 2009).

15.5 Role of sheep and goats during Neolithic period

The culling profile for the caprine, following Helmer's method (Helmer 1995, 2000), was established only in Unit 4, as sufficient maxillary and mandibular cheek teeth were available. The results for all specimens, including sheep, goat and sheep or goat teeth, reveal that both Stages C (6–12 months) and G (4–6 years) were the main ages at slaughter (Fig. 15.14). Hence, it is suggested that 1) exploitation of both primary (meat) and secondary products (milk and fleece) was important for Neolithic visitors to Damjili (Helmer et al. 2007: table 1); and 2) juvenile and adult caprine were included in the herd from the visitors' homebase. The latter would bias the culling profiles of settlements in the lowlands. However, when we establish culling profiles for sheep and goats separately, each profile displays different strategies: for sheep, focus was on juvenile to prime adults (Fig. 15.15), whereas for goats, focus was exclusively on adults (Fig. 15.16). This result is consistent with that from settlements in the lowlands, thus indicating the different roles of these two species during the Neolithic period.

This result also implicates the function and seasonality of site use because the assemblage lacks caprines at Stage B (2–6 months). Assuming spring (March to May) is the general birth season, at least based on the kill-off pattern of sheep and goats, there is no evidence of summertime occupation at the site. This indicates that Damjili was a temporary camp for herders from their villages between autumn to spring. This remains to be confirmed by the identification of other seasonal resources, such as birds, or by isotopic studies of caprine molars.

15.6 Concluding remarks

Analysis of faunal remains from Damjili provided a crucial dataset of past human-animal relations in the Southern Caucasus. The most important result indi-

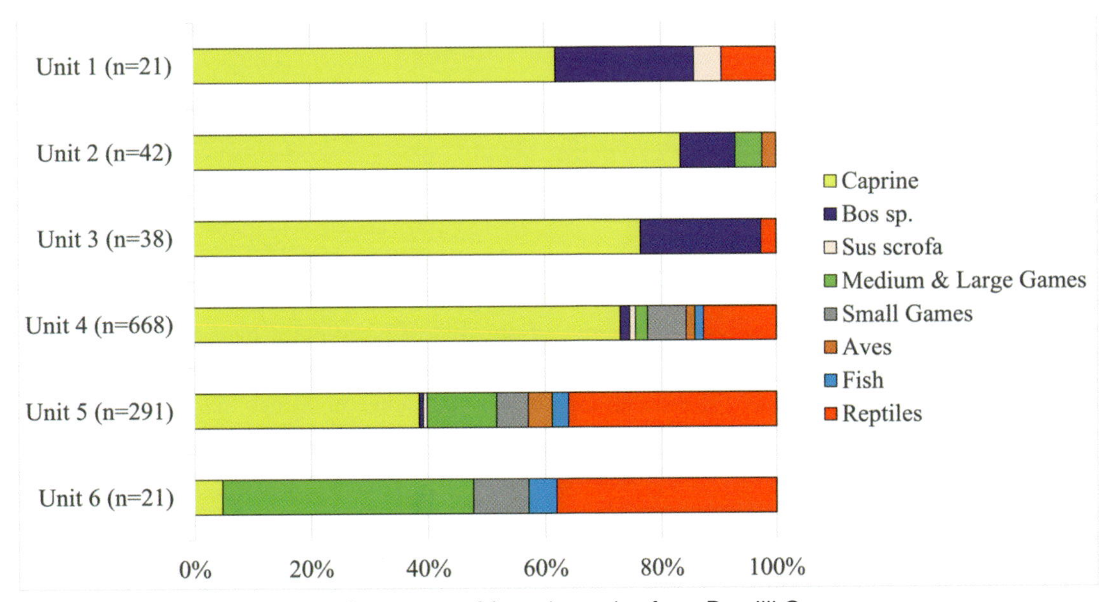

Fig. 15.12 Temporal change in frequency of faunal species from Damjili Cave.

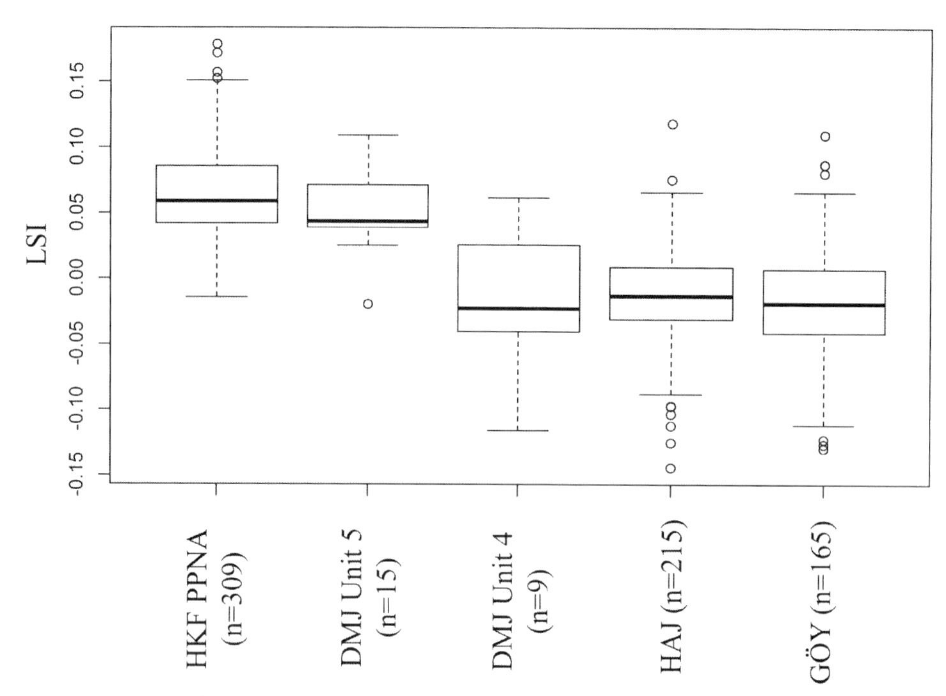

Fig. 15.13 Size comparison of *Ovis* sp. from Damjili Cave and other Neolithic sites. HKF=Hasankeyf (PPNA, Arai unpublished data), DMJ=Damjili, HAJ= Hacı Elamxanlı Tepe (Arai 2021), GÖY=Göytepe (Arai 2020).

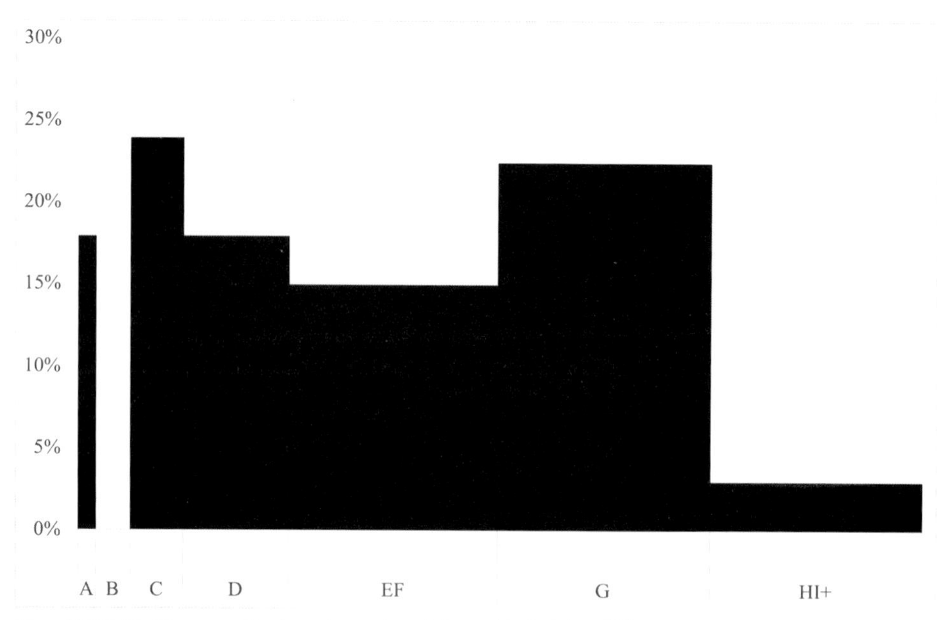

Fig. 15.14 Culling profile for sheep/goat from Unit 4 (n=40).

cated how quickly domestic ungulates were incorporated into the animal economy of local community. As indicated by the metrical analysis of sheep bones, there is no evidence of livestock use during the Late Mesolithic period, whereas the Neolithic economy fully relied on domestic ungulates, except for the additional use of small prey. The decline in large games may have involved a change in the role of hunting and hunters' positions within the community. These small games disappeared completely during the Chalcolithic period, suggesting the establishment of a mature livestock economy in the region. The function of Damjili may have changed from that of a temporal base for hunting to that of a pastoral camp. Wild fauna from the site indicate a strong focus on forest mammals; however, the Mesolithic people appear to have exploited a wider area since the steppic gazelle was frequently hunted. The results here should not be interpreted separately but together with the results from other sites because Neolithic settlements in the lowlands provide evidence for a wide range of animal exploitation, in-

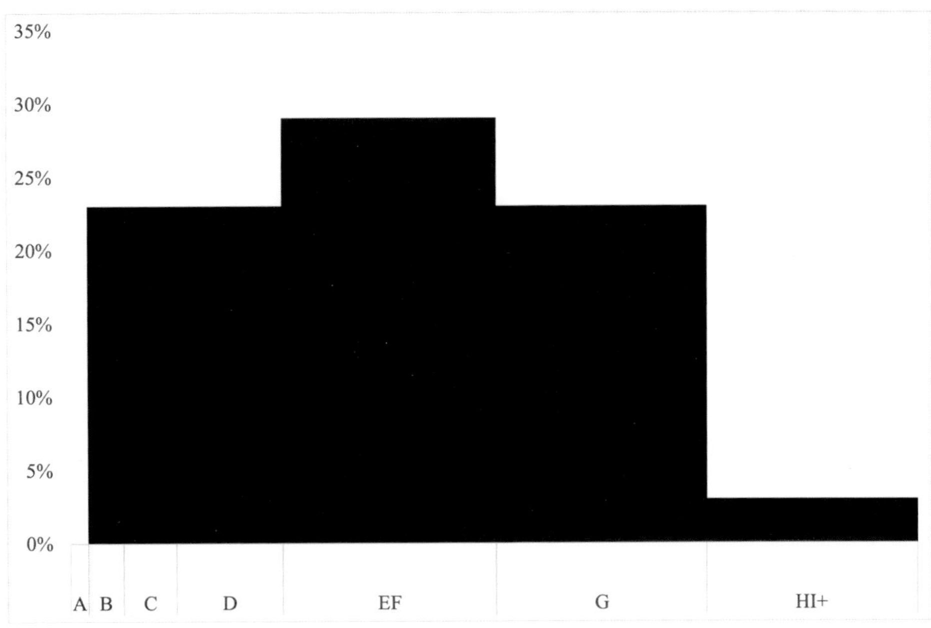

Fig. 15.15 Culling profile for sheep from Unit 4 (n=13).

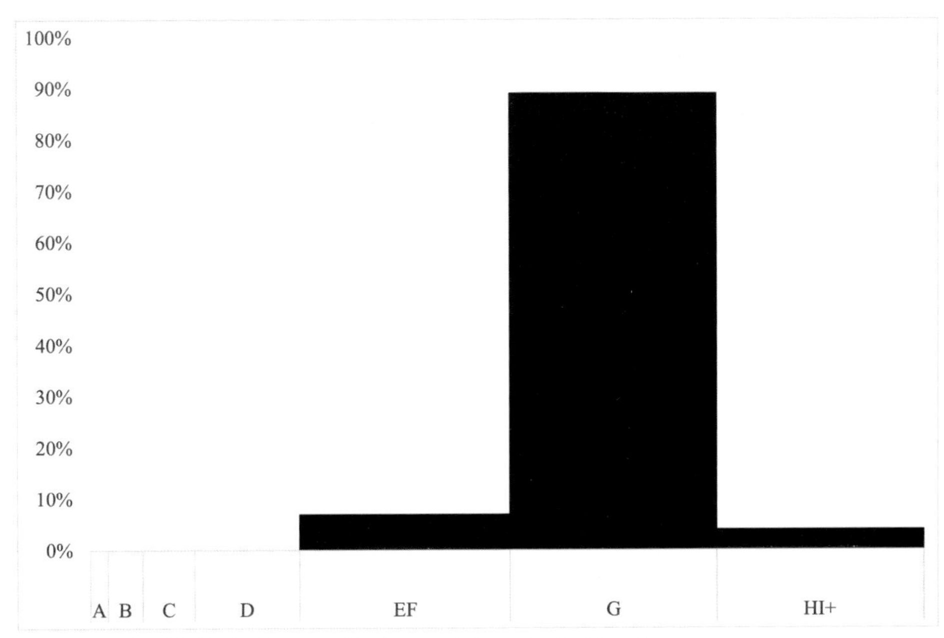

Fig. 15.16 Culling profile for goat from Unit 4 (n=8).

cluding both forest and steppic species (Arai 2020; Benecke 2017; Vila et al. 2017). Therefore, we need to integrate information from both settlements and camps to establish a complete understanding of the seasonality, mobility patterns, and livestock use of early farmers in the region.

The metric data for the faunal remains described in this chapter are presented in the Appendix (Tables 15A.1–15A.11).

References

Arai, S. (2020) Faunal remains from Göytepe. In: *Göytepe: Neolithic Excavations in the Middle Kura Valley, Azerbaijan*, edited by Y. Nishiaki and F. Guliyev, pp. 333–364. Oxford: Archaeopress.

Arai, S. (2021) Neolithic animal remains from Hacı Elamxanlı Tepe. In: *Hacı Elamxanlı Tepe: The Archaeological Excavations of An Early Neolithic Settlement in West Azerbaijan*, edited by Y. Nishiaki, F. Guliyev, and S. Kadowaki, pp. 195–224. Berlin: ex Oriente.

Bar-Oz, G., A. Belfer-Cohen, T. Meshveliani, N. Jakeli, Z. Matskevich, and O. Bar-Yosef (2009) Bear in mind: bear hunting in the Mesolithic of the Southern Caucasus. *Archaeology, Ethnology and Anthropology of Eurasia* 37(1): 15–24.

Benecke, N. (2017) Exploitation of animal resources in Neolithic settlements of the Kura Region (South Caucasia). In: *The Kura Projects: New Research on the Later Prehistory of the Southern Caucasus*, edited by B. Helwing, T. Aliyev, B. Lyonnet, F. Guliyev, S. Hansen, and G. Mirtskhulava, pp. 357–369. Archäologie in Iran und Turan 16. Berlin: Dietrich Reimer Verlag.

Boessneck, J., H. F. Müller, and M. Teichert (1964) Osteologische Unterscheidungsmerkmale zwischen Schaf (Ovis aries LINNE) und Ziege (Capra hircus LINNE). *Kühn-Archiv* 78: 1–129.

Helmer, D. (1995) Biometria i arqueozoologia a partir d'alguns exemple s del Proxim Orient. *Cota Zero* 11: 51–60.

Helmer, D. (2000) Étude de la faune mammalienne d'El Kowm 2 (Syrie), In: *Une île dans le désert: El Kowm 2 (Néolithique précéramique, 8000–7500 BP Syrie)*, edited by D. Stordeur, pp. 233–264. Paris: CNRS Éditions.

Helmer, D., L. Gourichon, and E. Vila (2007) The Development of the exploitation of products from Capra and Ovis (meat, milk and fleece) from the PPNB to the Early Bronze in the Northern Near East (8700 to 2000 BC Cal.). *Anthropozoologica* 42(2): 41–69.

Kushnareva, K. K. (1997) *The Southern Caucasus in Prehistory: Stages of Cultural and Socioeconomic Development from the Eighth to the Second Millennium B.C.* University Monograph 99. Philadelphia: The University Museum, University of Pennsylvania.

Stiner, M. C., N. D. Munro, and T. A. Surovell (2000) The tortoise and the hare: small-game use, the Broad-Spectrum Revolution, and paleolithic demography. *Current Anthropology* 41(1): 39–73.

Vila, E., A. Bălăşescu, V. Radu, R. Badalyan, and C. Chataigner (2017) Neolithic subsistence economy in the Plain of Aratat: Preliminary comparative analysis of the faunal remains from Aratashen and Khaturnarkh-Aknashen (Armenia). In: *Archaeozoology of the Near East 9, Vol. 1*, edited by M. Mashkour and M. Beech, pp. 98–111. Oxford/Philadelphia: Oxbow Books.

Appendix to Chapter 15

The abbreviations used below follow Driesch (1976).

Driesch, A. von den (1976) *A Guide to the Measurement of Animal Bones from Archaeological Sites*. Peabody Museum Bulletin 1. Cambridge: Peabody Museum of Archaeology and Ethnology, Harvard University.

Table 15A.1 Measurements of *Ovis* sp. * In the column for sector, "Pit" denotes one of the trenches opened in the 2016 season. The small shall apply hereafter.

Scapula

Sector*	Context	Unit	Fusion	Note	BG	LG	GLP	SLC
Pit 9	11	4	f		20.1	23.5	30.2	17.8
D2	13	5	f		22.92			19.98

Humerus

Sector	Context	Unit	Fusion	Note	Bd	BT	HTC	Ddm
D0	7	2	f		26.64	27.42	13.72	24.29
C2	4	4	f		27.69	26.53	14.11	
D1	10	4	f	BT estimated	28.61	26.24	13.82	26.17
B3	22	5	f		34.74	33.11	17.62	

Radius

Sector	Context	Unit	Fusion	Note	Bd	BFd	Dd
D0	10	4	f		34.68	28.55	23.49
D1	8	4	f		24.99	21.75	19.04

Intermediate carpal

Sector	Context	Unit	Fusion	Note	GB	GD	GL
B3	5	4			13.45	19.65	12.65

Metacarpal

Sector	Context	Unit	Fusion	Note	Bd	BT	Dd	DT
A0	7	4	f		28.86			
D0	9	4 + Mixed?	f		22.23	21.01	12.16	14.98
A0	14	5	f				19.94	

Femur

Sector	Context	Unit	Fusion	Note	Bp	DC
A0	15	5	f		51.52	23.98

Patella

Sector	Context	Unit	Fusion	Note	GB
A0	15	5			25.37

Tibia

Sector	Context	Unit	Fusion	Note	Bd	Dd
Pit 9	9	4	f		30.6	23.6
A2	8	4	u		25.97	23.54
C3	8	4	fusing		26.22	20.23

Astragalus

Sector	Context	Unit	Fusion	Note	GLl	Dl	GLm	Dm	Bd	LA
D0	5	2			32.62	18.7	31.19	19	20.36	
Pit 9	9	4				17.1		15		
A0	10	4			28.96	16.04	27.74	16.55	18.05	24.21
A1	7	4			24.98	13	23.4	12.97	15.39	20.75

Sector	Context	Unit	Fusion	Note					
C2	5	4			34.06	18.97			22.17
C2	4	4			28.42	15.99	28.1	17.29	17.21
C3	8	4		GLm estimated	30.04	17.08	28.33		18.64
D1	10	4			27.89	15.93	26.45	16.24	17.76
D1	10	4			27.99	15.29	26.49	15.15	16.46
D0	9	4 + Mixed?					25.54	15.43	16.22
Pit 9	19	5			33.3	19.1	32.7	20.9	20.5
B3	21	5			34.86	19.29	33.04		21.45
D1	17	5			35.52	19.33	33.32		21.45

Calcaneus

Sector	Context	Unit	Fusion	Note	GB	GD	GL	Bp	Dp	SD	DD
A3	5	4			18.62	22.29					
Pit 9	13	4?	f					17	20.1		
Pit 9	13	4?	f					17	19.7		
A0		5	fusing		21.73		68.68			9.24	16.64

Central tarsal

Sector	Context	Unit	Fusion	Note	GB
B3	21	5			31.63

Ph 1

Sector	Context	Unit	Fusion	Note	Bp	Dp	SD	Bd	Dd	GLpe
Pit 9	8	3	f		12.8		9.9	11		38.4
A2	7	4			11.2					
C1	2	4						13.43		
C1	2	4	f		14.28					
C1	4	4						12.37		
C2	4	4						10.96		
D1	11	4	f		15.22					
D1	12	4						13		
D2	9	4						13.37		
D2	10	4	f		13.24		10.96			
D0	9	4 + Mixed?	f		11.37		8.42	10.37		34.54
Pit 9	19	5			13.1					
B1	4	5		estimated	15.17					
B2	12	5	f	Bp estimated	13.23		9.44	10.29		
A0	15	5			13.05				11.35	
A0	16	5						13.86	12.46	
Pit 9	14	5?	f		14.6	17.4				

Ph 2

Sector	Context	Unit	Fusion	Note	Bp	SD	Bd	GL
Pit 9	11	4	f		12.3	8.7	9.5	24
A2	8	4	f		14.94	10.83	10.97	26.2
B3	4	4	f		13.08	9.53	9.85	25.24
C1	4	4					10.84	
C2	4	4	f		12.89	9.55	10.31	21.95
C2	4	4	f		12.05	8.08	8.66	22.52
C2	6	4	f		14.68	9.67	11.35	27.15
D0	10	4	f			9.65	11.09	24.12

Sector	Context	Unit	Fusion	Note				
D1	10	4	f		11.71	8.3	9.01	21.74
D3	8	4	f			9.62	10.13	23.68
D0	9	4 + Mixed?	f		9.88	6.98	7.66	20.15
A1	9	5	f		13.25	9.74	9.91	27.23
C2	10	5	f		13.87	9.77	11.14	25.82
D2	12	5	f		14	9.82	10.24	26.5

Ph 3

Sector	Context	Unit	Fusion	Note	Ld	DLS	MBS
B2	8	4			30.74	38.31	7.45
C1	2	4			26.29	30.78	
A3	12	5			30.48	37.55	6.84
C2	10	5			22.72	27.05	
A3	13	6			30.46	37.94	6.64

Table 15A.2 Measurements of *Capra* sp.

Radius

Sector	Context	Unit	Fusion	Note	Bd	BFd	Dd
B0	5	3	3		25.34	21.83	17.88

Metacarpal

Sector	Context	Unit	Fusion	Note	Bp	Dp	SD
Pit 9	9	4			24.6	16.4	17.3

Patella

Sector	Context	Unit	Fusion	Note	GB	Gd	GL
Pit 9	13	4?			9.5	18.1	30.4

Astragalus

Sector	Context	Unit	Fusion	Note	GLl	Dl	GLm	Dm	Bd
C1	7	5					26.45	15.17	17.32

Ph 1

Sector	Context	Unit	Fusion	Note	Bp	SD	Bd	GLpe
A1	8	4			11.01			
C2	4	4	f		11.91	9.28	11.05	37.07
C1	8	5	f		14.92			

Ph 2

Sector	Context	Unit	Fusion	Note	Bp	SD	Bd	GL
C2	10	5					9.71	

Table 15A.3 Measurements of *Bos* sp.

Radius

Sector	Context	Unit	Fusion	Note	Bd	BFd
Pit 9	9	4	u		65.2	63.4

Ph 1

Sector	Context	Unit	Fusion	Note	SD	Bd
C4	1	1	f		25.26	28.02

Ph 2 (Posterior)

Sector	Context	Unit	Fusion	Note	Bp	SD	Bd	GLpe
D1	4	1			28.97	22.69	22.85	39.78

Table 15A.4 Measurements of *Sus* sp.

Astragalus

Sector	Context	Unit	Fusion	Note	GLl	Dl	GLm	Dm	Bd
D1	10	4			46.75	25.38	42.11		29.78

Table 15A.5 Measurements of *Cervus elaphus*.

Central tarsal

Sector	Context	Unit	Fusion	Note	GB	GD
D2	12	4			43.54	40.49

Ph 1

Sector	Context	Unit	Fusion	Note	Bp	Dp	Bd
Pit 9	16	5					21
A3	11	5	f		22.31	26.51	
B3	12	5					21.64
C2	9	5			19.69		

Ph 1

Sector	Context	Unit	Fusion	Note	Bp
A0	18	5	f		20.65

Table 15A.6 Measurements of *Gazella subgutturosa*.

Scapula

Sector	Context	Unit	Fusion	Note	BG	LG	GLP	SLC
D0	5	2	f		20.51	22.5	28.98	17.47

Humerus

Sector	Context	Unit	Fusion	Note	Bd	BT	HTC	Ddm
D0	5	2	f		26.85	25.71	14.47	24.36

Astagaralus

Sector	Context	Unit	Fusion	Note	GLl	Dl	GLm	Dm	Bd	LA
A0	20	6			30.92	16.35	28.61	15.5	16.97	24.19

Ph 1

Sector	Context	Unit	Fusion	Note	Bp	Bd
Pit 9	22	5	f		10.5	
B3	20	5	f		10.5	
B3	20	5				9.73
B3	21	5				9.22
B0	15	6				9.52
A1	15	6				7.94

Ph 2										
Sector	Context	Unit	Fusion	Note	Bp	Dp	SD	Bd	Dd	GL
A0	15	5	f		12.97	12.57	9.21	9.94	11.87	27.19
Pit 9	14	5?	f		9.2		6.2	8.9		23.4

Ph 3						
Sector	Context	Unit	Fusion	Note	Ld	DLS
Pit 9	16	5			21.3	26.7
C0	12	5			21.17	25.7

Table 15A.7 Measurements of *Vulpes vulpes*.

Lower M1						
Sector	Context	Unit	Fusion	Note	L	B
B3	7	4			15.01	5.59

Scapula								
Sector	Context	Unit	Fusion	Note	BG	LG	GLP	SLC
A3	6	4	f		10.06	15.46	16.51	13.16

Calcaneus							
Sector	Context	Unit	Fusion	Note	GB	GD	GL
B3	7	4	f		7.45	7.37	16.54

Table 15A.8 Measurements of *Lepus europaeus*.

Radius					
Sector	Context	Unit	Fusion	Note	Bd
D1	11	4	f		8.22

Ulna								
Sector	Context	Unit	Fusion	Note	DPA	SDO	LO	BPC
A2	8	4	f		15.99	12.79		
D1	11	4	f		10.82	10.47	12.28	7.55

Femur					
Sector	Context	Unit	Fusion	Note	DC
C2	6	4			10.36

Astragalus						
Sector	Context	Unit	Fusion	Note	Bp	GLl
C2	6	4			10.24	19.08

Table 15A.9 Measurements of *Felis* sp.

Humerus								
Sector	Context	Unit	Fusion	Note	Bd	BT	HTC	Ddm
A3	6	4	f		21.16	15.8	7.39	
C1	4	4	f		15.39	10.71	4.6	8.45

Radius							
Sector	**Context**	**Unit**	**Fusion**	**Note**	**Bp**	**Dp**	**Bd**
D2	11	5	f		9.07	6.2	
D2	11	5	f				12.36

Table 15A.10 Measurements of *Panthera pardus*.

Ph 1								
Sector	**Context**	**Unit**	**Fusion**	**Note**	**Bp**	**SD**	**Bd**	**GL**
A1	13	5	f		13.49	9.53	9.05	33.99
A3	12	5	f		13.86			
C1	10	5	f	burnt	11.42			

Table 15A.11 Measurements of *Testudo graeca*.

Humerus					
Sector	**Context**	**Unit**	**Fusion**	**Note**	**SD**
B2	2		4		5.22
B2	4		4		4.41
B3	10		5		4.76
B2	11		5		4.78

Chapter 16

Bronze Age and medieval pottery, and other finds from Damjili Cave

Kazuya Shimogama

16.1 Introduction

This chapter discusses artifacts, particularly the pottery assemblage, recovered during excavations in various trenches and squares at Damjili Cave. During the excavation of the upper deposits lying directly on top of the Neolithic and Chalcolithic layer units, we encountered numerous pottery and glazed ceramic sherds dating back to the Bronze Age and medieval period.

We aim to (1) describe the basic information on the Bronze Age (Unit 2) and the medieval period (Unit 1) assemblage, (2) examine the relative chronology for each corpus of artifacts, and (3) define past human cave use during the Bronze Age and the medieval period by means of available archaeological evidence and interpret them in each archaeological context.

16.2 Bronze Age pottery from Unit 2

16.2.1 Description of the pottery sample

The Bronze Age pottery assemblage comprises ceramic sherds that were recovered from several contexts assigned to lithological Unit 2 in the primary excavation area located in the main rockshelter (Nishiaki et al. 2019). Intrusive sherds that clearly represent earlier Chalcolithic or Neolithic characteristics, amounting to 13 or more sherds, were not included in this analysis (see also Chapter 11 of this volume) because they are considered to be associated with disturbances, such as medieval period pits. Sherds from unreliable and mixed contexts were also excluded from analysis.

The assemblage was small in quantity and comprised four rim sherds and 62 body sherds (66 examples in total). The small number of sherds in Unit 2 can be explained by relatively thin stratigraphic deposits (c. 20–50 cm thick, only one or two layers from each square), probably reflecting ephemeral occupation during the Bronze Age.

16.2.2 Pottery ware types and vessel shape typology

The Bronze Age pottery sherds can be classified into the following four ware types:

Coarse Simple Ware

This was the most common ware type in Unit 2, accounting for 68% of the assemblage (Table 16.1). This ware is characterized by a coarse low-fired paste with fine sand and a mineral temper of less than 1 mm in size (Fig. 16.1: 1). Occasionally, it contains larger dark-colored grits (approximately 1–2 mm). The surface colors most often represented are brown and gray brown, while the core colors are invariably gray to dark gray brown or black in rare cases, indicating unoxidized low firing. The infrequent black soot observed on the exterior surface may have been caused by secondary firing, implying that these vessels were used as cooking pots. The irregularly smoothed surface and profile of the sherds indicate that they were all handmade, possibly using coiling and pinching techniques.

The majority of the sherds are undiagnostic, undecorated, and thick-walled body sherds (Fig. 16.2: 4); however, two examples of rim sherds from this ware type and their profiles indicate that they appear to belong to closed shapes, probably pot or jar forms. No clear traces of burnishing or slip were observed.

Orange Sandy Ware

This ware type represents the second most com-

Table 16.1 Pottery sherds from Unit 2 by ware types.

	Coarse Simple Ware	Orange Sandy Ware	Orange Vegetal Ware	Fine Simple Ware	Total
Unit 2 layers	45	11	4	6	66
	68.2%	16.7%	6.1%	9.1%	100.0%

Fig. 16.1 Representative ware types of the Bronze Age pottery from Unit 2. 1: Coarse Simple Ware; 2: Orange Sandy Ware; 3: Orange Vegetal Ware; 4: Fine Simple Ware.

mon group (Table 16.1). It is easy to recognize as it contains fine white sand temper abundantly (Fig. 16.1: 2). It is clearly distinguishable from the coarse red-brown ware frequently encountered in Unit 3 (Chalcolithic) layers because the orange sandy ware contains finer sand inclusions, whereas the angular and large-sized sand are common in the coarse red-brown ware (Chapter 11 of this volume, fig. 11.1: 3–5). The surface and core colors range from bright orange to light orange. Sherds with gray cores rarely occur, indicating relatively suitable firing conditions. The orange sandy ware is handmade, however, no vessel-forming traces can be observed among the sherds. The absence of large sherds also hinders the assessment of the vessel typology. Several sherds from the same context, which may be derived from an identical vessel, may form a medium-sized necked jar.

Orange Vegetal Ware

This group was macroscopically recognized by the apparent presence of an organic plant temper with a 5–10 mm-long straw fragment, leaving distinctive linear voids on the surface (Fig. 16.1: 3). The paste color varies from orange, which is similar to orange sandy ware, to light brown and grayish-brown. Most sherds are incompletely oxidized and often display a darker gray or even black core. As with other Bronze Age ware types, this ware was hand-formed.

Despite the extremely small quantity of this ware (only four sherds), one rim and body sherd belonged to a high-necked jar or pot with a slightly everted rim, without a sharp neck distinction, and a rim diameter of 15–20 cm (Fig. 16.2: 1, 2; also see Table 16.2).

Fine Simple Ware

This ware is also represented by small number of sherds: only six (9.1%). This ware is characterized by a finer paste composition than the other ware types. It usually contains fine sand inclusions of less than c. 0.5 mm, and is relatively well-fired. The surface color is a buff-oriented tint varying from pale brown to light orange, with a single sherd having a light gray core owing to incomplete oxidization. One of the rim fragments had a simple, everted rim (Fig. 16.2, 3). As the entire shape cannot be reconstructed, this may be part of an open-shaped bowl or the upper rim of a medium-sized jar. Again, there were no decorations on this type of pottery.

Overall, the four ware types are all handmade and, in most cases poorly fired. Although the assemblage includes a small number of finer wares (fine simple wares), other types of coarse- or orange-colored wares are predominant. Decorations of any sort have been virtually absent from the Bronze Age assemblage thus far, and notable surface treatments, such as burnishing and polishing, have not been identified on any sherd fragments. In addition, closed vessels such as jars and pots are more common than open bowl vessels in the assemblage.

16.2.3 Inter-site comparison and relative chronology

Considering the aforementioned ceramic features, we have insufficient materials for placing the Damjili Unit 2 assemblage within the established framework of regional chronology in the Southern Caucasus, because of the small quantity of our dataset and the scant occurrence of diagnostic sherds

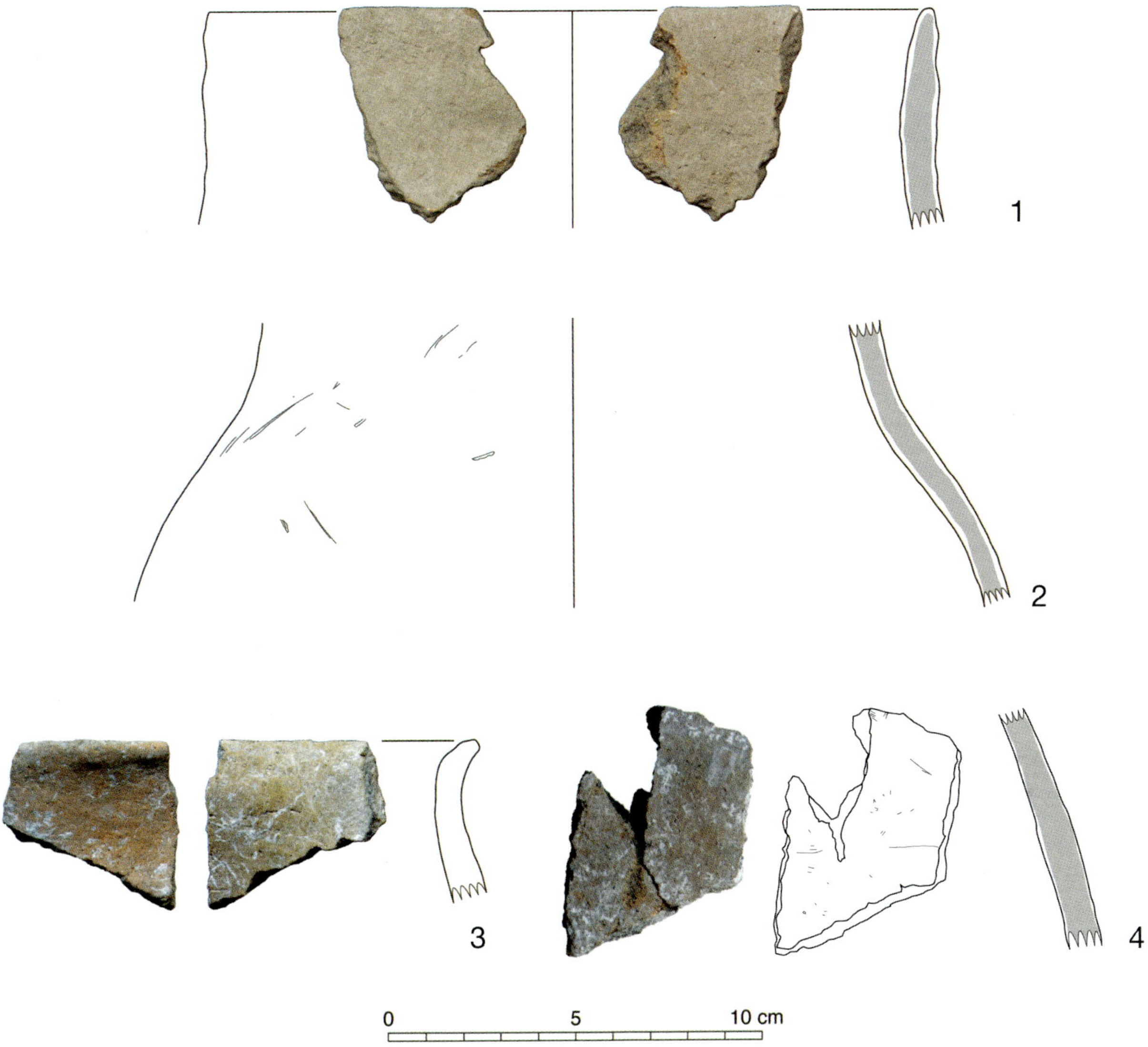

Fig. 16.2 Pottery from Unit 2. Unoxidized cores are indicated in gray for vessel wall profile.

Table 16.2 Pottery catalog for Fig. 16.2. * In the column for context, "Pit" denotes one of the trenches opened in the 2016 season. The same shall apply hereafter.

Fig.	Context*	Unit	Ware type	Color: exterior	Color: interior	Color: core	Temper	Forming	Remarks
16.2: 1	A1/3.4	2	Orange Vegetal Ware	10YR 7/2 dull yellow orange	7.5YR 7/2 light brownish gray	7.5YR 7/4 dull orange	sand; plant	handmade	gray core
16.2: 2	B0.3	2	Orange Vegetal Ware	7.5YR 5/3 dull brown	7.5YR 7/4 dull orange	N3/0 dark gray	plant	handmade	gray core
16.2: 3	Pit 7.8	2	Fine Simple Ware	5YR 6/4 dull orange	7.5YR 6/4 dull orange	7.5YR 6/4 dull orange	fine sand	handmade	
16.2: 4	Pit 7.8	2	Coarse Simple Ware	10YR 6/2-6/3 grayish yellow brown	7.5YR 6/3 dull brown	N2/0 black	fine sand	handmade	gray core

useful as chronological markers. Furthermore, despite recent discoveries across the region, the pottery chronology of the Bronze Age in Azerbaijan or neighboring areas has been a matter of debate because it continues to lack basic datasets from settlement sites with well-stratified sequences supported by reliable radiometric dating (Lyonnet 2014; Rova 2014; Sagona 2018: 213–378). This is particularly the case for undecorated wares such as those from Damjili Cave. However, here, we attempt to compare our pottery assemblage with currently available Bronze Age ceramic data from the central part of the Southern Caucasus.

Notably, the Damjili Bronze Age assemblage does not contain any later Bronze Age pottery traits; for instance, it is characterized by black-on-red elaborately painted ware of the Middle Bronze Age (for example, Trialeti complex, Sagona 2018: 352–353), and gray-black burnished wares of the Late Bronze-Early Iron Age (Khojaly-Gedebey, Lchashen-Tsitelgori ceramic traditions, and other contemporary related horizons) (Sagona 2018: 380, 403–410). Moreover, it is remarkable that it lacks the typical red black burnished ware from the Kura Araxes cultural tradition. Considering the general ceramic chronology, our evidence indicates clear ceramic affinity with what Palumbi calls "monochrome ware" described for the Kura Araxes cultural phenomenon (Palumbi 2008: 157–213). In contrast to red black burnished ware, monochrome ware is defined as wet-smoothed (not always burnished), primarily brown, gray to buff colored ware with coarse minerals or vegetal tempers (Palumbi 2008: 159, 162). At numerous settlement sites, including Kiketi, Samshvilde II, Kvatskhelebi, and Khizanaant Gora, the ceramic data consistently indicate the coexistence of red-black burnished ware with monochrome ware, with the latter being more common in most cases (Palumbi 2008: 157–213). However, a decline in the quantity of monochrome ware was observed in the later Kura-Araxes phase, particularly in the KA II phase, in the Shida Kartli region (Rova 2014: 53). The dominance of unburnished monochrome wares at Damjili Cave may be understood as either a local ceramic tradition or linked to a particular subsistence strategy or cultural behavior in the cave.

A similar coexistence pattern of two different ceramic groups has also been suggested at an Early Bronze Age site in Baba Dervish, one of the nearest mounded settlements to Damjili Cave, however, unfortunately, no quantitative data were provided. At this site, the red-black burnished Kura Araxes pottery, which was interpreted as food-serving wares, co-occurred with light-colored vessels as cooking pots (Ismailov 1977: 28–30).

Thus, the Damjili Bronze Age assemblage, which is exclusively composed of unburnished monochrome ware (our four ware types), cannot be easily compared with the reported pottery assemblages; however, a few typological links support their late Early Bronze Age (or early Middle Bronze Age) date. A jar or pot with a slightly everted rim (Fig. 16.2: 1, 2) can be more or less encountered in early Middle Bronze Age kitchen wares from Didi Gora and Tqisbolo-gora in eastern Georgia (Kastl 2008: fig. 2: 7–13), whereas an everted rim medium-sized vessel (Fig. 16.2: 3) offers us possible parallels to Kiketi monochrome jar (Palumbi 2008: fig. 5.3: 2), other types of monochrome ware vessels (Palumbi 2008: fig. 5.7: 2, 4, 6), Baba Dervish (Ismailov 1977: fig. VIII), and cooking pots from Mentesh Tepe Phase 2 (Lyonnet 2014: fig. 8: 2). As aforementioned, burnishing surface treatment and distinctive decorative elements, including incised or ribbed geometric motifs, relief and plastic applied motifs, and the so-called Nakhichevan lugs, all often associated with Early and Middle Bronze Age settlements, were completely absent in the Damjili pottery repertoire. Therefore, more precise and direct comparisons are difficult at the present stage of research.

With regard to absolute chronology, two radiocarbon dates from the Unit 2 samples at Damjili Cave (TKA-17144 and IAAA-160715) point to an early to late third millennium BC calibrated date (Nishiaki et al. 2019: table 1 and fig. 4; also see Chapter 5 of this volume). In terms of the archaeological phases in the Southern Caucasus, these ^{14}C dates fall from the late Early Bronze Age to the early Middle Bronze Age (Sagona 2018: 253–261, 299–303). However, based on quantitatively limited ceramic data, this chronological range does not contradict our techno-typological evaluation and inter-site comparisons. Therefore, it is suggested that the Damjili Bronze Age pottery can be dated to the late Early Bronze Age or early Middle Bronze Age in the middle of the third millennium BC.

16.2.4 Cave use at Damjili during the Bronze Age

Bronze Age pottery from Damjili Cave is interesting because it was recovered from unusual cave deposits. An example of the Areni-1 cave complex in Armenia, dated to the early to middle fourth millennium BC, provides one of the earliest and rarest pieces of evidence for cave use in wine making, storage, and/or ritual activities in the Late Chalcolithic period (Areshian et al. 2012; Wilkinson et al. 2012). In contrast, the Bronze Age use of caves and rockshelters is largely unknown, and their social and cultural significance has not been comprehensively studied. Nevertheless, increasing evidence suggests that caves and rockshelters were intermittently used as seasonal shelters by farmers, pastoralists, and various populations, at least as early as the Neolithic or Chalcolithic periods (Chapter 11 of this volume). For example, the Hovk-1 rockshelter and Yenokavan-2 cave located on the high-forested plateau of the Aghstev Valley, are considered short-term camps visited by hunters or transhumant pastoralists (Arimura et al. 2014: 263). The latter cave site revealed a more than 1 m thick Early Bronze Age layer that yielded a high density of pottery fragments and faunal remains associated with pits and hearths. Investigations of artificial rock-cut cave complexes in the Kvemo Kartli region of southeastern Georgia also offer valuable information on Bronze Age cave exploitation (Bakhtadze 2013a: 6–11). Excavations at some cave complexes in this region, including Samshvilde, Nakhiduri, and Zurtaketi, revealed the presence of both Late Bronze Age pottery and Early Bronze Age Kura Araxes burnished ware and Middle Bronze Age comb-impressed and incised sherds, although unfortunately they were found to be severely mixed without stratigraphical relationships (Bakhtadze 2013a: figs. 11–13). However, it has been suggested that rock-cut shelter habitation dates back to at least the Early Bronze Age. The Early Bronze Age pottery recovered from these cave complexes appears to be different from Damjili Unit 2 pottery in that the latter includes only coarse monochrome wares.

Based on ceramic evidence from Damjili Cave, the common occurrence of coarse cooking ware, the relatively small quantity of pottery sherds (possibly indicating a few complete vessels in the entire cave use episode), and the low frequency or absence of open vessels, such as bowls, related to household commensality are remarkable. Rather than the long-term use of the cave, it may be interpreted as a camp site used only seasonally and intermittently, perhaps visited by small groups equipped with a small set of undecorated pottery. It has also been suggested that the pottery vessels found at Damjili Cave may have been brought into the cave from a nearby settlement (possibly Baba Dervish), with a preferred and suitable vessel kit for cave activities.

16.3 Medieval period pottery and small finds from Unit 1

16.3.1 Description of the pottery sample

Contrary to the Bronze Age pottery samples, the medieval period pottery assemblage assigned to Unit 1 at Damjili Cave was recovered abundantly in the excavations. It comprises 680 sherd fragments (Table 16.3). It was divided into six pottery groups based on the excavated area and stratigraphic units/subunits.

The first group refers to pottery items recovered from six sounding trenches, Trenches 1–6, in Damjili Cave 2 during the 2016 season (192 potsherds). As analyzed below, this group exhibits remarkably similar techno-typological features and ware composition to other groups, suggesting that it should be ceramically and chronologically equated with the materials from the Main Cave. The pottery from these soundings in Damjili Cave 2 does not include the earlier Bronze Age sherds or Chalcolithic and Neolithic sherds.

The other five pottery groups, as indicated in Table 16.3, were derived from trenches in the Main Cave (c. 1.5 m-thick stratified deposits). The second group designated here as "Unit 1" contains the bulk of potsherds (216 in total) retrieved from the deposits immediately below the present surface; c. ~0.2 m topsoil. This "Unit 1" group also includes part of the pottery from the sounding trench (Trenches 7, 8, 9, and 10), which cannot be stratigraphically assigned to the subdivided phases within Unit 1 as explained below. Among these soundings, Trenches 8 and 10 yielded only a few dozen potsherds, primarily associated with stone cist graves. Unlike these trenches, the stratigraphic contexts of Trenches 7 and 9, which were incorporated into a much larger area in the following seasons, are easily compared with the lithological unit sequence of the main excavation area (Chapter 4 of this volume). Hence, the pottery from these contexts is included in each sub-unit group of Unit 1.

The three pottery groups of Units 1a, 1b, and 1c represent the sequences of different subunits or phases in Unit 1. They can be defined based on the stratigraphy of the main excavation areas. Unit

Table 16.3 Total counts of the medieval pottery by Unit 1 pottery groups.

	Rim	Body	Base	Others	Total
Damjili 2*	15	168	7	2	192
Unit 1**	23	181	8	4	216
Unit 1a	8	45	6	0	59
Unit 1b	7	75	1	2	85
Unit 1c	4	100	4	4	112
mixed layers	1	9	0	6	16
Total	**58**	**578**	**26**	**18**	**680**
	8.5%	85.0%	3.8%	2.6%	100.0%

* Counted here are the pottery specimens recovered from six sounding trenches (Trenches 1–6) in Damjili Cave 2, located to the northeast from the Main Cave in the distance of c. 100 m.

** "Unit 1" contains pottery sherds retrieved from the depoits immediately below the present surface, and those from the sounding trenches (Trenches 8 and 10) that cannot be stratigraphically assigned to each subdivided phase within Unit 1.

1a comprises the uppermost deposits of c. 0.5 m, containing pit graves found in Square D1 and E1. It is extremely likely that, according to their elevation, the stone cist graves in Trenches 8 and 10 are contemporaries with these pit graves, which are undoubtedly assigned to Unit 1a. Considering that these graves were dug by cutting them into lower deposits, the Unit 1a group may have contained earlier intrusive potsherds. A total of 59 sherds is included in the Unit 1a, none of which were contained as certain grave goods.

The deposits below Unit 1a belong to Unit 1b, representing a considerably thin gray-brown layer (c. 0.2–0.3 m thick). It is defined by a poorly preserved wall segment (D3.2) that runs parallel to the overhanging bedrock cliff to the north, a shallow hearth (E0.2; Pit 7.9), and a flask-shaped pit (D3.7); 85 pieces of potsherd associated with this subunit were not found *in situ* as well.

In the lowest Unit 1c with c. 0.6 m-thick soil deposit, another wall structure built with rectangular ashlar limestones (C4.3) was partly discovered. This wall fragment runs in the NE to SW direction, which is common in the stone wall of the upper Unit 1b. In addition, a round hearth (D4.6) filled with gray-black ash deposits immediately south of this wall and a cluster of sizeable stones (D0.3) to the north near the cave bedrock also belonged to Unit 1c. The pottery, totaling 112 sherds, indicated a similar recovery pattern to the other subunits, where neither complete vessels nor vessels with *in situ* contexts were recovered.

To the last pottery group, we added finds (16 sherds), which can definitely be related to mixed contexts caused by disturbances of pit and grave digging. Pottery data from this group were excluded from quantitative analysis elaborated below.

There was limited opportunity to examine the spatial distribution of pottery materials in the exposed trenches, largely because a small number of pottery was found in extremely fragmented and random condition. The sherd counts in Table 16.3 reveal that there are no differences in vessel part distribution between the pottery groups, demonstrating consistently a higher ratio of undiagnostic body sherds (85%) than other ceramic parts (rims, bases, handle fragments). The absence of near-complete and complete vessels and the high breakage pattern in the Damjili Cave assemblage suggest that the pottery vessels may have been used carelessly within the rockshelters, implying an expedient nature of cave use. Further, a gradual shift in the ceramic composition for successive Units 1a–c subunits indicates no abrupt change in cave use patterns throughout the Unit 1 occupation (Fig. 16.4).

16.3.2 Pottery ware types and vessel shape typology

After classification into the aforementioned stratigraphic or provenance groups, every sherd of the Unit 1 pottery assemblage could be analyzed macroscopically according to the following variables related to ceramic techno-typology: paste, inclusion or tempering materials, surface treatment (slip, painting and incising decoration, burnishing, and glaze application), firing conditions, colors (exterior, interior, and core), vessel shape and typology, manufacturing technique (handmade, wheel-thrown, or finished on a wheel), and secondary use wear (such as cooking, sherd reuse). Based on sherd observations, the assemblage was divided into a series of ware types distinguished by a set of shared characteristics. Examination of the recovered pottery sherds allowed us to define seven different wares for the Unit 1 pottery, as follows:

Coarse Kitchen Ware (Fig. 16.3: 1)
This is one of the most common ware types for the Unit 1 assemblage, totaling 186 sherds of all the pottery groups (27.4%) (Table 16.4). This coarse ware mostly contains fine sand and dark-colored angular minerals of less than a few millimeters (most probably crushed basalt grit); occasionally, fine white lime and mica are also visible. Generally, no plant tempers were observed. Surface color usually displays a contrasting effect between the exterior (darker gray or orange, gray-brown) and interior surfaces (lighter orange, buff, and brown). This coloring effect appears to have been caused by secondary firing, as many sherds of this ware reveal mottled black soot, particularly on their exterior surfaces. It is often argued that carbonate sand and other grit tempers in the clay matrix, as in the case of the Damjili coarse kitchen ware and other coarse wares, can produce a vessel that is resistant to repeated thermal shocks (Rice 1987: 228–230). This appearance and sand-abundant fabric features suggest that the coarse kitchen ware vessels were used as cooking pots. The core color ranges from gray-brown to orange to reddish-brown, among which incompletely oxidized gray cores are frequently observed, suggesting that the ware was fired at a low temperature.

The pottery sherds assigned to this fabric typically exhibit crudely manipulated and undulated surfaces. This indicates that they were always made by hand without the use of rotational devices. An example of a jar rim sherd (Fig. 16.7: 4) presents rough horizontal striations on both surfaces, possibly made when the surfaces were smoothed on a wheel, being an exception for this type of ware.

Fig. 16.3 Representative ware types of the medieval period pottery from Unit 1. 1: Coarse Kitchen Ware; 2: Coarse Common Ware; 3: Orange Coarse Ware; 4: Fine Common Ware; 5: Fine Orange Ware; 6: Red Slipped Ware; 7: Glazed Ware.

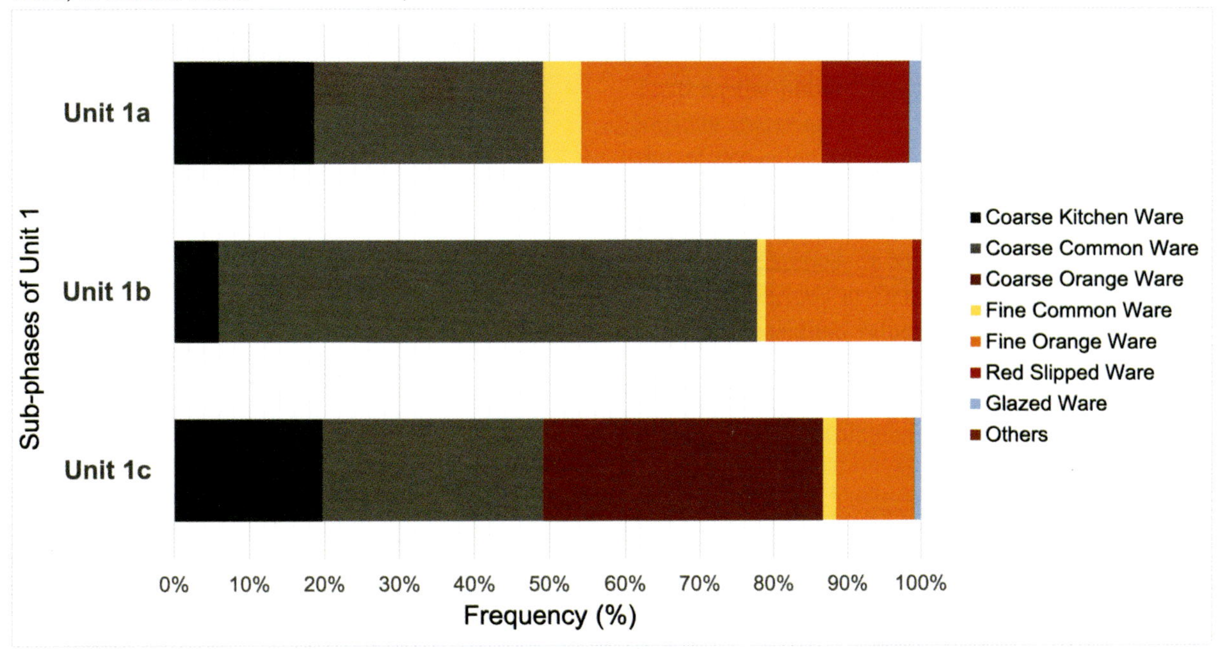

Fig. 16.4 Frequency of ware types from Unit 1 by subunits.

Table 16.4 Pottery sherds from Unit 1 by ware types.

	Coarse Kitchen Ware	Coarse Common Ware	Coarse Orange Ware	Fine Common Ware	Fine Orange Ware	Red Slipped Ware	Glazed Ware	Others	Total
Damjili Cave 2	**93**	**18**	**9**	**8**	**35**	**24**	**5**	**0**	**192**
	43.1%	8.3%	4.2%	3.7%	16.2%	11.1%	2.3%	0.0%	100.0%
Unit 1*	**54**	**48**	**43**	**20**	**38**	**6**	**6**	**1**	**216**
	25.0%	22.2%	19.9%	9.3%	17.6%	2.8%	2.8%	0.5%	100.0%
Unit 1a	**11**	**18**	**0**	**3**	**19**	**7**	**1**	**0**	**59**
	18.6%	30.5%	0.0%	5.1%	32.2%	11.9%	1.7%	0.0%	100.0%
Unit 1b	**5**	**61**	**0**	**1**	**17**	**1**	**0**	**0**	**85**
	5.9%	71.8%	0.0%	1.2%	20.0%	1.2%	0.0%	0.0%	100.0%
Unit 1c	**22**	**33**	**42**	**2**	**12**	**0**	**1**	**0**	**112**
	19.6%	29.5%	37.5%	1.8%	10.7%	0.0%	0.9%	0.0%	100.0%
mixed layers	**1**	**9**	**0**	**1**	**2**	**2**	**1**	**0**	**16**
	6.3%	56.3%	0.0%	6.3%	12.5%	12.5%	6.3%	0.0%	100.0%
Total	**186**	**187**	**94**	**35**	**123**	**40**	**14**	**1**	**680**
	27.4%	27.5%	13.8%	5.1%	18.1%	5.9%	2.1%	0.1%	100.0%

It is difficult to reconstruct the entire vessel shape for this ware, however, the recovered sherd fragments suggest that most of them typologically represent short-necked jars with globular bodies and everted rims (Fig. 16.5: 1–4; Fig. 16.6: 4–6; Fig. 16.7: 1–7). Various hand-pinched rims are found including simple, thin, flattened, and beaded. No base shreds have been identified to date. Some well-preserved sherds were equipped with handles with flat sections (Fig. 16.7: 6, 7). With respect to decoration, it is remarkable that horizontal crudely incised lines (Fig. 16.5: 2–4; Fig. 16.6: 4; Fig. 16.7: 3, 6, 7), incised geometric motifs including garland-like semi-circles (Fig. 16.6: 4), shallowly excised pendent semi-circles (Fig. 16.6: 6), a row of dotted impressions on rim top (Fig. 16.5: 1; Fig. 16.6: 5), and circular thumb impressions on the body or handle (Fig. 16.7: 6, 7) are commonly found. They appear to be applied only to the upper part of the vessel.

Coarse Common Ware

This ware, being the largest group (27.5%) (Table 16.4), represents another coarse ware type in the assemblage (Fig. 16.3: 2). This ware is much less coarse than coarse kitchen ware. This ware is characterized by a relatively thick-walled (more than 10 mm) mineral-tempered fabric. The particle size of the inclusions is generally homogeneous on any sherd, but sometimes reveals a greater density of sand and grit, exhibiting a coarse texture in the paste. No organic tempers were present. The color variation was relatively wide, ranging from light buff, brown, and orange to darker gray and brown. Unoxidized gray cores are uncommon, indicating good firing conditions. This ware is also handmade and probably formed first using the coiling technique.

The dominant presence of thick-walled sherds illustrates that the vessels belonging to this group were predominantly large jars or pots. One such example is wide-mouthed pots with multiple horizontal incisions on the body (Fig. 16.7: 8). A rim sherd may be a fragment of a short-necked jar with a globular body (Fig. 16.7: 12). The overall black soot on the exterior surface indicates that this jar was a cooking pot, similar to coarse kitchen ware jars. Pot or jar fragments decorated with incised and impressed motifs also indicate some affinity with coarse kitchen ware (Fig. 16.7: 8–11). A body sherd with an impressed band (Fig. 16.7: 13) may have been used for cooking or storage of foodstuffs. Despite the clear differences in ceramic ware fabrics, it is possible that coarse common ware has functions similar to those of the coarse kitchen ware.

Coarse Orange Ware (Fig. 16.3: 3)

The bulk of orange-colored fabric is frequently encountered in the Damjili Cave Unit 1 assemblage. Based on the fabric and other technical features, these examples can be classified into two subgroups: coarse and fine orange ware (see below). The coarser variety of orange ware exhibits a much higher density of inclusions, such as sand, mineral grit, mica, and rarely organic chaff (Fig. 16.7: 15). In contrast to other coarse ware types, coarse orange ware appears on bowls (Fig. 16.7: 16) and medium-sized jars (Fig. 16.7: 17), which are all handmade. Regardless of the large quantities of sherds from this ware recovered (19.6% of the assemblage), it is difficult to define the typological profiles from the available samples, such as rims or bodies.

Surface colors are largely restricted to orange-

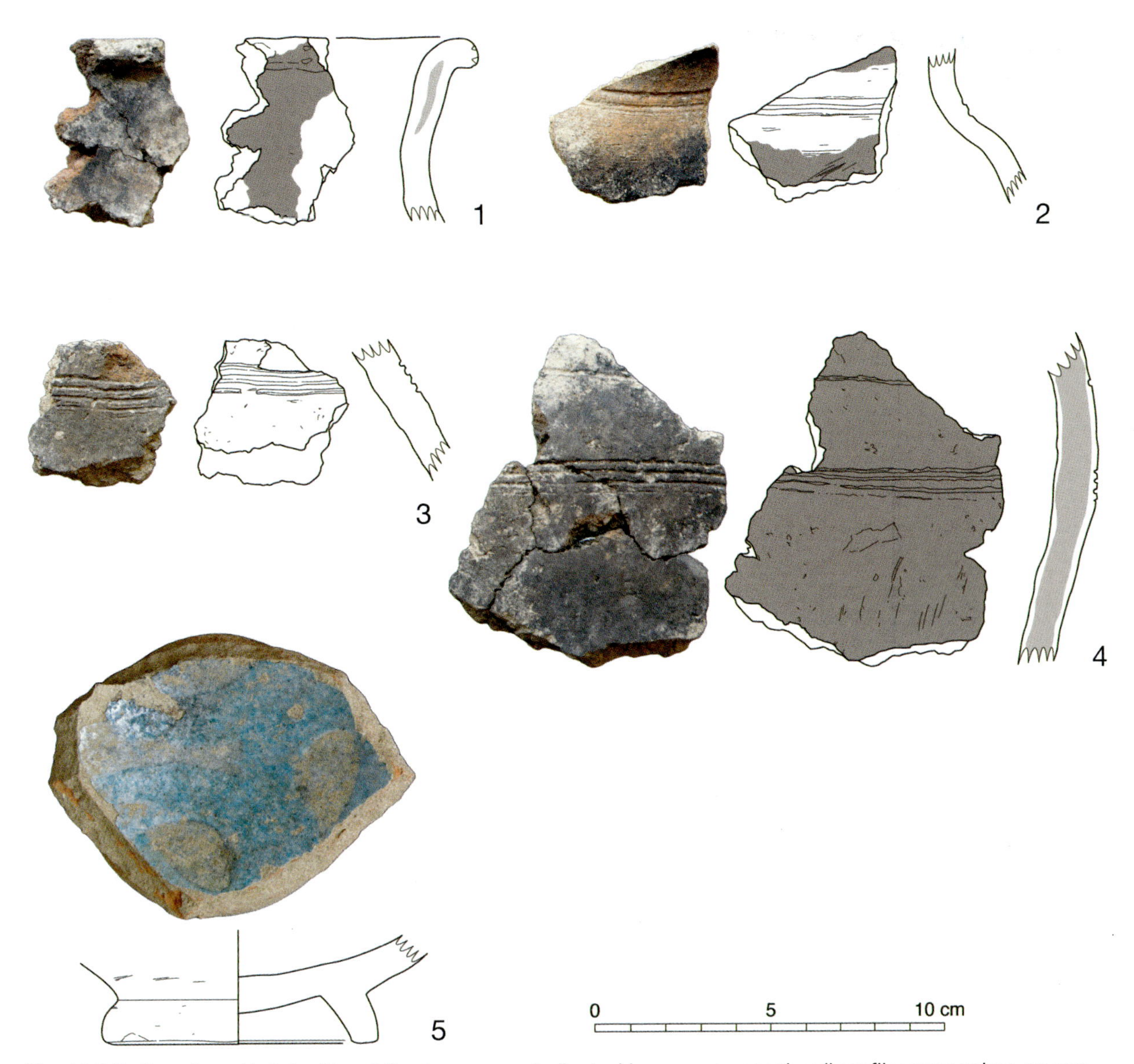

Fig. 16.5 Pottery from Unit 1c. Unoxidized cores are indicated in gray on vessel wall profile; gray colors on surface are soot from secondary firing.

Table 16.5 Pottery catalog for Fig. 16.5.

Fig.	Context	Unit	Ware type	Color: exterior	Color: interior	Color: core	temper	forming	remarks
16.5: 1	Pit 7.5	1c	Coarse Kitchen Ware	7.5YR 4/2 grayish brown	5YR 5/4 dull reddish brown	2.5YR 5/6 bright reddish brown	sand	handmade	gray core
16.5: 2	Pit 7.5	1c	Coarse Kitchen Ware	5YR 5/4 dull reddish brown	5YR 6/4 dull orange	7.5YR 4/2 grayish brown	fine sand	handmade	
16.5: 3	Pit 7.5	1c	Coarse Kitchen Ware	10YR 4/1 brownish gray	5YR 5/4 dull reddish brown	5YR 4/4 dull reddish brown	sand; grit	handmade	
16.5: 4	Pit 7.5	1c	Coarse Kitchen Ware	10YR 2/1 black	5YR 6/4 dull orange	5YR 5/6 bright reddish brown	sand	handmade	gray core
16.5: 5	C3.2	1c	Glazed Ware	2.5YR 6/4 dull orange	turquoise-blue (glaze)	2.5YR 6/8 orange	fine sand; lime	wheelmade	carved decorations

based colors, varying from light, vivid orange to darker dull orange or grayish-orange. This may reflect the consistent firing conditions and similar clay types used for the vessel manufacturing.

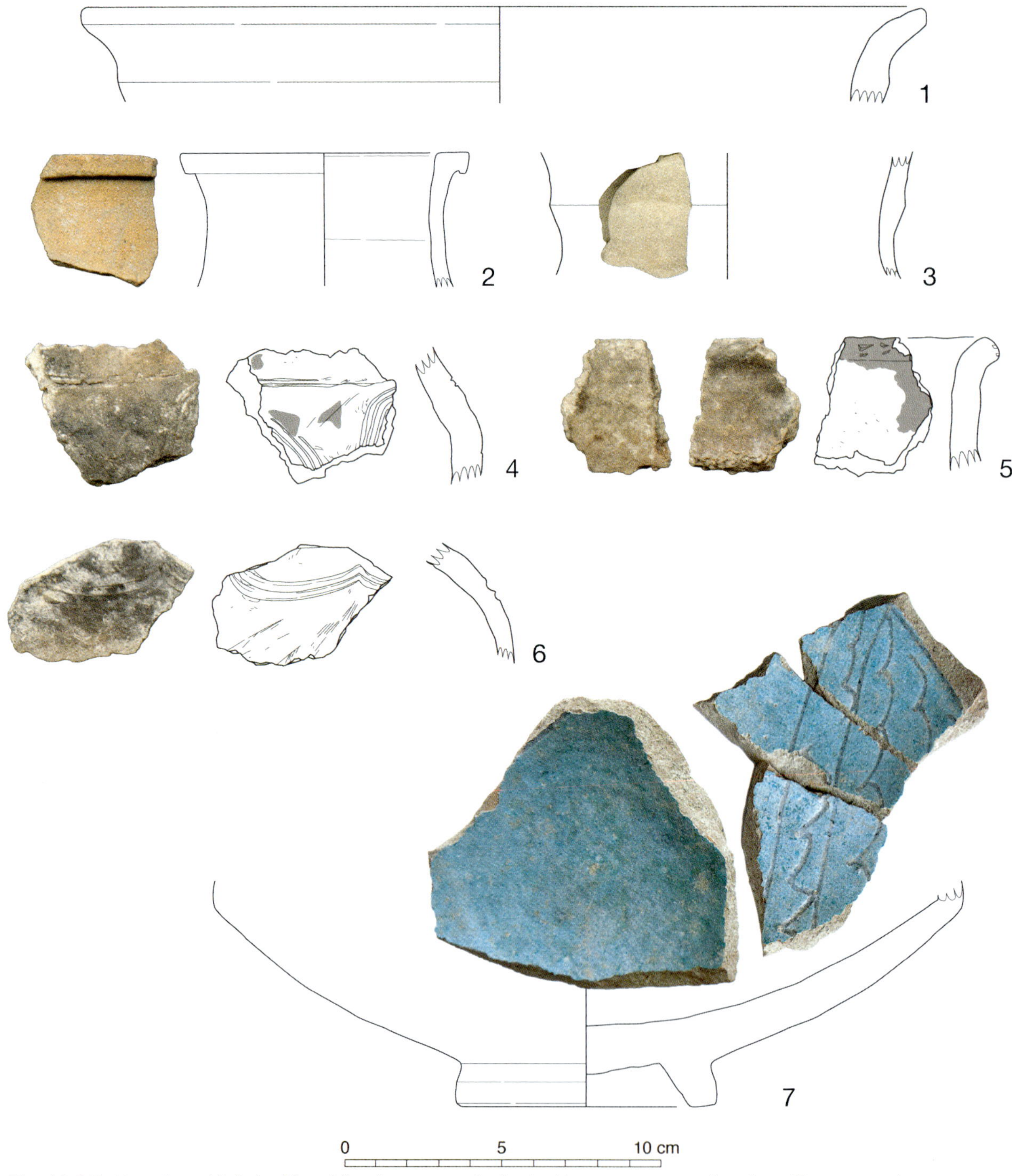

Fig. 16.6 Pottery from Unit 1a. Unoxidized cores are indicated in gray on vessel wall profile; gray colors on surface are soot from secondary firing.

Simultaneously, most of the sherds do not exhibit incompletely fired core.

Fine Common Ware

This is one of the minor ware types, comprising only 5.1% of the entire assemblage (Table 16.4). This ware is defined by a fine sand-tempered fabric, buff to brown surface color, and wheel-finished surfaces (Fig. 16.3: 4; Fig. 16.7: 20). As tempering materials, small fragments of shiny micaceous material are occasionally mixed with fabrics other than well-selected fine sand of ~1 mm in size. Sometimes, well-levigated clay with no visible inclusions is observed (Fig. 16.6: 3). In addition, minute horizontal striations are commonly caused by the smoothing using rotating wheel. Generally, this ware includes medium-fired sherds, however, sherds with gray cores also exist.

Various pottery forms are attested as fine common ware. In contrast to coarse ware, open-shaped bowls are commonly produced by this ware (Fig. 16.7: 20). A jar neck with a carination (Fig. 16.6: 3), min-

Table 16.6 Pottery catalog for Fig. 16.6.

Fig.	Context	Unit	Ware type	Color: exterior	Color: interior	Color: core	Temper	Forming	Remarks
16.6: 1	Pit 9.2	1a	Fine Orange Ware	5YR 6/6 orange	5YR 6/6 orange	5YR 6/6 orange	fine sand; mica	wheelmade	
16.6: 2	Pit 7.2	1a	Fine Orange Ware	7.5YR 7/4 dull orange	7.5YR 7/4 dull orange	5YR 6/6 orange	fine sand	wheelmade	
16.6: 3	Pit 9.2	1a	Fine Common Ware	10YR 8/3 light yellow orange	2.5Y 8/4 pale yellow	10YR 8/3 light yellow orange	none	wheelmade	carinated jar neck?
16.6: 4	Pit 7.2	1a	Coarse Kitchen Ware	7.5YR 5/3 dull brown	7.5YR 6/4 dull orange	7.5YR 5/1 brownish gray	fine sand	handmade	incised decorations
16.6: 5	Pit 7.2	1a	Coarse Kitchen Ware	5YR 4/3 dull reddish brown	7.5YR 5/3 dull brown	7.5YR 4/4 brown	fine sand	handmade	impressions on the lip
16.6: 6	Pit 7.2	1a	Coarse Kitchen Ware	10YR 6/2 greyish yellow brown	7.5YR 5/3 dull brown	7.5YR 5/3 dull brown	fine sand	handmade	incised decorations
16.6: 7	E3.2	1a	Glazed Ware	2.5YR 6/6 orange	turquoise-blue (glaze)	2.5YR 6/6 orange	very find sand	wheelmade	

iature jar originally attached to a larger vessel (Fig. 16.7: 18), and a possibly disc-shaped lid with four finger impressions (Fig. 16.7: 19) are also found in this category. Incised grooves, although rarely appearing, are typical decorations in fine common ware.

Fine Orange Ware

Finer variety of orange-colored ware, that is fine orange ware (Fig. 16.3: 5) constituted a relatively high ratio among the assemblages (18.1%). Most of this ware bears wheel-finished traces as faint horizontal striations on the exterior and interior surfaces, indicating the final wet smoothing treatment on a fast-rotating wheel. In most cases, the inclusions are very fine sand less than 0.5–1 mm, to which rare white lime particles and mica are added. The clay paste used for pottery-making may have been selectively prepared by potters with such a small quantity of tempering materials. The surface color is restricted to light orange or orange (sometimes with gray shades), implying a highly controlled firing atmosphere in pottery kiln. No unoxidized cores were observed.

Few sherds allowed us to reconstruct the entire vessel shape. The few available profiles suggest that small jugs with tall handles (Fig. 16.7: 21) and carinated bowls with straight rims (Fig. 16.7: 22) form part of the vessel typology. The handle most likely originates from a water jug (*bardaq*) with an upright neck-to-shoulder handle. Decorations are extremely rare for this ware compared with the various types of coarse wares described earlier.

Red Slipped Ware

One of the most distinctive and easily recognizable wares is the red slipped ware (Fig. 16.3: 6). It constitutes only 5.9% of the entire assemblage (Table 16.4). Red to dark red slip featuring this ware is, overall, applied on different fabrics, however, the large proportion of the sherds exhibits fine wheel-finished ware with minor sand inclusions, similar to fine common ware and fine orange ware. Although less common, but the red slip was found on handmade coarse wares with numerous sand-mineral tempers, resembling coarse orange wares (Fig. 16.7: 26, 28). Interestingly, the red-slipped surfaces, which are more likely to be on the exterior, are occasionally burnished.

Red slipped ware is produced in several typological categories. The most common are small- to medium-sized closed forms such as jars or pots (Fig. 16.7: 23–25). Large pots or deep bowls with thicker profiles are formed on the coarse fabric variants (Fig. 16.7: 27, 28). A red slipped ware lid (Fig. 16.7: 26) is an uncommon item, which probably covers medium or large pots, similar to Fig. 16.7: 28.

Glazed Ware

This ware (Fig. 16.3: 7) was subdivided according to glaze color patterns and decorative techniques. In this study, glazed ware was classified into three major types: monochrome-glazed, underglaze-incised, and polychrome-splashed ware.

Monochrome-glazed ware refers to green lead-glazed ware (Fig. 16.7: 29, 33) and dark brown (possibly manganese) glazed ware (Fig. 16.7: 30) on reddish or orange earthenware fabric. This type of glazed ware is not decorated with incised designs.

Underglaze-incised ware is characteristic for two different types of decorative technique under a

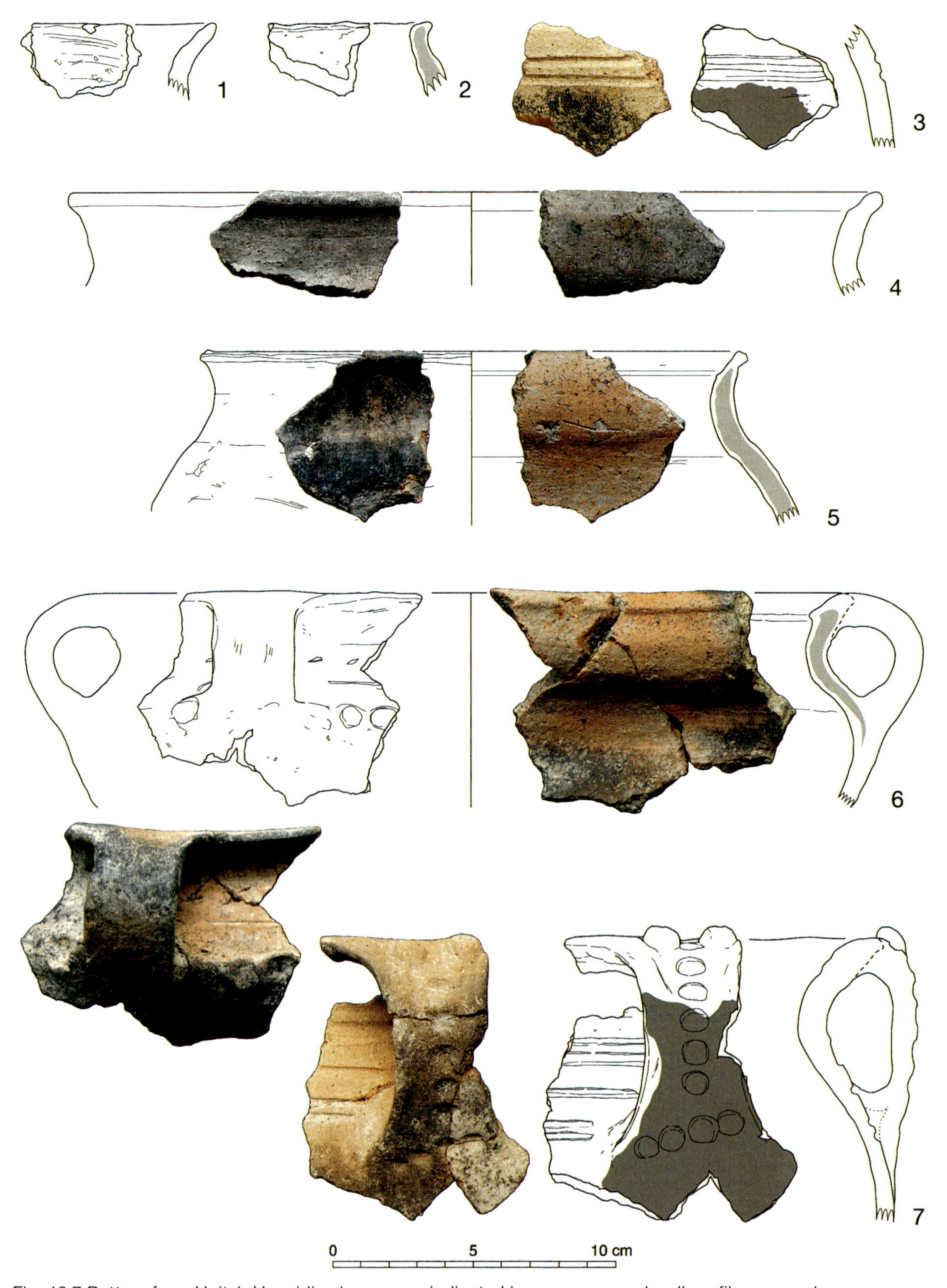

Fig. 16.7 Pottery from Unit 1. Unoxidized cores are indicated in gray on vessel wall profile; gray colors on surface are soot from secondary firing.

vivid turquoise-blue glaze; carved designs of leaves and a "pedestal" (Fig. 16.5: 5; Fig. 16.7: 31) and incised geometric or floral designs (Fig. 16.6: 7; Fig. 16.7: 32).

Polychrome-splashed ware were less common than in the other groups. One of the best-preserved sherds is characterized by alternating splashed patterns of green, yellow, and dark brown glaze on a

Table 16.7 Pottery catalog for Fig. 16.7.

Fig.	Context	Unit	Ware type	Color: exterior	Color: interior	Color: core	Temper	Forming	Remarks
16.7: 1	Pit 1.11	1	Coarse Kitchen Ware	2.5Y 7/1 light gray	2.5Y 7/2 grayish yellow	2.5Y 5/1 yellowish gray	sand	handmade	
16.7: 2	Pit 1.11	1	Coarse Kitchen Ware	2.5Y 8/2 light gray	2.5Y 8/2 light gray	2.5Y 6/1 yellowish gray	grit	handmade	
16.7: 3	Pit 9.8	1	Coarse Kitchen Ware	10YR 7/3 dull yellow orange	5YR 6/6 orange	5YR 6/6 orange	sand; mica	handmade	incised lines
16.7: 4	Pit 1.11	1	Coarse Kitchen Ware	10YR 7/2 dull yellow orange	7.5YR 7/3 dull orange	10YR 8/3 light yellow orange	sand	wheelmade?	
16.7: 5	Pit 1.11	1	Coarse Kitchen Ware	7.5YR 7/4 dull orange	5YR 7/6 orange	2.5Y 6/1 yellowish gray - 7.5YR 7/4 dull orange	sand; grit	handmade	gray core
16.7: 6	Pit 5.2	1	Coarse Kitchen Ware	7.5YR 7/6 orange	7.5YR 7/4 dull orange	7.5YR 5/6 bright brown - 7.5YR 5/1 brownish gray	sand; grit	handmade	gray core; incised lines and finger impressions
16.7: 7	Pit 9.1	1	Coarse Kitchen Ware	5YR 6/4 dull orange	5YR 6/4 dull orange	5YR 6/4 dull orange	sand	handmade	incised lines and finger impressions

white slip (Fig. 16.7: 34). This fragment, derived from a base, displayed a radiating pattern of polychromatic glazes. A single tiny fragment with green glaze on a white slip (Fig. 16.7: 35) was reused, presumably as a pendant or spindle whorl, because half of the broken edges were abraded to form a rounded shape.

Overall, various glazed wares account for only a small percentage (2.1%) of the assemblage. Notably, the ware fabrics consistently contain orange or reddish clay with fine sand inclusions. The vessel shapes were, without exceptions, thrown on a fast-rotating wheel. In this sense, the fabrics of glazed wares appear to be equivalent to those of fine orange wares.

Regarding vessel forms, all glazed ware sherds represented bowls with ring bases. An incised glazed ware bowl displays a carinated part on the upper body (Fig. 16.6: 7), probably having a vertical rim, according to the complete specimens from neighboring sites (for example, Bakhtadze 2013b: figs. 53, 54). The two rim fragments (Fig. 16.7: 29, 30) illustrate various rim profiles in the glazed-ware type, one of which may have originated from a small cup (Fig. 16.7: 30).

16.3.3 Chronological change in the Unit 1 assemblage

For the Damjili Unit 1 assemblage, the limited number of diagnostic sherds with well-preserved profiles prevents us from fully comprehending the diachronic development in terms of ceramic typology. Nevertheless, some diagnostic sherds recovered from secure subunit contexts reveal minor chronological differences in typological variability (Figs. 16.5–16.7; Unit 1b lacks any good pottery sherds with recognizable profiles, with the exception of a single carinated bowl sherd presented in Fig. 16.7: 22). Typologically common vessel shapes occur in both Units 1c and 1a, for example, a coarse kitchen ware jar with everted rims (Fig. 16.5: 1; Fig. 16.6: 5). Turquoise-blue glazed ware having different decorative styles were encountered in Units 1c and 1a (Fig. 16.5: 5; Fig. 16.6: 7), however, both types were found in an identical context (Pit 9.1), demonstrating no chronological gap in the different decorative styles (Fig. 16.7: 31, 32). Yet, a much wider variety of decoration patterns in coarse kitchen ware is likely to appear only in Unit 1a, where garland-like semicircular motifs (Fig. 16.5: 4, 6) are striking. In short, there are few chronological differences between Units 1c and 1a from a typological perspective.

Interestingly, the frequency of ware types by subunit in Unit 1 allowed us to observe a significant pattern in the assemblage (Table 16.4; Fig.16.4). In Unit 1c, the earliest medieval phase, coarse ware groups, including coarse kitchen ware, coarse common ware, and coarse orange ware are predominant (more than 80%). Among them, coarse orange ware is the most frequent ware type in this subunit (37.5%). In contrast to coarse wares, fine wares are attested with rather less frequency (less than 15%), with fine orange ware as the most common ware type (10.7%). Only one example of glazed ware was included, suggesting a limited use of highly valued glazed pottery in Unit 1c.

In the succeeding Unit 1b, coarse common ware was much more common, with over 70% of all the

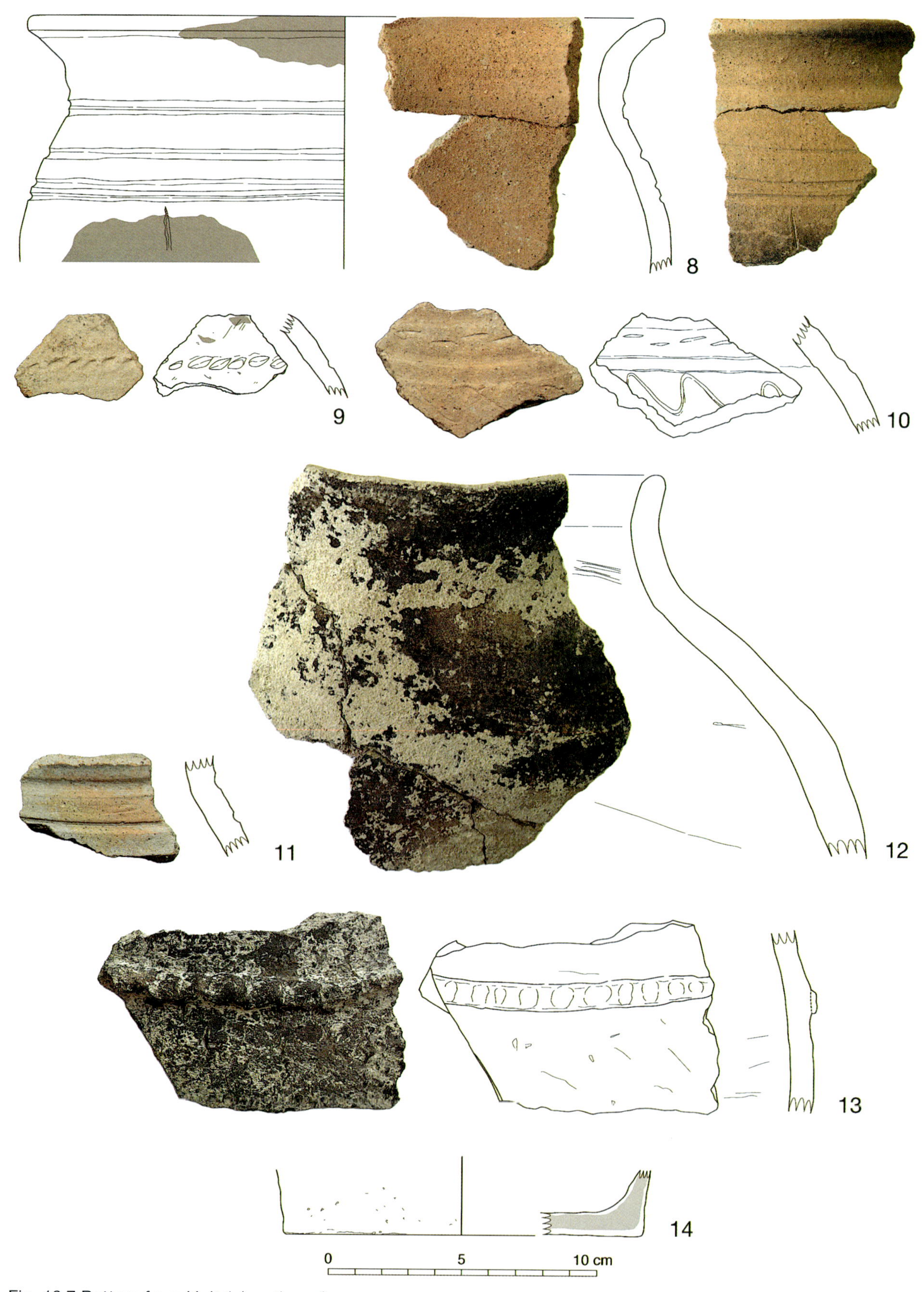

Fig. 16.7 Pottery from Unit 1 (continued).

sherds, whereas coarse orange ware is no longer found from this subunit onward. In contrast, coarse kitchen ware is marked by a decrease in frequency. This abrupt replacement observed in coarse ware types may explain a technological shift in local coarse ware production or a change in favored ware for culinary practices; however, to prove this hypothesis, it is necessary to examine more ceram-

Table 16.7 Pottery catalog for Fig. 16.7 (continued).

Fig.	Context	Unit	Ware type	Color: exterior	Color: interior	Color: core	Temper	Forming	Remarks
16.7: 8	C0.1	1	Coarse Common Ware	5YR 7/6 orange	5YR 6/6 orange	5YR 5/4 dull reddish brown	fine sand; mica	wheelmade	incised decoration; potter's mark
16.7: 9	Pit 9.1	1	Coarse Common Ware	10YR 6/2 grayish yellow brown	7.5YR 6/3 dull brown	5YR 6/6 orange	sand; lime grit	handmade	impressed dots
16.7: 10	Pit 9.1	1	Coarse Common Ware	5YR 6/4 dull orange	5YR 5/3 dull reddish brown	5YR 5/4 dull reddish brown	sand	handmade	incised decorations
16.7: 11	Pit 9.1	1	Coarse Common Ware	5YR 7/6 - 7.5YR 7/4 dull orange	5YR 6/6 orange	5YR 6/6 orange	fine sand	wheelmade	
16.7: 12	A4.1	1	Coarse Common Ware	gray brown	light brown	dark brown	sand	handmade	
16.7: 13	n/a	1	Coarse Common Ware	gray	dark gray - black	dark gray - black	sand; mica	handmade	applied band with impressions
16.7: 14	Pit 4.2	1	Coarse Common Ware	5YR 6/6 orange	7.5YR 7/4 dull orange	7.5YR 5/1 brownish gray	sand; grit	handmade	

ic datasets quantitatively and techno-typologically. It must also be indicated that a new ware type, red slipped ware, first appeared in this subunit. Fine orange ware continued to persist, with a significant increase from the preceding phase.

The uppermost medieval subunit, Unit 1a, is marked by a further increase in the frequency of fine common ware, fine orange ware, and red slipped ware, corresponding to a relative decline in coarse wares. Noticeably, an increasing number of glazed wares were introduced in Unit 1a. Most of the fine wares, including red slipped ware and glazed ware, share similar characteristics in fabric, inclusion, and wheel-forming methods with fine orange wares, as mentioned earlier. This suggests that fine orange ware may have been produced as the principal ware type for all the fine-ware groups. Red slipped ware, possibly a variant of fine orange ware with the application of red slip, is found in high numbers in this subunit.

It is difficult to interpret the bulk of the unstratified Unit 1 pottery indicated here simply as "Unit 1" (Fig. 16.7). The frequency pattern by ware type features a substantial quantity of coarse orange ware (c. 20%), followed by coarse kitchen wares (25%) and coarse common wares (22%) (Table 16.4). Considering that several coarse orange ware sherds derived from Trench 10 were related to cist graves and other contexts that were disturbed by pits, their common presence should be underestimated. The other ware types demonstrate a pattern largely similar to that of Unit 1a. In terms of stratigraphy, the "Unit 1" assemblage should be placed to a phase immediately following Unit 1a. Thus, despite some intruded sherds (for example, earlier sherds and post medieval materials, including modern square tile pieces), such a frequency pattern may suggest that there is no clear chronological gap between them.

Further, the pottery from Damjili Cave 2 indicates a similar frequency by ware type with this bulk of "Unit 1" (Table 16.4). The most common are coarse kitchen ware sherds (Fig. 16.7: 1, 2, 4, 5) (over 40%), followed by fine orange ware and red slipped ware. The close similarity in ware frequency and shared typological traits to the "Unit 1" bulk and the Unit 1a assemblage leads us to date Damjili Cave 2 occupation to the contemporaneous medieval period, which is, the final use phase at Damjili Cave.

16.4 Other medieval finds

Small finds, in contrast to pottery vessel fragments, are extremely rare. All non-pottery artifacts dating to the medieval period from Damjili Cave are catalogued in this section. They comprised 14 items of diverse materials, including burnt clay, iron, and glass.

Four small pieces of fired clay lumps with partly smoothed white surfaces were recovered from Trench 6 in Damjili Cave 2 (Fig. 16.8: 1, 2). They are usually small fragments measuring 2–5 cm and are occasionally tempered with angular lime grit and sand. The core section showed no unoxidized gray core which are often observed in coarse ware pottery sherds. Irregularly wet-smoothed surface of these clay objects resembles traces of striations

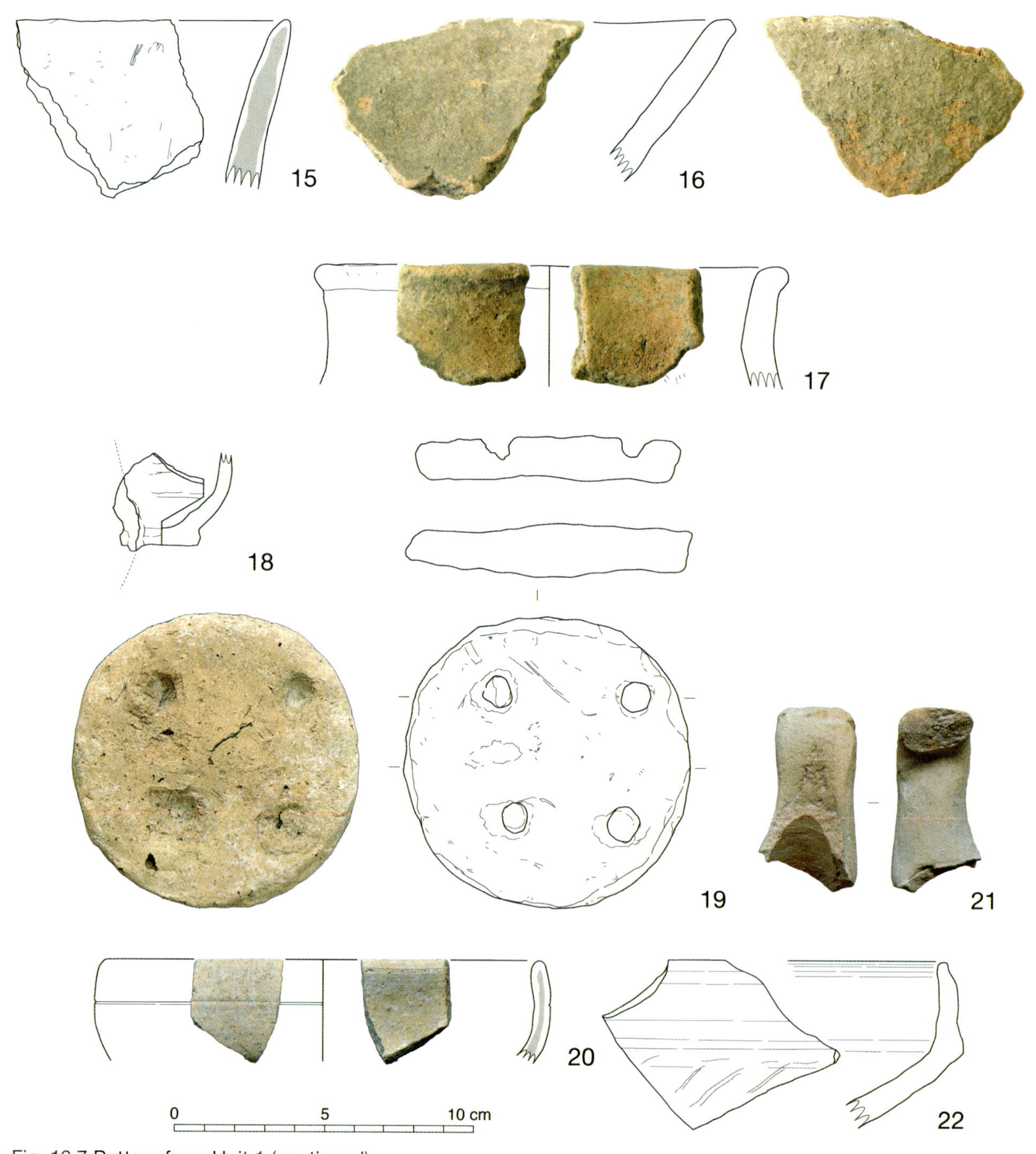

Fig. 16.7 Pottery from Unit 1 (continued).

left by wheel finishing, which makes us believe that they are vessel fragments at first glance; however, the reverse sides always exhibit a weathered broken appearance. They are considered to be fragments of mud plaster, probably from cylindrical bread ovens (*tandir*).

Tile fragments were found in Trenches 7 and 10, both assigned to Unit 1 in the main cave (Fig. 16.8: 3, 4). Although similar to sand-tempered coarse orange ware pottery in fabric, they have a well-smoothed, parallel-sided, and relatively thick profile (20–27 mm thick), indicating that they are roof tile fragments. The presence of a massive tile-roofed structure under the rockshelter, however, is highly dubious, because they are extraordinarily few in number for architectural tool.

One of the iron objects is represented by an arrowhead with a V-shaped split tip (Fig. 16.8: 5). This was found in the topsoil deposits of Trench 9. The arrowhead is exceptionally elongated in shape with 16.2 mm in extant length and 29.7 g in weight, comprising a complete tang (7.3 cm long) and a V-shaped two-pronged tip that is separated from the tang part by a ridge-like rest. Both distal parts of the tip were broken. According to Akhmedov's arrowhead classification system, this type of large arrowhead be-

Table 16.7 Pottery catalog for Fig. 16.7 (continued).

Fig.	Context	Unit	Ware type	Color: exterior	Color: interior	Color: core	Temper	Forming	Remarks
16.7: 15	Z1.1	1	Coarse Orange Ware	7.5YR 7/4 dull orange	5YR 6/6 orange	10YR 4/1 brownish gray	fine sand; chaff	handmade	gray core
16.7: 16	Z1.1	1	Coarse Orange Ware	5YR 6/6 orange	5YR 6/6 orange	5YR 6/6 orange	fine sand	handmade	
16.7: 17	Z1.1	1	Coarse Orange Ware	10YR 6/4 dull yellow orange	5YR 6/6 orange	5YR 4/1 brownish gray	fine sand	handmade	
16.7: 18	Pit 9.1	1	Fine Common Ware	7.5YR 7/4 dull orange	5YR 6/4 dull orange	5YR 7/4 dull orange	none	wheelmade	a miniature vessel for a larger pot
16.7: 19	B3.3	1	Fine Common Ware	7.5YR 6/4 dull orange	5YR 6/4 dull orange	5YR 6/4 dull orange	fine sand; mica	handmade	lid; four round impressions
16.7: 20	Pit 6.2	1	Fine Common Ware	7.5YR 7/4 dull orange	7.5YR 7/6 orange	10YR 6/2 grayish yellow brown	fine sand	wheelmade	an incised line; gray core
16.7: 21	Pit 1.14	1	Fine Orange Ware	2.5YR 7/4 pale reddish orange	2.5YR 7/4 pale reddish orange	2.5YR 6/4 dull orange	fine sand	handformed/ wheelmade	handle
16.7: 22	D4.2	1b	Fine Orange Ware	orange	orange	orange	fine sand	wheelmade	

longs to Type 4 (Akhmedov 2017: 97). Although detailed typological studies based on arrowhead morphology fail to identify the chronological development of this class of weaponry, many studies on Eurasian medieval iron arrowheads indicate that a large variety of arrowhead types co-occurred during the same time period, most likely because of their functions (Medvedev 1966; Kaminsky 1996; Akhmedov 2017; Bozer et al. 2020). The earliest occurrence of two-pronged arrowheads was reported in East and Central Asia (Miyashita 1980). However, notably in the North Caucasian context, the two-pronged arrowheads are interpreted as coming into use after the tenth century (Kaminsky 1996: 102). Other chronological assessments from excavated examples across Western Eurasia to Eastern Europe indicate a later introduction to these areas, placing this arrowhead type within the time range of the ninth to twelfth century (Medvedev 1966: 72–73; Bonzer et al. 2020: 349). Major urban settlements in early medieval Azerbaijan also yielded similar arrowheads, although rarely, among iron objects (Dostiyev 2008: 118, fig. 36: 32). With respect to the function of the two-pronged arrowheads, written historical records, including the early thirteenth century diplomat Joannes de Plano Carpini, explain that this particular type of arrowhead was used for bird shooting or hunting, rather than for war weapons against humans (Świętosławski 1999: 62).

Two iron nails were also found (Fig. 16.8: 6, 7). One was a large nail that was 10 cm long with a round head 8 mm in diameter. Its double-bent profile suggests that it was once used for different (possibly wooden) material, and then discarded. The isolated recovery from Trench 10 did not provide any functional interpretation. The closest parallels are known from various urban centers such as Beylaqan and Shamkir (Dostiyev 2008: fig. 34: 13, fig. 40: 19–23; Dostiyev 2010: 54). The other nail specimen, 3.6 cm long (with the lower part broken), displays a curved shape with a folded swirl-like tip. Similar iron nails are attested widespread across Azerbaijan, such as Shamakhı, Beylaqan, Qabala (Selbir area) and other medieval cities (Dostiyev 2008: fig. 33: 16, fig. 39: 15; Babayev 2016: 179, n.162).

A blue glass bangle fragment with a rounded section (Fig. 16.8: 8) was recovered from the upper deposit of Damjili Cave 2 (Trench 4). Only a quarter of the entire circular bangle was extant, which was originally approximately 7 cm in diameter. This bangle is made from a simple glass string rather than twisted strings, which was also common in early medieval glass bangles. Such glass bangles, with a wide variety of colors and forms, are frequently attested from the early medieval urban settlements in various parts of Azerbaijan and the Southern Caucasus (Dostiyev 2008: fig. 96). The bangles from Shamkir dated from the ninth to twelfth century (Dostiyev 2010: 219, 221, 223) and from the Selbir area in Qabala dated from the eighth to the tenth century (Babayev 2016: 177, n. 134–135), represent the most comparable finds.

16.5 Dating the Unit 1 assemblage

Based on the pottery assemblage of Unit 1 and the

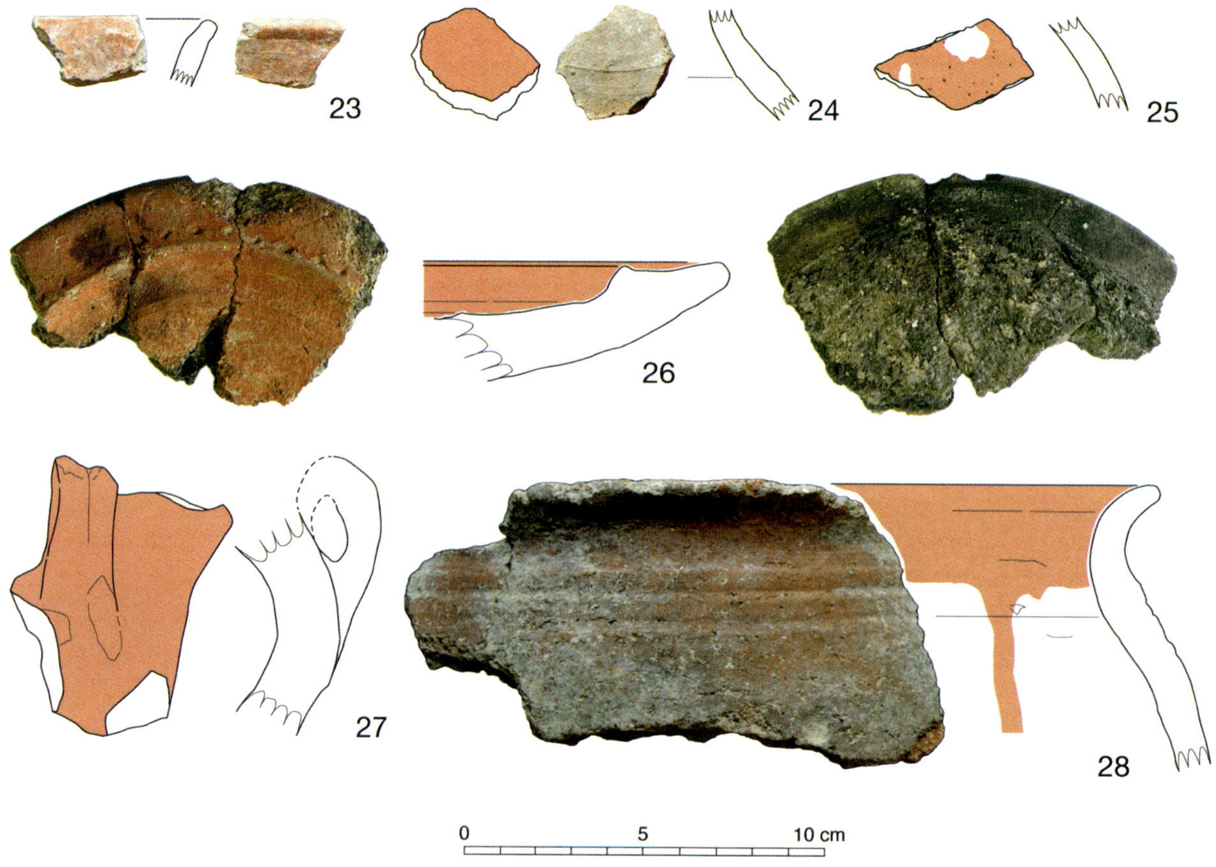

Fig. 16.7 Pottery from Unit 1 (continued).

Table 16.7 Pottery catalog for Fig. 16.7 (continued).

Fig.	Context	Unit	Ware type	Color: exterior	Color: interior	Color: core	Temper	Forming	Remarks
16.7: 23	Pit 1.11	1	Red Slipped Ware	2.5YR 5/6 bright reddish brown	2.5YR 5/6 bright reddish brown	7.5YR 7/4 dull orange	fine sand	wheelmade	
16.7: 24	Pit 6.3	1	Red Slipped Ware	10R 5/6 red	5YR 6/6 orange	5YR 5/6 bright reddish brown	fine sand	wheelmade	
16.7: 25	B1.3	1	Red Slipped Ware	10YR 5/6 yellowish brown	5YR 5/6 bright reddish brown	7.5YR 6/2 grayish brown	fine sand; mica	wheelmade	
16.7: 26	B4.1	1	Red Slipped Ware	10R 5/8 red	10YR 6/2 grayish yellow brown	10YR 6/2 grayish yellow brown	white sand	handmade	coarse paste; shallow wavy incision and a row of impressions on the outer rim
16.7: 27	E0.1	1	Red Slipped Ware	red slip; burnished	-	-	fine sand	handmade	
16.7: 28	Pit 9.1	1	Red Slipped Ware	7.5R 4/6 red	7.5R 4/4 dusky red - 2.5YR 5/6 bright reddish brown	2.5YR 4/6 reddish brown	sand; mica	handmade	horizontal grooves

small finds presented above (Sections 16.3–16.4), we attempted to determine the relative and absolute chronologies of Unit 1 by comparing our materials with related assemblages in the Southern Caucasus.

Chronologically, much attention has been paid to glazed wares in the medieval archeology of the Middle East, primarily because of their noticeable aesthetic value and the early collection acquired through archaeological excavations. However, it has been argued that the established framework of glazed ware chronology and art historical evaluations should be reconsidered with a more systematic

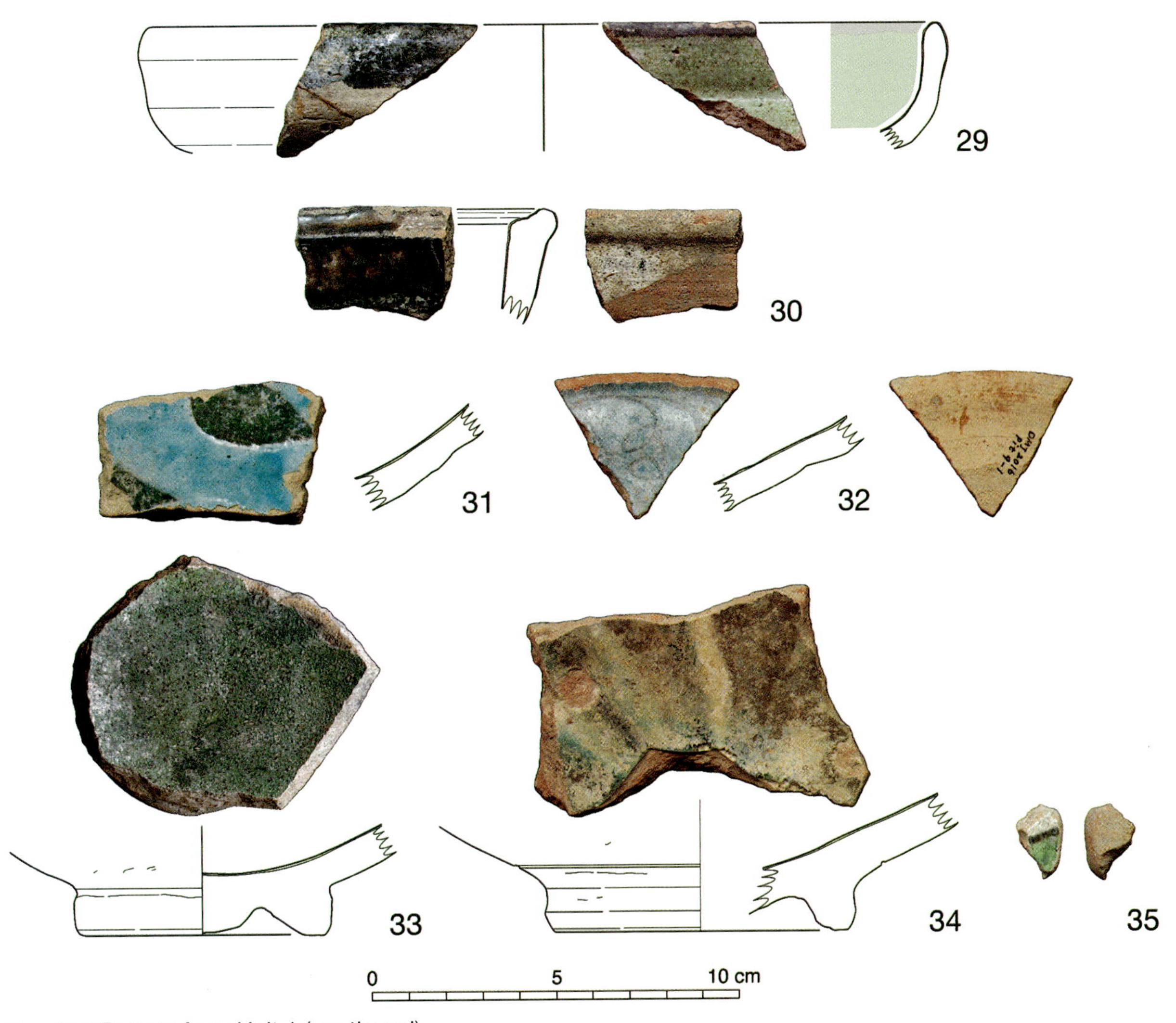

Fig. 16.7 Pottery from Unit 1 (continued).

Table 16.7 Pottery catalog for Fig. 16.7 (continued).

Fig.	Context	Unit	Ware type	Color: exterior	Color: interior	Color: core	Temper	Forming	Remarks
16.7: 29	Pit 1.11	1	Glazed Ware	dark green (glaze); 7.5YR 8/3 light yellow orange	light green (glaze)	7.5YR 7/6 orange	fine sand	wheelmade	monochrome
16.7: 30	Pit 9.1	1	Glazed Ware	2.5YR 6/6 orange; faded pale glaze	dark brown (glaze)	2.5YR 6/6 orange	fine sand	wheelmade	monochrome
16.7: 31	Pit 9.1	1	Glazed Ware	5YR 7/6 orange	turquoise-blue; green (glaze)	5YR 6/6 orange	fine sand	wheelmade	underglaze incised
16.7: 32	Pit 9.1	1	Glazed Ware	5YR 7/4 dull orange	turquoise-blue (glaze)	5YR 6/6 orange	fine sand	wheelmade	underglaze incised
16.7: 33	Pit 1.11	1	Glazed Ware	10YR 7/3 dull yellow orange	light green (glaze)	7.5YR 7/4 dull orange	fine sand	wheelmade	monochrome
16.7: 34	Pit 9.1	1	Glazed Ware	2.5YR 5/4 dull reddish brown	green, yellow, brown on white slip (under transparent glaze)	2.5YR 6/8 orange	fine sand	wheelmade	polychrome glazed under a transparent glaze
16.7: 35	Pit 5.4	1	Glazed Ware	2.5YR 5/5 dull reddish brown	green glazed and paint? on white slip (under transparent glaze)	2.5YR 6/9 orange	fine sand	wheelmade	glazed under a transparent glaze; reused as a pendant?

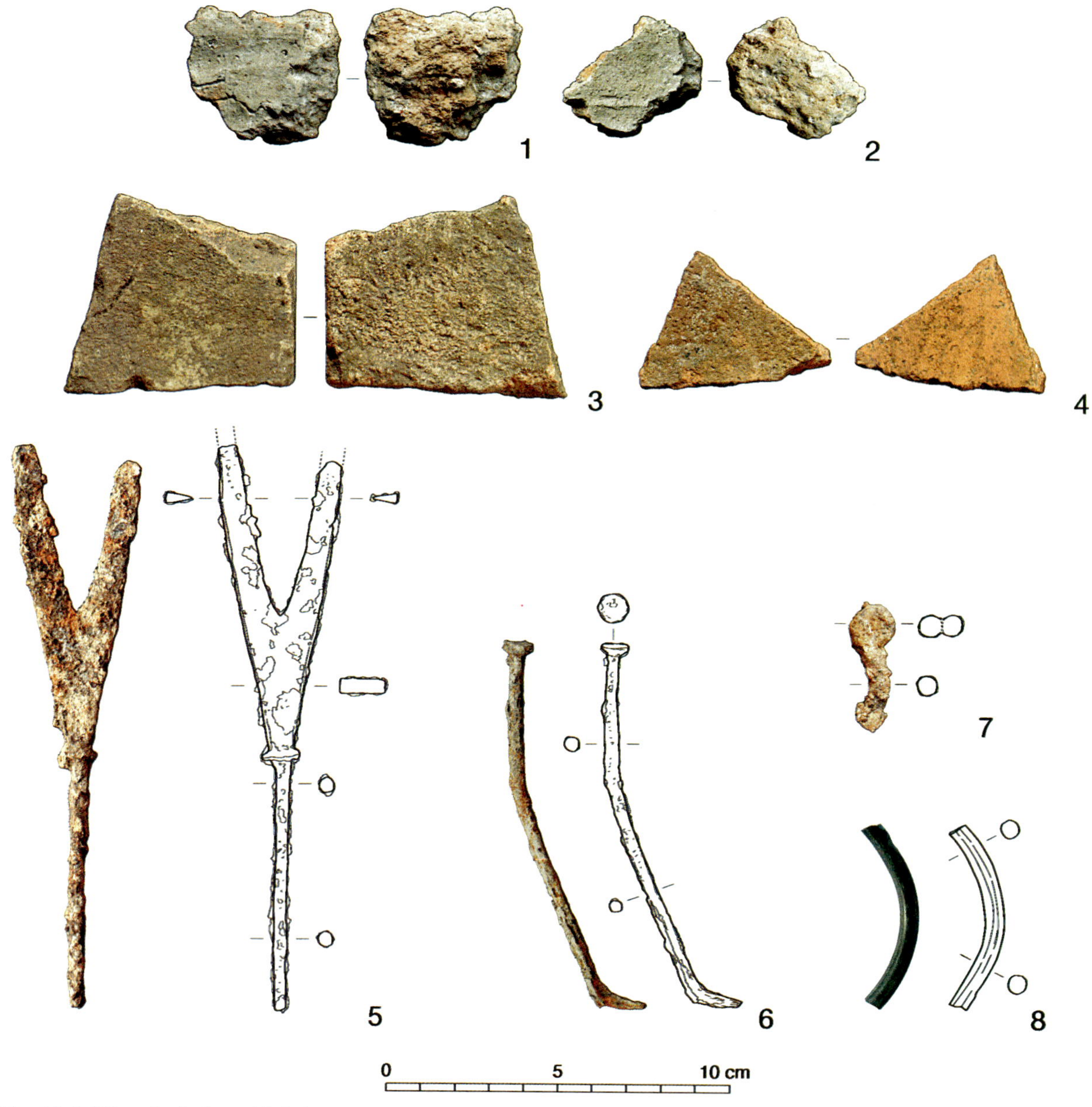

Fig. 16.8 Other finds from Unit 1.

Table 16.8 Pottery catalog for Fig. 16.8.

Fig.	Context	Unit	Material	Object	Color	Remarks
16.8: 1	Pit 6.2	1	bunrt clay	*tandir* fragment?	ext. light gray; core: orange	one irregularly smoothed surface
16.8: 2	Pit 6.2	1	bunrt clay	*tandir* fragment?	ext. light gray; core: light orange	one irregularly smoothed surface
16.8: 3	Pit 7.2	1	bunrt clay	unglazed tile	ext. yellowish brown; core: orange	both surfaces smoothed
16.8: 4	Pit 7.2	1	bunrt clay	unglazed tile	ext. orange brown; core: orange	both surfaces smoothed
16.8: 5	Pit 9.1	1	iron	arrowhead	-	weight 29.7 g
16.8: 6	Pit 10.6	1	iron	nail	-	weight 8.6 g
16.8: 7	Pit 5.2	1	iron	nail?	-	swirled tip
16.8: 8	Pit 4.1	1	glass	bracelet	blue	

collection of archaeological ceramic data (Watson 1999). Nonetheless, for the Southern Caucasus of concern in this section, recent accumulations of published data from excavated sites provide useful keys for chronologically assigning ceramic examples.

Polychrome-splashed wares are one of the most widely attested glazed wares in this region. The base fragment from Damjili (Fig. 16.7: 34) can be best compared, for example, with underglazed polychrome painted ware from Shamkir (Dostiyev

2010: 138, 140; Dostiyev 2017: 648–652, fig. 9: 2), and from Dmanisi (Bakhtadze 2013b: figs. 24, 26), all dated to the early medieval period (prior to the Mongolian invasion), namely the ninth to tenth century. Additionally, the monochrome ware provides a similar date in the early medieval period. Clear parallels for the green-glazed ware with no underglaze-incised decorations (Fig. 16.7: 29, 33) can be found again at Shamkir (Dostiyev 2010: 150), whereas the dark-brown glazed ware (Fig. 16.7: 30) is comparable with a similar vessel from the Selbir area at Qabala (Babayev 2016: 181, n.103). Both of these examples date to around the tenth to eleventh century.

A similar date was also provided by the incised or often called *sgraffiato* glazed ware under a turquoise-blue glaze (Fig. 16.5: 5; Fig. 16.6: 7; Fig. 16.7: 31, 32). Comparable glazed ware examples are known from early medieval settlements in eastern Georgia such as Dmanisi (Maysuradze 1954: plate 32, nos. 283, 368, 2476), Nekresi (Bakhtadze et al. 2010: 40), Tbilisi, and Rustavi (Bakhtadze 2013b: figs. 43, 44, 49, 51, 53, 54). These turquoise-blue glazed incised ware sherds are dated to the eleventh–thirteenth century.

More interestingly, the twelfth to thirteenth century artificial rock-cut caves (hermitages) at Nakhiduri in southeastern Georgia yielded a similar combination of glazed wares to the Damjili examples: monochrome green and dark brown glazed wares and underglazed-incised ware with a turquoise-blue glaze (Bakhtadze 2013a: fig. 15).

However, apart from these glazed wares, virtually absent within the Damjili glazed wares are early glazed ware with slip-painted decorations under colorless or light yellow transparent glazes (Dostiyev 2017: 640–643), underglazed ware with manganese and copper oxide geometric painted motifs (Dostiyev 2017: 643–645), and green-glazed slip-painted ware (Dostiyev 2017: 645–648). These early glazed wares are assigned to the eighth to tenth century, suggesting that the Damjili assemblage may belong to a period when the early glazed ware series may no longer have been in common use.

Similarly, to define the *terminus ante quem* date for Damjili glazed ware, the absence of glazed fritware is of particular significance. Fritware production, followed by its first development in late eleventh-century Egypt, was introduced to Syria and Iran in the twelfth century (Watson 2017: 494–495). Potters in the Southern Caucasus may also have adopted this ceramic industry as late as the twelfth century, in the Seljuk era. Fritware vessels with turquoise and light blue glaze (Dostiyev 2010: 167–169, 172–175) and lustre fritware (Dostiyev 2010: 176–178) have been found from Shamkir, mainly in the layers belonging to the twelfth–thirteenth century, until the last destruction by the Mongolian troops in the early decades of the thirteenth century. This indicates that either the medieval dwellers in Damjili Cave abandoned cave use before the development of glazed fritware vessels, or they simply did not know sophisticated white fritware fabrics but reddish orange ones for glazed ware. However, it should be emphasized that there may well be some overlaps in the estimated absolute dating for glazed wares, depending on the production technology and predilection of glazed ware consumers in each area. This chronological problem requires further refinement using more excavated materials and radiometric dates.

Unlike glazed wares, unglazed wares provide limited reference data because they are difficult to date owing to the scarcity of securely dated pottery types. Notably, coarse kitchen ware (locally called *qazan*) demonstrates a wide typological variant in the rim profile, handle shapes, and decoration methods, possibly reflecting a chronological change in this vessel category. For example, cooking pot ware of the ninth to the tenth century displays globular jar form and sometimes with crescent-shaped handles on shoulder (Dostiyev 2020: 33, plates XVI, XVII), whereas hemispherical hole-mouth pots and more elaborated applied decorations including triangular handles become more common later in the eleventh–twelfth century (Dostiyev 2020: 41). Considering these typological variants, the Damjili coarse kitchen ware pots appear to exhibit a simpler vessel form of the globular body type (Fig. 16.7: 6, 7), probably dating to an earlier phase, the ninth to tenth century (however, note that crescent-shaped handles are not present).

Several red slipped wares are often found in the early medieval context in the fortified urban settlement of Shamkir (Dostiyev 2020). They are observed in diverse types of pottery vessels (bowls, jars, jugs, large pithoi, and lids) at Shamkir, which is also the case at Damjili. Comparisons with the Shamkir sequence, an incised lid (Fig. 16.7: 26), and jars with outwardly curved rims (Fig. 16.7: 28) can be dated to the tenth–eleventh century.

In summary, the Damjili Unit 1 assemblage can be dated to some point between the ninth century and the end of the eleventh century. In particular, comparable evidence from early medieval sites across the region may indicate a more restricted pre-Mongolian period between the tenth and the eleventh century. This chronological evaluation corresponds well with the two radiocarbon dates obtained from human bones in Trench 10 (Chapter 5 of this volume).

However, chronological examination of the glazed and unglazed pottery does not confirm, to a lesser extent, the stratigraphic distinction of the three subunits, Units 1a–1c. For instance, glazed ware sherds belonging to the same group are present in different subunits (in the case of underglazed-incised ware with a turquoise-blue glaze). This might be principally because the subunits defined here covered a short period, possibly within a few centuries. Furthermore, it remains difficult to link our data on ware composition to the relative chronology because comparable quantitative ceramic data are not available for other sites in the region. The changing patterns in the relative frequencies of coarse wares and a gradual increase in red slipped ware and glazed ware discussed in Section 16.3.3 (Fig. 16.4) can be interpreted not as commonly shared trends in the regional ceramic tradition, but more likely as peculiarities observed in the Damjili Unit 1 assemblage.

16.6 Interpreting the Unit 1 assemblage and cave use at Damjili Cave during the medieval period

Following its short-lived use during the Early Bronze Age, Damjili Cave was reoccupied during the early medieval period. This recurring occupation does not necessarily appear to be similar to earlier cave use.

Damjili Unit 1 revealed that during the medieval period, the cave was intensively occupied (possibly one or two centuries). The recovery of the well-made stone walls, pits, and graves in Units 1c–1a (indicating domestic activities including building, storing foods, cooking, waste disposal, and mortuary practices, burial of the human body), as well as the thick accumulation of ashy deposits, also reinforces the argument that human occupations in the medieval period were completely different in nature, perhaps more diversified in socio-economic behavior from the earlier prehistoric ones. The more abundant quantity of pottery sherds compared with the Bronze Age assemblage confirms that the cave was much more intensively occupied and exploited as a domestic living place in the medieval period. The pottery from three successive subphases (Units 1c–1a) revealed a wide variety of fine and coarse wares, including glazed wares, suggesting that medieval cave dwellers required a combination of ordinary tableware, storage, and culinary vessels, as city dwellers did.

Notably, human use and occupation of Damjili Cave became more spatially extensive, reaching its maximum area during the early medieval period. In Unit 1a, the latest phase of Unit 1, sporadic use began in the previously unoccupied at Damjili Cave 2. This possibly provided sufficient space for cooking, storage, and waste-discarding activities, as evidenced by the hearths and flask-shaped pits in the sounding trenches. Additionally, medieval activities would have extended to the western part of the cave, where pits with medieval glazed ware were discovered during the 1956–1957 excavations (Hüseynov 2010: 54).

As there is no archaeological evidence of craft production or workshops with pottery firing kilns to date, it is highly probable that none of these pottery vessels were produced in the cave itself but were largely imported as required from nearby settlements or urban centers. The presence of iron and glass objects demanding specialized craft technology also explains why virtually all of the material culture was required to be supplied from outside the cave. Contemporary medieval settlement sites in the vicinity on which Damjili Cave dwellers depended have been documented sporadically (some exceptions include Göyezen Qalası and Taglarbeyli, but not excavated sufficiently) (Aveydagh State Historical-Cultural Reserve of the Ministry of Culture and Tourism); the daily needs, whether material or provisions, for the most part, must have been brought into the cave probably from these settlements or from some distance.

In this regard, the sporadic but increasing evidence of cave/rockshelter dwellings in the late first millennium is noteworthy. For instance, the Mashavera and Khrami valleys in the Kvemo Kartli region of eastern Georgia, adjacent to the Qazax Plain, experienced several rock-cut occupations, including Nakhiduri, Muguti, and Samshvilde (Bakhtadze 2013a). Keshikchidagh and Davit Gareja monastic complexes were also hallmarks of the early medieval religious tradition in northwestern Azerbaijan and the Kakheti region, that, according to the archaeological records, experienced one of the most prosperous age in the eleventh to the twelfth century (Nasibli 2010).

It is possible that Damjili Cave, located in the Aveydagh rural landscape, played a socio-economic role similar to these cave complexes when placed in broader regional development in the late first millennium. The reason why these rock-cut dwellings were established in the, occasionally remote, harsh ecological settings, can be connected with the development of monastic institutions linked to the Christian eremitism in the territories of medieval Caucasian Albania and Georgia (Bakhtadze 2013a: 15; Baumer 2021: 198–199). Although we have no archaeological signature expressing the ritual practices or religious identity of the Damjili dwellers, a group of artificial rock-cut hollows at a higher elevation around Damjili Cave may have been suitable for hermitages. The particular set of glazed ware types from Damjili Cave, which finds strong affinity with that of glazed ware, in particular, from sites related to monastic and religious communities (for example, Dmanisi, Nekresi, and Nakhiduri) is noteworthy.

Caves and rockshelters were not only occupied as hermitages by religious communities. They may have also offered refuge or asylum for outlaws, soldiers, and displaced people who were driven out of urban life owing to political conflicts, constant wars, or economic failures, however, managed to take advantage of the emergent international trade network in the early medieval period. This aspect can be hinted at by the presence of material culture, such as highly valued glazed vessels compared with contemporary urban centers.

Considering the archaeological evidence, the renewed occupation of Damjili Cave in the early medieval period would also be closely connected to the urbanization process in the Southern Caucasus, which culminated in the twelfth century (Dostiyev 2008: 30–183; Rayfield 2012: 73). An accumulating corpus of archaeological records across the region proves that numerous urbanized centers, such as Barda, Beylaqan, Ganja, Shamkir, Tbilisi, Dabil, and Qabala, developed under the auspices of local kingdoms and polities (Dostiyev 2008: 30–32). In addition to early medieval urbanism in some major settlements, intensive land use in rural areas was in full swing, as suggested by a marked increase in settlement number and density in the Tovuz-Qovlar area in western Azerbaijan (Shimogama and Alakbarov 2020). These settlement dynamics in the late first millennium were triggered by an increasing population throughout the region.

In the context of urbanism and unstable political circumstances when intense rivalry between ruling elites sought independence, and foreign powers attempted to impose sovereignty over the region (Abbasid caliphs, Byzantine rulers, and northern nomadic groups of the Khazars and Alans) (e.g., Bünyadov 2007; Baumer 2021), supra-regional interaction network connecting urban centers and rural villages enabled an unprecedented scale of the movement of people and goods. The subsistence activities performed by cave dwellers undoubtedly depended on this ever-growing network. The occupation of Damjili Cave ended only when the urban network was devastated by the Mongol invasion in the early thirteenth century.

At present, it remains early to relate archaeological evidence to any specific historically documented event (e.g., infiltration of the Seljuk Turks or the emergence of the unified Georgian kingdom). Further detailed studies on excavated materials will certainly contribute to elucidating the cultural and historical significance of cave occupations in the Southern Caucasus and a dynamic understanding of the medieval period in its entirety.

16.7 Conclusions

This chapter attempted to investigate how Damjili Cave was exploited in the periods after the Mesolithic, Neolithic, and Chalcolithic periods, primarily based on excavated materials.

Generally, archaeological data from cave sites suggest that human use of caves was not homogeneous but diverse during the Holocene period across Western Eurasia (e.g., Bonsall and Tolan-Smith 1997; Bergsvik and Skeates 2012; Bergsvik and Dowd 2018). At Damjili Cave, the analysis of the material remains from Unit 2 and Unit 1 deposits (largely focused here on the pottery assemblages) clearly reveals different cave use patterns during these different occupation periods. During the Early Bronze Age, the expedient and homogeneous character of archaeological materials indicated less investment in caves. This suggests that small groups of farmers or herders visited the cave periodically and used it as a shelter or place for seasonal camps.

In contrast, more intense domestic exploitation and long-term occupation may have entailed more large-scale modifications of the cave environment during the early medieval period. A wide range of human activities, such as building, pit digging, and burials at Damjili Cave, represent one such modification that is archeologically recognizable. Despite the absence of evidence of repeated religious activities, the development of monasticism in the region may be linked to and stimulated by the reoccupation of Damjili Cave in the late first millennium. Further, the pottery assemblage and other finds of various materials highlight strong connectivity to the region-wide interaction network, which must have sustained the lifestyles of the past community in the cave.

The differences in cave use intensity during the two later periods at Damjili Cave suggest that the culturally assigned values and meanings to the cave differed for each period. To clarify this aspect of the cave, we await more comprehensive fieldwork and systematic archaeological analyses.

References

Akhmedov, S. A. (2017) Classification of 9th–13th century arrowheads found in Azerbaijan. *Archaeology, Ethnology & Anthropology of Eurasia* 45(4): 93–101.

Areshian, G. E., B. Gasparyan, P. S. Avetisyan, R. Pinhasi, K. Wilkinson, A. Smith, R. Hovsepyan, and D. Zardaryan (2012) The Chalcolithic of the Near East and south-eastern Europe: Discoveries and new perspectives from the cave complex Areni-1, Armenia. *Antiquity* 86: 115–130.

Arimura, M. (2014) Forest exploitation during the Holocene in the Aghstev Valley, Northeast Armenia. In: *Stone Age of Armenia: A Guide-book to the Stone Age Archaeology in the Republic of Armenia*, edited by B. Gasparyan and M. Arimura, pp. 261–281. Kanazawa: Center for Cultural Resource Studies, Kanazawa University.

Aveydagh State Historical-Cultural Reserve of the Ministry of Culture and Tourism. Official webpage. https://avey-heritage.az/content/3 (accessed on 19 December 2022) (in Azerbaijani).

Babayev, I. (2016) *The Gabala Archaeological Expedition 2014: Reports, Findings*. Baku: CBS.

Bakhtadze, N. (2013a) The results of archaeological research of the rock-cut monuments in the Kvemo Kartli region (Georgia). *Światowit* XI(LII)/A: 1–20.

Bakhtadze, N. (2013b) *Ceramics in Medieval Georgia*. Tbilisi: Georgian National Museum.

Bakhtadze, N., N. Tevdorashvili, and G. Bagrat'ioni (2010) *Nekresi: Tsnobari Momlotsveta da Mogzaurtat'vis*. Nekresi: Diocese of Nekresi, Partriarchate of Georgia (in Georgian).

Baumer, C. (2021) *History of the Caucasus: At the Crossroads of Empires*. Volume One. London: I.B. Tauris.

Bergsvik, K. A., and M. Dowd (2018) Caves and rockshelters in medieval Europe: Religious and secular use. In: *Caves and Ritual in Medieval Europe, AD 500–1500*, edited by K. A. Bergsvik and M. Dowd, pp. 1–9. Oxford: Oxbow Books.

Bergsvik, K. A., and R. Skeates (2012) Caves in context: an introduction. In: *Caves in Context: The Cultural Significance of Caves and Rockshelters in Europe*, edited by K. A. Bergsvik and R. Skeates, pp. 1–9. Oxford and Philadelphia: Oxbow Books.

Bonsall, C., and C. Tolan-Smith (1997) *The Human Use of Caves*. British Archaeological Reports, International Series 667. Oxford: BAR Publishing.

Bozer, R., A. Yavaş, and Ü. Güder (2020) Arrowheads (temren) found from the excavations at the Sultan Giyaseddin Keyhüsrev-II caravanserai in Eğirdir, Isparta. *Sanat Tarihi Dergisi* 29/2: 333–369.

Bünyadov, Z. (2007) *Azerbaijan in the Seventh–Ninth Centuries*. Baku: Shargh-Gharb (in Azerbaijani).

Dostiyev, T. (2008) *Azerbaijan Archaeology. Vol. VI: Middle Ages*. Institute of Archaeology and Ethnography, National Academy of Sciences of Azerbaijan. Baku: Shargh-Gharb (in Azerbaijani).

Dostiyev, T. (2010) *Middle Age City of Shamkir: Archaeological Investigations in 2009* (in Azerbaijani).

Dostiyev, T. (2017) Glazed ceramics of medieval Shamkir city. In: *Glazed Pottery of the Mediterranean and the Black Sea Region, 10th–18th Centuries, Vol. 2* edited by S. Bocharov, V. François, and A. Sitdikov, pp. 639–674. Kazan: A. Kh. Khalikov Institute of Archaeology, Academy of Sciences of the Republic of Tatarstan (in Russian).

Dostiyev, T. (2020) *Ceramics from Medieval Period City of Shamkir, Vol. I: Unglazed Ceramics*. Baku: Institute of Archaeology and Ethnography, Azerbaijan Academy of Sciences (in Azerbaijani).

Huseynov, M. (2010) *The Lower Paleolithic of Azerbaijan*. Baku: National Academy of Sciences of Azerbaijan (in Russian).

Ismailov, G. S. (1977) *Arkheologicheskoe Issledovanie Drevnego Poseleniya Baba-Dervish*. Baku: Elm (in Russian).

Kaminsky, V. N. (1996) Early medieval weapons in the North Caucasus — A preliminary review. *Oxford Journal of Archaeology* 15(1): 95–105.

Kastl, G. (2008) Didi Gora and Tqisbolo-gora: Two Middle Bronze Age settlements in the Alazani Valley, Kakheti, eastern Georgia. In: *Ceramics in Transition: Chalcolithic through Iron Age in the Highlands of the Southern Caucasus and Anatolia*, edited by K. S. Rubinson and A. Sagona, pp. 185–198. Ancient Near Eastern Studies Supplement 27. Leuven: Peeters.

Lyonnet, B. (2014) The Early Bronze Age in Azerbaijan in the light of recent discoveries. *Paléorient* 40(2): 115–130.

Maysuradze, Z. (1954) *Gruzinskaya Khudozhestvennaya Keramika XI–XIII vv.* Tbilisi: Izdatel'stvo Akademii Nauk Gruzinskaya SSR (in Russian).

Medvedev, A. F. (1966) *Ruchnoe Metatel'noe Oruzhie: Luk i Strely, Samotstrel VIII–XIV vv.* Moscow: Izdatel'stvo Nauka (in Russian).

Miyashita, S. (1980) Considération sur pointe de flèche à la forme de croissant représenté dans la scène de chasse de Bahram Gûr. *Bulletin of the Ancient Orient Museum* II: 83–97 (in Japanese).

Nasibli, Y. M. (2010) The Keshikchidagh Monastic Complex. *Visions of Azerbaijan* July-August 2010: 86–90.

Nishiaki, Y., A. Zeynalov, M. Mansrov, C. Akashi, S. Arai, K. Shimogama, and F. Guliyev (2019) The Mesolithic-Neolithic Interface in the Southern Caucasus: 2016–2017 Excavations at Damjili Cave, West Azerbaijan. *Archaeological Research in Asia* 19: 100140.

Palumbi, G. (2008) *The Red and Black: Social and Cultural Interaction between the Upper Euphrates and the Southern Caucasus Communities in the Fourth and Third Millennium BC.* Studi di Preistoria Orientale Volume 2. Roma: Sapienza Università di Roma.

Rayfield, D. (2012) *Edge of Empires: A History of Georgia.* London: Reaktion Books.

Rice, P. (1987) *Pottery Analysis: A Sourcebook.* Chicago: University of Chicago Press.

Rova, E. (2014) The Kura-Araxes culture in the Shida Kartli region of Georgia: An overview. *Paléorient* 40(2): 47–69.

Sagona, A. (2018) *The Archaeology of the Caucasus: From Earliest Settlements to the Iron Age.* Cambridge: Cambridge University Press.

Shimogama, K. and V. Alakbarov (2020) Chapter 8: Archaeological reconnaissance survey around Göytepe, Tovuz-Qovlar Tovuz region. In: *Göytepe: Neolithic Excavations in the Middle Kura Valley, Azerbaijan*, edited by Y. Nishiaki and F. Guliyev, pp. 137–166. Oxford: Archaeopress Publishing.

Świętosławski, W. (1999) *Arms and Armour of the Nomads of the Great Steppe in the Times of the Mongol Expansion (12th–14th Centuries).* Łódź: Oficyna Naukowa MS.

Watson, O. (1999) Museums, collecting, art-history and archaeology. *Damaszener Mitteilungen* 11: 421–432.

Watson, O. (2017) Ceramics and circulation 800–1250. In: *A Companion to Islamic Art and Architecture*, vol. I, edited by F. B. Flood and F. Necipoğlu, pp. 478–500. Oxford: John Wiley & Sons.

Wilkinson, K. N., B. Gasparian, R. Pinhasi, P. Avetisyan, R. Hovsepyan, D. Zardaryan, G. E. Areshian, G. Bar-Oz, and A. Smith (2012) Areni-1 Cave, Armenia: A Chalcolithic–Early Bronze Age settlement and ritual site in the southern Caucasus. *Journal of Field Archaeology* 37(1): 20–33.

Chapter 17

Damjili Cave in the context of Neolithization in the South Caucasus

Yoshihiro Nishiaki, Azad Zeynalov, and Yagub Mammadov

17.1 Research background

Food production forms the basis of our life in modern society. Given that the norm in prehistory was hunting and gathering, the emergence processes of food production socio-economy, that is Neolithization continues to attract scholarly effort. The case in the South Caucasus has also been studied for a long time from multiple perspectives (Chapter 1). As a result of the intensive international collaboration research in the last decades, the old theory claiming independent origin of prehistoric food production has been almost declined. Many researchers have now agreed that food production technology was introduced around 6000 BC from the "Fertile Crescent" of Southwest Asia, where the Neolithic socio-economy occurred much earlier. Our investigations at Göytepe (2008–) and Hacı Elamxanlı Tepe (2012–2016) in West Azerbaijan have greatly contributed to defining the first Neolithic societies of the South Caucasus. The excavations at Hacı Elamxanlı Tepe provided the evidence of the first Neolithic settlement on the Middle Kura (c. 5950–5800 cal BC), while the work at Göytepe has been revealing how the first farming society developed in the following period (c. 5650–5460 cal BC). An important conclusion of these sets of field and laboratory investigations was that the farming was introduced from the Fertile Crescent of Southwest Asia. However, the research also revealed that it was not a wholesale import of the West Asian Neolithic. Although the cultigens and livestock were definitely introduced from West Asia, the cultural elements of the South Caucasian Neolithic were not always considered as direct imports. For example, in our study region, the production of pottery vessels became popular well after the introduction of the Neolithic economy (Nishiaki and Guliyev 2020; Nishiaki et al. 2021).

Then, the central question is how the local Mesolithic hunter-gatherers involved in the Neolithic process. Were they suddenly replaced by farming and pastoral groups incoming from Southwest Asia, or did the interaction between the two groups lead to the establishment of a new Neolithic society? This question has remained unsolved for long due to the lack of referable Mesolithic sites in the region. The excavation of Damjili Cave in west Azerbaijan from 2016 to 2022 filled this gap. Moreover, it provided not only the latest Mesolithic but also the early Neolithic cultural layers at a single site. In other words, it was found to be an invaluable site to argue the transition from the Mesolithic to the Neolithic in the South Caucasus region and the relationship between the two communities (Chapter 1).

17.2 Investigations at Damjili Cave

Damjili is a cave site on Mount Avey. There were claims that assert the presence of many prehistoric caves including Paleolithic ones at the eastern foot of this mountain. However, as a result of our surveys in 2015 and 2016, we concluded that many of those caves and rockshelters are from the Bronze Age or later (Chapter 2). Accordingly, Damjili Cave is currently the only prehistoric site known to us at the eastern foot of the mountain that dates back to before the Bronze Age. The cave, located at the basin of a rainwater catchment at the mountain (Chapter 3), was excavated for three seasons in the 1950s (Appendix). A large amount of prehistoric stone artifacts were recovered during those excavation at the basin of the waterfall, but the excavators at the time concluded that those artifacts were from stratigraphically mixed contexts due to significant later disturbance. Our re-excavations in 2016–2022 partially support this statement: the water flow (Trench 10) near the basin of the waterfall, and the medieval cemetery (Trench 8), and constructions (Trenches 1–5) widely disturbed the underlying prehistoric deposits (Chapter 4).

Our excavation area (originally defined as Trenches 7 and 9), located about 60 m away from the basin of the waterfall, demonstrated that it was a rare location where prehistoric remains were preserved stratigraphically. Although there was evidence of stratigraphic mixing at the time of the Middle Paleolithic to the earlier Mesolithic, probably caused by water flow, the cultural layers after the late Mesolithic were identified in order. In excavations in the 1950s, Middle Paleolithic, Upper Paleolithic, Mesolithic, and Neolithic stone tools were said to have been found intermingled beneath medieval sediments (Appendix), whereas our excavations found no Upper Paleolithic elements. On the contrary, Chalcolithic and Bronze Age deposits were recovered.

The cultural stratigraphy identified from our excavations is as follows (Chapters 4–6).

Unit 6: Sterile, containing Middle Paleolithic artifacts on top,
Unit 5: Mesolithic, 6500–6000 BC,
Unit 4: Neolithic, 6000–5300 BC,
Unit 3: Chalcolithic, 4500–3700 BC,
Unit 2: Bronze Age, 2800–2200 BC, and
Unit 1: medieval ages, 6th century AD and after.

In the 1950s, settlements in the medieval ages were reported to date from the 9th to 13th centuries (Chapter 4), but our radiocarbon dating and pottery analysis indicate the existence of earlier occupations, dating from at least in the 6th century (Chapters 5 and 16). It is also to be noted that the interface between Units 6 and 5 exhibited stratigraphic disturbance containing both Middle Palaeolithic and Mesolithic artifacts (Chapter 7). Since no *in situ* charcoals were obtained from the layer, OSL dating was applied, yielding a Mesolithic age (Chapter 6). Although the disturbance in the re-excavated area was minor compared to the 1950s, it is clear that it was not irrelevant.

17.3 Mesolithic and Neolithic at Damjili Cave

Archaeological evidence for the Mesolithic (Unit 5) and Neolithic (Unit 4) is paramount to the study of the South Caucasus' Neolithization. None of the remains from both periods at Damjili contain mud walls. The main Mesolithic features are rows of limestone cobbles and rubble that seem to have formed artificial structures of an unknown shape. The features from the Neolithic levels were different (Chapter 4). First, the Neolithic levels contained cobbles and rubble that displayed a circular alignment. Secondly, well-constructed fireplaces were recovered. In the Mesolithic period, the fire-place was nothing more than a concentration of ashes and charcoal, whereas those of the Neolithic represented cobble-filled pits. Cobble-filled pits have also been found in the Hacı Elamxanlı Tepe, the oldest Neolithic settlement in the South Caucasus. A similar type is common in the Neolithic of the Fertile Crescent.

Most common artifacts were flaked lithics in both periods (Chapter 8). Raw materials were mainly glassy sedimentary rocks such as flint, or andesite and obsidian. The use of obsidian shows remarkable stratigraphic changes, from less than 10% at the bottom of the Mesolithic (Unit 5.3) to over 70% at the beginning of the Neolithic Age (Unit 4.4). Our analysis revealed that this change did not occur suddenly, but rather began as a transitional period at the end of the Mesolithic Age (Unit 5.1). It was also found that pressure debitage was the main technique for exfoliation of obsidian cores in both periods (Chapter 9), but some typological changes may have occurred also in the late Mesolithic period (Unit 5.1), when typologically unique Chok points appeared. This suggests that the North Caucasian lithic tradition has emerged in the South Caucasus. Since the use of obsidian is assignable to a southern tradition, both northern and southern elements intermingled in the late Mesolithic period. This point is important evidence for considering the Neolithization of this region.

Grinding and ground stones were scarce (Chapter 10). Compared to the mound settlements of Göytepe and Hacı Elamxanlı Tepe in the plain, the scarcity is obvious. This is not surprising when considering the geographical setting of the cave. Nevertheless, a chronological change was detected that the use of ground stones increased during the Neolithic period in Damjili. Hammers and pounders were characteristic of the Mesolithic levels, while ground/grounding tools increased in the Neolithic, although it is not clear whether they were related to grain processing within the cave.

More than 200 pottery sherds were unearthed from the Neolithic levels (Chapter 11). Previous studies at Göytepe and Hacı Elamxanlı Tepe have demonstrated that little pottery was utilized in the early Neolithic period of the MIddle Kura Valey, the South Caucasus. The stratigraphic analysis at Damjili Cave confirmed this estimate in a single cultural sequence (Fig. 17.1). At the beginning of the sixth millennium BC, when agriculture and livestock were brought to the South Caucasus, painted pottery was abundantly used in the Fertile Crescent Mesopotamia. Nevertheless, pottery had no practical significance at the beginning of the Neolithic period in the Southern Caucasus.

Bone tools were not recovered at Damjili (Chapter 12). Since they are found in large quantities at mound sites such as Hacı Elamxanlı Tepe and Göytepe, this finding suggests that bone tools were not frequently used in caves. However, finds from the latest Mesolithic level of Damjili contained an ornamental piece. It is a large boar tusk ornament. Similar works have been found at the earliest

Neolithic sites like Hacı Elamxanlı Tepe, but none at the Late Neolithic Göytepe. This finding reinforces our model that surmises a gradual formation of the Shomutepe Neolithic culture by stages.

In terms of small finds, there was an important discovery from the Mesolithic period. It is a human figurine made of stone (Chapter 13). Its stylistic features differ considerably from the human clay figurines popular in the Shomutepe culture of the Neolithic. Given that there has been no reliable report of human figurines form the early phase of the Neolithic in the Middle Kura Valley, this discovery suggest a break between the Mesolithic and Neolithic in the realm of symbolic sphere, contrary to the case of the above ornamental findings.

We already know that a fully agro-pastoral economy was established at the earliest Neolithic settlement of Hacı Elamxanlı Tepe. Evidence from Damjili, a cave site at the mountain, may not be sufficient to monitor the development of food production economy. For example, it is expected that even during the Neolithic period, the subsistence at temporary occupations at a cave might have been similar to that of the Mesolithic.

However, the present research provides pieces of important evidence from Damjili Cave that could distinguish the Mesolithic and Neolithic in terms of both plant and animal use. Charred grains or grain lumps were found only in strata after the Neolithic age (Unit 4) (Chapter 14). On the other hand the macro-botanical study revealed evidence of the use of Artemisia throughout the Mesolithic and Neolithic periods. This plant is rarely used in the Fertile Crescent of Southwest Asia, the homeland of Neolithic culture. In other words, its steady use since the Mesolithic may indicate a part of the continuity of plant use in the indigenous South Caucasus population. The animal remains from the Damjili excavations also provided remarkable evidence about animal use in the Mesolithic to the Neolithic (Chapter 15). The composition and measurements of the excavated bone remains indicate that a complete set of livestock (goats and sheep) was introduced at the beginning of the Neolithic period (Unit 4.4).

17.4 Conclusions

Damjili Cave provides rare archaeological evidence allowing us to examine socio-economic changes from the Mesolithic to Neolithic in a single sequence. This evidence contains two important elements to understand the Neolithization of the South Caucasus. First, Neolithic culture was formed by both Mesolithic traditions and newly introduced traditions. It is clear that the basic plants and animals for food production economy, such as cereal grains and livestock animals, were brought from Southwest Asia. Also, the necessary techniques for food production, such as sickle blades for harvesting crops and grinding stones for processing, rarely seen in the Mesolithic levels, are thought to have been introduced from Southwest Asia. Cooking techniques using cobble-filled pits would also be of a foreign origin. At the same time, the Damjili study revealed a series of features suggesting continuation of indigenous technology. First of all, life without pottery in the early Neolithic can be mentioned. The Damjili records showed the very rare use of pottery for a few centuries even after the start of food production, confirming our view presented from the investigations at Hacı Elamxanlı Tepe and Göytepe. The frequent use of obsidian in flaked stone tools and the employment of pressure debitage techniques, as well as the boar tusk ornaments in bone artifacts,

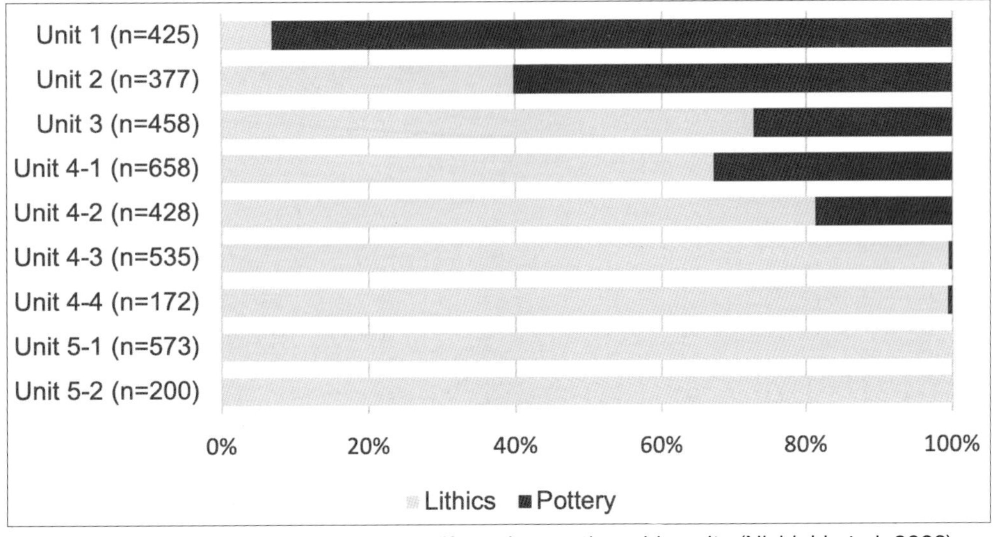

Fig. 17.1 Ratio of pottery sherds to flaked stone artifacts by stratigraphic units (Nishiaki et al. 2022).

in the early Neolithic also show continuity from the Mesolithic. Furthermore, there is a high possibility that the use of Artemish also suggests the continuity of local plant use.

These pieces of evidence suggest that the transition from Mesolithic to Neolithic was not a cultural and/or population turn-over. Additionally, the Damjili research revealed an important point about when the transition occurred. The transformation did not occur at the onset of the Neolithic but seems to have started at the late Mesolithic (Unit 5.1). Good examples for this phenomenon are seen in the flaked stone industry. First, the heavy use of obsidian characterizing the Neolithic industry began in the late Mesolithic. Second, the type of flaked stone tools also showed a unique feature: the northern-type microliths of the Chok-type intruded at the end of the Mesolithic period. When regarding the obsidian use as a southern element, the introduction of northern elements suggest complex cultural dynamics occurring at the end of the seventh millennium BC. The possiblility of their relation to cultural responses to the 8.2ka event, the short episode of abrupt climate changes, would deserve closer examination in the future. According to radiocarbon dates, there is almost no recognizable time difference between Mesolithic and Neolithic transitions in the middle Kura River valley where Damjili Cave is located (Fig. 17.2). The emergence of a complete Neolithic economy around 6000 cal BC would have been based on the ground leveling brought about by the cultural contacts that were taking place on a regional scale at the end of the Mesolithic (Nishiaki 2021).

Our resumed investigations of Damjili Cave thus produced significant data on not only the Mesolithic-Neolithic transitions as mentioned above, but also the longer cultural history starting from the Middle Paleolithic period. We believe that Damjili Cave is worthy of further investigations as a unique site in defining the prehistoric cultural occurrences in the Southern Caucasus.

References

Nishiaki, Y. and F. Guliyev (2020) *Göytepe – The Neolithic Excavations in the Middle Kura Valley, Azerbaijan*. Oxford: Archaeopress.

Nishiaki, Y. (2021) Hunter-gatherers and famers in the Mesolithic-Neolithic contact period of the Southern Caucasus. In: *Hunter-Gatherers in Asia: From Prehistory to Present*, edited by K. Ikeya and Y. Nishiaki, pp. 109–123. Osaka: National Museum of Ethnology.

Nishiaki, Y., F. Guliyev, and S. Kadowaki (2021) *Hacı Elamxanlı Tepe – The Archaeological Investigations of an Early Neolithic Settlement in West Azerbaijan*. Berlin: ex oriente.

Nishiaki, Y., A. Zeynalov, M. Mansurov, and F. Guliyev (2022) Radiocarbon chronology of the Mesolithic-Neolithic sequence at Damjili Cave, Azerbaijan, Southern Caucasus. *Radiocarbon* 64(2): 309–322.

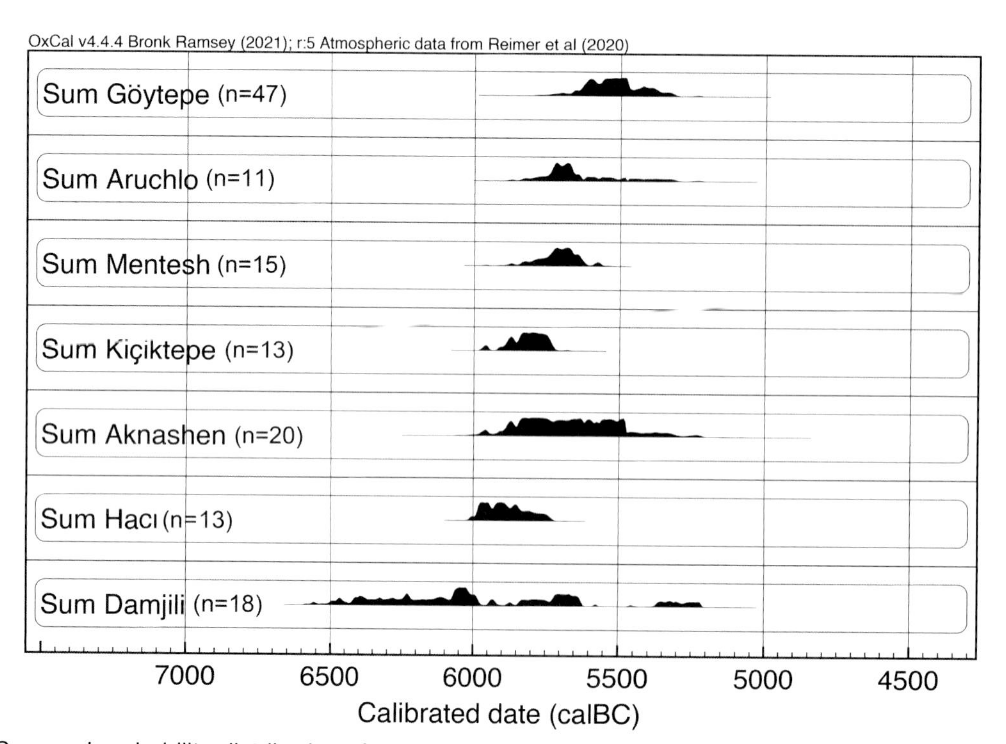

Fig. 17.2 Summed probability distribution of radiocarbon dates for late Mesolithic and Neolithic sites with sufficient radiocarbon dating in the South Caucasus (Nishiaki et al. 2022).

Xülasə

Qida istehsalı müasir cəmiyyətdə həyatımızın əsasını təşkil edir. Tarixdən əvvəlki dövrlərdə qida istehsalının sosial-iqtisadi inkişafı forması ovçuluq və yığıcılıq olduğunu nəzərə alsaq, neolitləşmə prosesləri elmi səyləri cəlb etməkdə davam edir.

Cənubi Qafqazdakı bu məsələ uzun müddətdir ki, müxtəlif aspektlərdən öyrənilir. Son onillikdə aparılan intensiv beynəlxalq tədqiqatları nəticəsində, tarixdən əvvəlki qida istehsalının müstəqil mənşəli olduğunu iddia edən köhnə nəzəriyyə demək olar ki, öz təstiqini tapmadı.

Bir çox tədqiqatçılar indi yekdilliklə razılaşıblar ki, qida istehsalı texnologiyası təxminən e.ə. VI minillikdə neolit sosial-iqtisadiyyatının daha əvvəl yarandığı Qərbi Asiyanın "Bərəkətli Aypara"sından qaynaqlanmışdır. Azərbaycanın qərbində Göytəpə (2008–2015) və Hacı Elamxanlı Təpə (2012–2016) abidələrində apardığımız araşdırmalar Cənubi Qafqazın ilk neolit cəmiyyətlərinin müəyyənləşdirilməsinə böyük töhfə verib.

Hacı Elamxanlı Təpə qazıntıları Cənubi Qafqazda ilk neolit məskənin olduğunu sübut etmiş (e.ə. 5950–5800), Göytəpədə tədqiqatlar erkən əkinçi cəmiyyətlərinin sonrakı dövrlərdə (e.ə. 5650–5460) necə inkişaf etdiyini aşkar etmişdir. Bu tədqiqatların vacib nəticəsi laborator analizlər əsasında müəyyən olmuşdur ki, əkinçilik Qərbi Asiyanın "Bərəkətli Aypara"sından qaynaqlanmışdır. Bununla belə, tədqiqatlar neolit iqtisadiyyatının bütövlükdə Qərbi Asiya neolitindən ixrac olunmadığını göstərdi.

Baxmayaraq ki, mədəni bitki növləri və mal-qara Qərbi Asiyadan gətirilmişdir, Cənubi Qafqaz neolitinin mədəni elementləri bir başa idxal olmayıb. Misal üçün, saxsı qabların istehsalı neolit iqtisadiyyatının tətbiqindən sonra geniş yayılıb.

Bundan əlavə, yerli yığıcı-ovçuların neolitləşmə prosesində rolu əsas elmi məsələdir. Onlar Qərbi Asiyadan gələn əkinçi və çoban qrupları ilə qəflətən əvəz olundular, yoxsa iki qrup arasında qarşılıqlı əlaqədə yeni neolit cəmiyyəti yarandı? Regionda istinad edilə bilən Mezolit məskənlərinin olmamasına görə məsələ uzun müddət həll olunmamış qaldı. Qərbi Azərbaycanda yerləşən Damcılı mağarasında 2016–2019 və 2022–2023-cü illərdə aparılmış tədqiqatlar bu boşluğu doldurdu. Bu bir məskəndə həm son mezolit həm də erkən neolit təbəqələrinin olduğunu sübut etdi. Başqa sözlə, burada Cənubi Qafqaz regionunda mezolitdən neolitə keçidi və iki icma arasında əlaqələri təsbit etmək üçün mükəmməl bir məskən olduğu müəyyən edildi (I Fəsil).

Damcılı Avey dağında bir mağara düşərgəsidir. Dağın şərq ətəklərində paleolit dövrü da daxil olmaqla çoxlu mağaraların olması iddia olunurdu. Lakin 2015–2016-cı illərdə aparılmış kəşfiyyat işləri nəticələri göstərdi ki, bu mağaraların və qaya sığınacaqlarının çoxu tunc dövrünə və daha gec dövrlərə aiddir (II Fəsil). Damcılı mağarası dağın şərq ətəklərində tunc dövründən əvvəlki dövrə aid olunan bizə məlum olan hazırda yeganə tarixdən əvvəlki dövr mağaradır. Dağdan tökülən yağış sularının hövzəsində yerləşən mağara (III Fəsil), 1950-ci illərdə üç mövsüm tədqiq edilmişdir (IV Fəsil). Qədim alətlərin çoxu tədqiqatlar zamanı şəlalənin hövzəsində aşkar olunmuşdur, lakin tədqiqatçılar belə qənatə gəlmişdirlər ki, bu artefaktlar sonradan baş vermiş dəyişikliklərə görə stratiqrafik cəhətdən qarışıq təbəqələrdəndirlər. Bizim 2016–2019 və 2022–2023-cü illərdə bərpa edilən tədqiqatlarımız qismən bu fikiri dəstəklədi: şəlalə hövzəsi yaxınlığında su axını (şurf 10) orta əsr qəbirstanlığı (şurf 9) və tikililər (şurf 1–5) alt tarixdən əvvəlki dövr təbəqələri (V Fəsil) geniş dağıtmışdır.

Şəlalənin hövzəsindən təxminən 60 m kənarda yerləşən, bizim qazıntı sahəmiz (şurf 7 və 9), göstərdi ki, ora tarixdən əvvəlki dövr qalıqların statiqrafik qorunduğu nadir yerdir. Bununla belə, orada da, çox güman ki, su axını səbəbi ilə, orta paleolitin erkən mezolitə statiqrafik qatışması sübütü var. Son mezolitdən sonrakı mədəni təbəqələr ardıcıl müəyyən olunmuşdur. 1950-ci illərdəki qazıntılarda orta paleolit, üst paleolit, mezolit və neolit daş alətləri orta əsr çöküntülərinin altında qarışmış halda tapıldığı qeyd olunmuşdur (Əlavə). Bizim təqdiqatlarda üst paleolit elementləri aşkar olunmamışdır. Əksinə, xalkolit və tunc dövrü təbəqələri aşkar olunmuşdur.

Bizim qazıntılardan müəyyən olunmuş mədəni təbəqələrin statiqrafiyası aşağıdakı kimidir (V və VI Fəsillər).

Təbəqə VI: Sterildir

Təbəqə V: Mezolit, e.ə. 6500–6000

Təbəqə IV: Neolit, e.ə. 6000–5300

Təbəqə III: Xalkolit, e.ə. 4500–3700

Təbəqə II: Tunc dövrü, e.ə. 2800–2200

Təbəqə I: Orta əsrlər, VI əsr və sonrası

1950-ci illərdə aparılmış tədqiqatlarda məskənin orta əsrlər dövrü IX–XII əsrlərə aid edilmişdir (IV Fəsil), lakin bizim radiokarbon nəticələrimiz məskəndə erkən məskunlaşmanın ən azı VI əsrdə olduğunu göstərir (XVI Fəsil). Həmçinin qeyd etmək lazımdır ki, VI–V təbəqələr arasında torpaq orta paleolit və mezolit artefaktlarından ibarət stratiqrafik pozuntunu nümayiş etdirdi (VII Fəsil). Təbəqədən heç bir *in situ* kömür aşkar olunmadığından OSL təyini tətbiq olundu və nəticələr mezolit dövrünü müəyyən etdi (VI Fəsil). Yenidən qazılan ərazidə müşahidə olan bu pozuntu 1950-ci illərlə müqayisədə cüzi olsa da, bunun əhəmiyyətsiz olmadığı aydındır.

Mezolit (V təbəqə) və neolit (IV təbəqə üçün) üçün arxeoloji sübutlar Cənubi Qafqazın neolitləşməsinin tədqiqi üçün mühüm əhəmiyyət kəsb edir. Damcılıda hər iki dövrə aid qalıqların heç birində palçıq divarları yoxdur. Əsas xüsusiyyətlər əhəngdaşı daşları və naməlum formalı süni strukturlar əmələ gətirən fraqmentlərdir. Neolit təbəqələrinə aid xüsusiyyətlər fərqlidir (IV Fəsil).

Birincisi, Neolit təbəqələrində dairəvi düzülüş nümayiş etdirən daşlar və fraqmentlər var idi. İkincisi, yaxşı tikilmiş ocaq yerləri aşkar olunmuşdur. Mezolit dövründə ocaq yeri daşların cəmləşməsindən başqa bir şey olmasa da, neolitdəkilər isə daşla dolu çuxurlarla təmsil edilmişdir. Qeyd etmək lazımdır ki, onlar daş cərgəsi ilə əhatə olunmuşdur. Daşla doldurulmuş çuxurlar Cənubi Qafqazda ən qədim neolit məskəni olan Hacı Ələmxanlı təpə də aşkar olunmuşdur. Bənzər tiplər Bərəkətli Aypara neolitində də, çox yayılmışdır.

Hər iki dövrdə əksər artefaktlar qəlpələnmiş daşlar olmuşdur (VIII Fəsil). Xam mal əsasən çaxmaqdaşı, andezit və dəvəgözü kimi şüşəşəkilli çöküntü suxurlarındandır. Obsidiandan istifadə nəzərəçarpan stratiqrafik dəyişiklik nümayiş etdirir. Belə ki, mezolitin dibində (V təbəqənin III horizontu) 10 % dən az olsa da, neolit dövrünün başlanğıcında 70 % dən çoxdur (IV təbəqənin IV horizontu). Təhlillər göstərdi ki, bu dəyişiklik qəfildən baş verməmişdir, lakin keçid dövrü daha doğrusu mezolit dövrünün sonunda (V təbəqə I horizontu) başlamışdır. Müəyyən edilmişdir ki, sıxma üsulu ilə qoparma hər iki dövrdə (IX Fəsil) obsidıan nüvələrinin istifadəsinin əsas üsulu olmuşdur, lakin bəzi texnoloji dəyişikliklər son Mezolit dövründə də meydana gələ bilərdə, tipoloji cəhətdən nadır Çox itiuçlular kimi (V təbəqə I horizontu). Bu, Şimali Qafqaz ənənəsinin yarandığını deməyə əsas verir. Obsidianın istifadəsi cənub ənənəsinə aid olduğundan, həm şimal, həm də cənub elementləri son Mezolit dövründə qarışmışdır. Bu məqam ərazinin neolitləşməsini nəzərdən keçirmək üçün mühüm sübutdur.

Üyütmək və əzmək üçün istifadə edilən daş əmək alətləri azlıq təşkil edir (X Fəsil). Düzənlikdə yerləşən Göytəpə və Hacı Elamxanlı təpə yaşayış yerləri ilə müqayisədə qıtlıq nəzərə çarpandır. Mağaranın coğrafi mövqeyini nəzərə alsaq, bu, təəccüblü deyil. Bununla belə, xronoloji dəyişiklik göstərmişdir ki, Damcılıda neolit dövründə üyütmə üçün daş alətlərdən istifadə artmışdır. Zərb və iri zərb alətləri mezolit təbəqələri üçün xarakterik olduğu halda, neolitdə isə üyütmə alətləri artmışdır. Bununla belə, onların mağara daxilində taxıl emalı ilə bağlı olub-olmadığı aydın deyil.

Neolit və xalkolit təbəqələrindən (XI Fəsil) 200 dən artıq keramika parçaları aşkar olunmuşdur. Göytəpə və Hacı Elamxanlı təpədə aparılan əvvəlki tədqiqatlar, Cənubi Qafqazın erkən neolit dövründə az miqdarda saxsı qabdan istifadə edildiyini göstərmişdir. Damcılı mağarasında aparılan stratiqrafik təhlil bu təxminləri vahid mədəni ardıcıllıqla təsdiqlədi (şək. 17.1). E.ə. VI minilliyin əvvəllərində Cənubi Qafqaza əkinçilik və heyvandarlıq gətirildiyi zaman, Bərəkətli Aypara Mesopotamiyasında boyalı saxsı qablardan çox istifadə olunurdu. Bu yerlərdən əldə edilən məlumatlar göstərir ki, Cənubi Qafqazda neolit dövrünün əvvəllərində dulusçuluq heç bir praktik əhəmiyyət daşımırdı.

Damcılıda sümük materialları son dərəcə azdır (XII Fəsil). Onlar Hacı Elamxanlı Təpə və Göytəpə kimi məskənlərdə çoxlu miqdarda tapılmışdır. Bu tapıntı onu deməyə əsas verir ki, sümük alətləri naməlum funksional məqsədlər üçün mağaralarda tez-tez istifadə olunmamışdı. Lakin parçaların üstünlük təşkil etdiyi bərpa edilmiş alətlər az sayda da olsa, təpə yaşayış yerlərində yayılmış alətlərə bənzəyir. Bundan əlavə, Damcılının ən son mezolit səviyyəsinə aid tapıntılar arasında erkən neolit dövrünün davamlılığını göstərən bəzək əşyası parçası da var. Bu, iri qaban dişindən olan bir bəzəkdir (kulon). Çox saylı sümük və buynuz alətlərinin tapıldığı son neolit Göytəpə məskənində olmasa da, bənzər əşyalar erkən neolit məskəni olan Hacı Ələmxanlı təpədə tapılmışdır. Bu, son neolitdə deyil, mezolitdən erkən neolitə simvolik dəyərlərdə davamlılığın olduğunu göstərən mühüm sübutdur.

Kiçik tapıntılara gəlincə, mezolit dövrü təbəqəsindən əhəmiyyətli bir tapıntı aşkar olunmuşdur. Bu daşdan hazırlanmış insan heykəlciyidir (XIII Fəsil). Onun üslub xüsusiyyətləri neolit dövrünə aid Şomutəpə mədəniyyətində məşhur olan insan gil heykəlciklərindən xeyli fərqlənir. Nəzərə alsaq ki, Orta Kür vadisində neolitin ilkin mərhələsinə aid insan heykəlcikləri haqqında etibarlı məlumat yoxdur, bu tapıntı yuxarıdakı ornamental tapıntıların əksinə olaraq simvolik sferada mezolit və neolit arasında fasilə olduğunu göstərir.

Bildiyimiz kimi, ən erkən neolit yaşayış yeri olan Hacı Elamxanlı təpədə tam əkinçi-çoban təsərrüfatı qurulmuşdur. Damcılı mağarasından əldə edilən dəlillər qida istehsalı iqtisadiyyatının inkişafını izləmək üçün yetərli olmaya bilər. Məsələn, güman edilir ki, hətta neolit dövründə də, mağarada müvəqqəti məskunlaşan yaşayış məskənləri mezolit dövrünə bənzəyirdi.

Bununla belə, bu tədqiqat Damcılı mağarasından həm bitki, həm də heyvan istifadəsi baxımından mezolit və neoliti fərqləndirə biləcək mühüm dəlillər təqdim edir. Biz taxıl, kömürlənmiş taxıl və ya taxıl parçaları tapacağımızı gözləməsək də, onlar neolit dövründən sonrakı təbəqələrdə tapılmışdır (XIV fəsil) Bununla belə, iri miqyaslı makrobotanik tədqiqatlar, mezolit və neolit dövrlərində yavşandan (*Artemisia*) istifadə olduğunu göstərmişdir. Bu bitki, neolit mədəniyyətinin vətəni olan, Bərəkətli Ayparada nadir hallarda istifadə olunmuşdur. Başqa sözlə desək, mezolitdən bəri onun davamlı istifadəsi yerli Azərbaycanın və ümumən Cənubi Qafqaz əhalisinin bitki istifadəsinin davamlılığının qismən göstərə bilər. Damcılı mağarasından əldə olunmuş heyvan qalıqları mezolitdən neolitə kimi heyvanlardan istifadə ilə bağlı nəzəçarpan sübutlar vermişdir (XV Fəsil). Əldə olunmuş sümük qalıqlarının tərkibi və ölçüləri göstərir ki, neolit dövrünün əvvəllərində mal-qaranın tam dəsti (keçi və qoyun) gətirilmişdir (IV təbəqənin IV horizontu).

Beləliklə, Damcılı mağarası mezolitdən neolitə qədər sosial-iqtisadi dəyişiklikləri ardıcıllıqla araşdırmağa imkan verən nadir arxeoloji sübutlar təqdim edir. Bu sübut Cənubi Qafqazın neolitləşməsini başa düşmək üçün iki mühüm elementi ehtiva edir. Birincisi, neolit mədəniyyəti həm mezolit ənənələri ilə, həm də yeni gətirilən ənənələr tərəfindən formalaşmışdır. Aydındır ki, taxıl və mal-qara kimi qida istehsalı iqtisadiyyatı üçün əsas bitki və heyvanlar Qərbi Asiyadan gətirilmişdir. Həmçinin, qida istehsalı üçün zəruri texnikalar, məsələn, məhsulu yığmaq üçün oraq bıçaqları və üyüdülməsi üçün daşlar, mezolit səviyyələrində nadir hallarda müşahidə olunur, buna görə də onların Qərbi Asiyadan gətirildiyi güman edilir. Daş ilə doldurulmuş çuxurlardan istifadə edərək bişirmə üsulları da xarici mənşəli olardı. Eyni zamanda, Damcılının öyrənilməsi yerli texnologiyanın davam etdiyini göstərən bir sıra sübutlar ortaya qoydu. İlk növbədə saxsısız həyatı qeyd etmək olar. Damcılı qeydləri göstərdi ki, qida istehsalının başlanmasından bir müddət sonra çox nadirən keramikadan istifadə olundu. Bu, Hacı Ələmxanlı və Göytəpə tədqiqatlarından təqdim olunan fikirləri təsdiqləyir. Erkən neolitdə obsidianın daş alətlərdə tez-tez istifadəsi və sixma üsulu ilə qoparma üsullarının tətbiqi, həmçinin sümük alətlərdəki heyvan dişlərindən bəzəklər də mezolitdən davamlılığı göstərir. Bundan əlavə, yavşandan istifadənin yerli bitki istifadəsinin davamlılığını göstərməsi ehtimalı yüksəkdir.

Bu dəlillər açıq şəkildə təsdiq edir ki, mezolitdən neolitə keçid mədəniyyətin və ya əhalinin dəyişməsi deyildi. Bundan əlavə, Damcılı araşdırması keçidin nə vaxt baş verdiyi ilə bağlı vacib bir məqamı ortaya qoydu. Keçid neolitin başlanğıcında baş verməmişdir, lakin görünür, son mezolitdə başlamışdır (V təbəqənin I horizontu). Bu məsələ üçün yaxşı nümunə daşı qoparma üsulunda görünür. Birincisi, neolit sənayesini xarakterizə edən obsidiandan sıx istifadə mezolit dövründə başlamışdır. İkincisi, pullu daş alətlərin növü də özünəməxsus xüsusiyyətə malik idi: mezolitin sonunda şimal tipli Çox mikrolitlər daxil olmuşdur.

Cənub elementi kimi obsidian istifadəsinə gəldikdə, şimal elementlərinin meydana gəlməsi eramızdan əvvəl VII minilliyin sonunda kəskin iqlim dəyişikliklərinin qısa epizodu olan 8200 hadisəsinə reaksiyalarla əlaqəli ola biləcək bir tip mədəni dinamika təklif edir. Radiokarbon tarixlərinə görə Damcılı mağarasının yerləşdiyi Kür çayının orta vadisində mezolit və neolit keçidləri arasında demək olar ki, zaman fərqi yoxdur (şək. 16.2). Tam neolit iqtisadiyyatının təxminən e.ə. VI minillikdə qəfil ortaya çıxması, mezolitin sonunda baş verən mədəni təmasların əsasında yaranmış platformaya arxalanır.

Damcılı mağarası ilə bağlı qeyd olunanan araşdırmalarımız yuxarıda qeyd etdiyimiz kimi təkcə mezolit-neolit keçidlərini deyil, həm də orta paleolit dövründən başlayaraq daha uzun bir mədəniyyət tarixini ortaya çıxarmışdır. Damcılı mağarası Cənubi Qafqazda tarixdən əvvəlki mədəni hadisələrin müəyyənləşdirilməsində unikal məkan kimi əlavə tədqiqatlara layiqdir.

Azad Zeynalov
Azərbaycan Milli Elmlər Akademiyası, Arxeologiya və Antropologiya İnstitutunun aparıcı elmi işçisi, tarix üzrə fəlsəfə doktoru

Yaqub Məmmədov
Azərbaycan Milli Elmlər Akademiyası, Arxeologiya və Antropologiya İnstitutunun böyük elmi işçisi, tarix üzrə fəlsəfə doktoru

Yoshihiro Nishiaki
Tokio Universiteti, Universitet Muzeyinin direktoru, Professor

Appendix:

Research at Damjili Cave in 1953–1957

Introduction to Appendix: Research at Damjili Cave in 1953–1957

Yagub Mammadov, Ulviya Safarova, and Yoshihiro Nishiaki

The first scientific investigations of Damjili Cave were carried out by Azerbaijani archaeologists in the 1950s. Although they demonstrated the potential significance of this cave for the prehistoric research, especially of the Paleolithic to the Neolithic periods, the results have not been fully published. One of the best-known publications for English readers might be the principal investigator's narrative presented in a synthetic book dedicated to the Lower Paleolithic of Azerbaijan (Huseynov 2010). However, it spares only a few pages for Damjili Cave, which obviously do not convey the details of the 1950s investigations. It is also obvious for the Azerbaijan-Japanese team who has been carrying out re-investigations and more to come in the future to understand the achievements of the past studies. Accordingly, we decided to make three useful reports written in Azerbaijani and Russian on the 1950s investigations available in English in this volume.

One is the first report of the excavation of this cave by Sergei Nikolayevich Zamyatnin in 1953 (Chapter A1: Zamyatnin 1958). The other two are the unpublished reports of the 1956 and 1957 seasons' work in the archives of the Institute of Archaeology, Ethnography, and Anthropology, Azerbaijan, signed by Mammadali Huseynov: one report (Chapter A2: Huseynov 1957a) presents detailed descriptions on the recovered lithic artifacts, whereas the other one (Chapter A3: Huseynov 1957b) provides general conclusions of the 1950's field investigations.

These reports sufficiently revealed that Damjili Cave had cultural deposits dating back to the Paleolithic period. They also provide invaluable information about how Damjili Cave and its terraces, now covered with asphalt and rest areas, looked like more than half a century ago. From a modern perspective, the estimates of the age of the stone artifacts excavated at that time, as mentioned in those reports, inevitably need to be reconsidered. Likewise, the stratigraphy also requires re-examination. Nevertheless, Zamyatnin and Huseynov's research reports served the best guide for our re-excavation.

We tried to translate the original Russian texts as precisely as possible but readers are asked to understand a number of difficulties, such as the terminologies used in the texts unfamiliar to us today and the presence of hand writings in the unpublished reports which are not easily readable. Further, we have not been accessible to the figures referred to in the report of Chapter A3. This deficiency is hopefully compensated by a number of figures included in Chapter A2, whose contents are partially overlapped with the report of Chapter A3. One more note: for the sake of readability, we inserted appropriate section headings to all these reports and figure captions, which were not included in the original reports.

Therefore, the following translation of the reports should be rather regarded as an edited version of the original reports. We are most grateful to the Authorities of the Institute of Archaeology and Anthropology, the National Academy of Sciences of Azerbaijan, who allowed us to publish the English version of these important articles of the 1950s.

References

Huseynov, M. (1957a) *Paleolithic Site in Azerbaijan: Report of 1957.* Manuscript on file of the Institute of Archaeology, Ethnography, and Anthropology, Baku (in Russian).

Huseynov, M. (1957b) *Paleolithic Station at Damjili Cave: Report of 1957.* Manuscript on file of the Institute of Archaeology, Ethnography, and Anthropology, Baku (in Russian).

Huseynov, M. (2010) *The Lower Paleolithic of Azerbaijan.* Baku: National Academy of Sciences of Azerbaijan (in Russian with English summary).

Zamyatnin, S. N. (1958) Scientific research on the Stone Age in Azerbaijan. *Bulletin of the Institute of History* 13: 5–18 (in Russian).

Chapter A1

Scientific research on the Stone Age of Azerbaijan, autumn 1953*

* Zamyatnin, S. N. (1958) Scientific research on the Stone Age of Azerbaijan, Autumn 1953. *Bulletin of the Institute of History* 13: 5–18 (in Russian).

A1.1 Introduction

In November 1953, on behalf of the Institute of History and Philosophy of the Academy of Sciences of the Azerbaijan Soviet Socialist Republics (SSR), I made a trip to search for the remains of human culture of the Paleolithic period. Although the locations, or even isolated finds of this period, were not previously found on the territory of the Azerbaijan SSR, the possibility of their discovery is beyond doubt. Evidence confirming the possibility of human settlement in the Quaternary period of present-day Azerbaijan can be, on the one hand, paleogeographic data, and on the other hand, finds in neighboring territories.

The paleogeography of Azerbaijan in the Quaternary period has not been sufficiently studied from the point of view that occupies us. It is known that changes in the natural environment, up to the location of land and sea, reached a significant scale. Nevertheless, the characteristic present features of the country, namely the extraordinary variety of landscape conditions, were preserved throughout this vast period of time. A variety of natural conditions ensured the presence on the territory of Azerbaijan in the Quaternary period of areas favorable for their settlement by primitive man. The best confirmation of this is the fairly impressive number of finds of bone remains of large Quaternary mammals recorded so far in Azerbaijan. The presence of a rich fauna of mammals such as horses, different types of bulls and deer, rhinoceros, and elephants, was an indispensable condition for the existence of primitive man. Where these animals lived, there could exist, and it is often possible to be convinced that there were people who hunted them.

The same nature of indications is given by archaeological finds in neighboring territories. It suffices to point to the locations of Armenia and Georgia, which demonstrate the existence of populations in Transcaucasia during the early Quaternary time. Those ancient sites of human activity have been discovered both on the volcanic highlands of the South Caucasus (the sites of Satanidar on Aragats and Arzni near Yerevan) and the central part of the southern slope of the Greater Caucasian Range (Lashe-Balta and other localities of South Ossetia), and finally, further, on the Black Sea coast (Yashtukh, Kyurdere, and other finds on the high terraces of Abkhazia). The most numerous and richest finds of the Upper Paleolithic period came from the caves of western Georgia: Motsameti caves near Kutaisi, caves and rockshelters in the vicinity of Chiatura, and caves of Sagvardzhile and Devis-Khvreli in smaller numbers. We mentioned above only the most important sites. In recent years, information about the presence of Paleolithic sites in Dagestan has also begun to accumulate. To the south, abroad, the first indications of the occupation of the caves of Iran in the Quaternary times were delivered by American excavations of recent years.

Thus, starting the search for the Paleolithic in Azerbaijan, one has to think not about whether this can be done but only about how to do it faster and easier. If in the plains the only possible way to explore Paleolithic sites is to record and systematically check on the site the bones of extinct large Quaternary mammals, then in a mountainous country, a simpler and more promising way is digging caves and rockshelters. In both cases, we are talking about sites of the Upper Paleolithic and Mousterian age, with an undisturbed cultural layer, where excavations are possible. The earlier occupational traces usually occur in a redeposited state and the searching methodology is different.

It was this last task that was set during the 1953 reconnaissance trip. The trip lasted from 1 to 22 November. For its implementation, the Institute provided an equipped truck. M. M. Huseynov, a researcher at the Museum of the History of Azerbaijan, took part in the work throughout their entire length. The route of the trip was planned on the basis of information available to the Institute about the caves and taking into account their availability in late autumn. At the suggestion of Department of History of Material Culture of the Institute, a tour of the surroundings of the village, Krakhkesaman, was included in the route, where several roughly split basalt flakes were discovered shortly before.

The work was started by a series of inspections in the vicinity of the village Maraza, near which there are widely known artificial caves, carved into the rock. In places, in the same loose Tertiary limestones in which artificial caves are located, there are also caves and rockshelters of natural origin. Some of them were examined, but turned out to be shallow, usually not containing deposits, and, as far as one can judge, should be assigned to the Holocene by the time of occurrence: i.e., they cannot contain the cultural remains that were the subject of our research.

Inspections in the vicinity of Khanlar (now Göy-Gol) also gave negative results. Here, opposite the city, on the left bank of the Ganjacay, a small cave known as "Cave of Beck" was visited, as well as several long and narrow caves and rockshelters located nearby. In the Beck's Cave, excavations were carried out by Y. I. Hummel, who discovered the remains of the Eneolithic period in the pit which he dug to the bedrock. The rockshelters that exist in a number of places in the vicinity are currently mostly devoid of sediments and have an exposed bedrock. Some of them were nevertheless inhabited in antiquity (probably also in the Eneolithic), because on the slopes, below some of the shelters, several flint flakes and fragments of ancient ceramics were collected, apparently originating from the destroyed deposits of caves. It was not possible to identify any sites where it would be possible to search for cultural remains of an earlier time in the vicinity of Kanlar.

A1.2 Research of the Dash Salahli region

A more favorable environment for searches is available in the vicinity of the village, Dash Salahli of the Gazakh region, where in dense Mesozoic limestone there is a fairly large number of various kinds of caves, rockshelters and rocky walls, sometimes with deposits on the bottom. A number of such caves and rockshelters were examined, and pits were opened in some of them. Whereas without dwelling evidence on the list of these localities examined in the vicinity of Dash Salahli, which did not provide positive data, it will be necessary to consider in some detail the results of reconnaissance in the Damjili area, where, although modest in quantity, but undoubted indications of the existence of a human settlement in the Upper Paleolithic and earlier periods were obtained.

For example, Damjili Cave ("dripping") is located approximately 1 km south of the village Dash Salahli. Here, at the foot of the limestone mountain that forms sheer cliffs, and at the cliffs' base covered with talus, there are two caves. One of these caves, smaller in size and distinguished by the feature that water continuously seeps from the cracks in its ceiling, apparently gave rise to the name of the entire region. This cave had been worshiped as a mazar until relatively recently. Water dripping from the ceiling of the cave was considered healing and was carefully collected. Two recesses arranged on the floor for collecting water, enclosed with stone tiles and located in those places where drops fell from the ceiling, have survived to this day. On the other hand, the second cave, which is much larger and located somewhat higher, is apparently used as a paddock for sheep.

There are sediments at the bottom of both caves. When inspecting the caves in their vicinity, including in the immediate vicinity of them, down the slope, a certain number of flakes and blades of flint and obsidian were collected, undoubtedly knapped by a human hand. Both caves were subjected to test sounding and yielded evidence of ancient human settlement.

Damjili Cave 1

The first, smaller, cave, which has the shape of an irregular semicircle, is distinguished by an uneven bedrock, which in some places comes to the surface. There are few deposits inside the cave and they have the greatest thickness at the entrance in the terrace. An exploratory pit was opened here, which showed that the bedrock in this part of the cave is at a depth of 0.80–1.00 m and has a descending to the north-east direction, towards the entrance. The upper part of the deposits consisted of an intense colored layer of humus with cultural remains from the late medieval period (including fragments of glazed pottery and bones of domestic animals). The thickness of the layer is 0.50–0.60 m.

Below, to the bedrock, there was a light yellow clay layer containing cultural remains of the Paleolithic period in the form of flint and obsidian tools and fragments of split tubular bones, differing in the nature of fossilization from the finds in the upper layer. In this layer, there were small fragments of charcoal, apparently bone, and pieces of ocher. Unfortunately, the lower layer with Paleolithic remains was extremely badly damaged during the subsequent occupation of the cave. Numerous later pits filled with sediments of the upper, medieval layer, and often brought to the bedrock. The lower layer is so damaged that only its insignificant portions have been preserved. The same ancient damage to the Paleolithic layer apparently explains the fact that individual characteristic chipped flints were picked up on the surface, below the cave along the slope and in its immediate vicinity.

Although the stone artifacts from the pit are not numerous, they allow us to outline a fairly accurate chronological framework. First of all, a small cutting tool made on an elongated blade of transparent black obsidian should be noted (Fig. A1.1: 1). Its secondary retouch consists of an ordinary notch in the upper part of the tool, made by abrupt retouch to form a back for the finger, and several flat facets along the opposite edge (possibly traces of use). The possible way the tool was used is shown in Fig. A1.1: 2. Similar specimens are well known from finds in the Upper Paleolithic sites of Georgia, mainly the later ones, namely in the Gvardzhilas-klde Cave near Chiatura (excavations by S. A. Krukovsky) and the Barmaksyz Cave, the valley of the River Khrami (excavations by B. A. Kuftin).

S. A. Krukovsky, who first discovered this type of tools in Transcaucasia, proposed the name "Rgani's knife" for it. This characteristic tool is sometimes erroneously described as an atypical dart tip. The presence of a flint microblade with a blunt edge (Fig. A1.1: 3) and a rounded scraper (Fig. A1.1: 4), also made of dark pink flint, is in good agreement with the Rgani knife. The blank for the latter was a transverse flake from the reduction of a massive scraper or a small core, but the retouching that formed the working edge of the tool was undoubtedly applied after the flake had been detached from the parent flint.

In addition to these three tools, the following

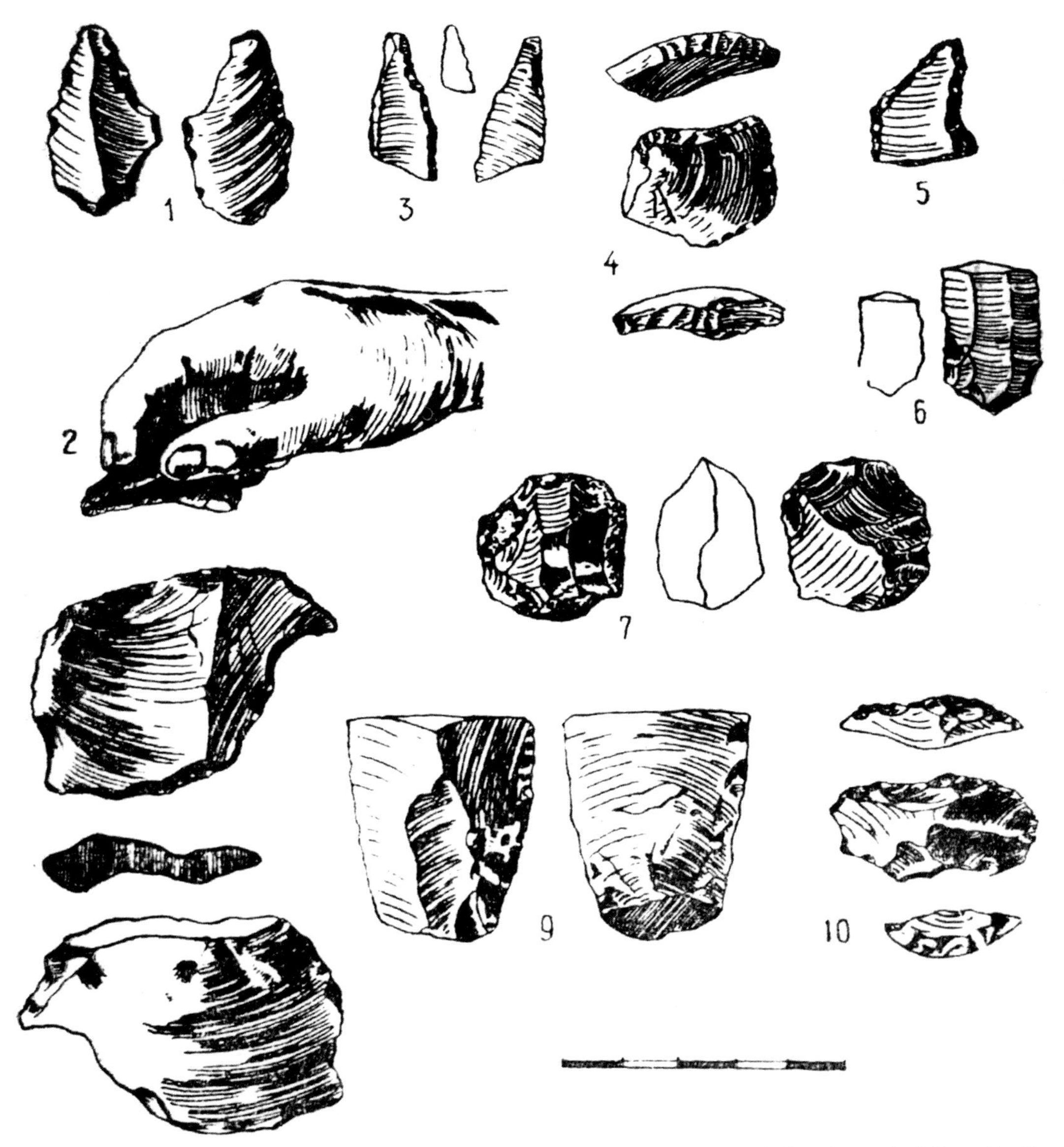

Fig. A1.1 Lithic artifacts from the village of Dash Salahli of the Gazakh region. 1: An elongated Rgani knife; 2: A way to use the Rgani knife; 3: Microlithic blade; 4: Scraper; 5: Obsidian tool; 6: Flint knife blade; 7: Core-shaped piece of obsidian; 8: Large wide flake of flint; 9: Thick knife-blade of flint; 10: Double scraper of obsidian.

come from the excavation pit in the cave: a fragment of a tool made of dark obsidian with retouching along the entire edges (Fig. A1.1: 5), a small knife-like bladelet of pink flint (Fig. A1.1: 6), a massive core-shaped piece of dark translucent obsidian (Fig. A1.1: 7), and a large wide flake with a long winding striking platform, made of excellent tobacco-brown flint. This last flint gives the impression of a more primitive technique (Fig. A.1: 8) and seems to be an accidental addition, although, of course, it is impossible to speak with certainty from a single find. In addition to the above, there were several inexpressive fragments of obsidian and flint. I present here a drawing of a massive knife-like blade made of opaque dark pink flint with careful opposite retouching, picked up on the surface near the cave (Fig. A1.1: 9). Its origin from the cultural deposits of the cave is probable.

The findings in the first cave are exhausted by the foregoing.

Damjili Cave 2

The second cave in Damjili is located next door, significantly larger in size. Rather, it is not a cave, but a rockshalter. Its length is 70 m and its depth is 20 m. It used to be much deeper, but then the outer edge of the roof collapsed. Huge blocks of collapsed rock are visible right next door, a little lower down the slope. On the surface of one of the collapsed rocks, a trough of a medieval water pipe was laid. The thickness of sediments in the second cave is much larger than in the first and apparently has a thickness of at least 12–15 m.

The sounding pit was laid at the middle part of the cave in the form of a narrow trench (1 m) perpendicular to the rock. From the surface to a depth of 0.30–0.40 m, there was a humus layer with recent remains, starting from a depth of 0.60 m, a number of fragments of obsidian were found, undoubtedly knapped by a human hand and one piece of knapped pink flint was also discovered. Of the tools, only one small double scraper made of black obsidian was found (Fig. A1.1: 10). In addition, several fragments of a split tubular bone and a bull's tooth were found. No coloration of the cultural layer was noted. The sounding pit was brought to a depth of 2.00 m in the same homogeneous clay.

Unfortunately, due to lack of time and due to weather conditions, further work in the Damjili region had to be stopped. Continuation of work in the vicinity of Dash Salahli and, in particular, completion of the sounding pit in the second cave deserve attention.

A1.3 Research of the Agstafa Region

Krakhkesaman

Examination of the site of finds in the vicinity of Krakhkesaman provided additional data to clarify their age, which cannot be as ancient as it was drawn on the basis of the initially obtained information. The village Krakhkesaman of Agstafa region is located on the right bank of the Kura. The finds in question come from the Shorsu area, located opposite the village, on the high left bank of the river. The first collections in the Shorsu tract were made by the engineer B. A. Kantor, who delivered his collection to the Institute of History and Philosophy of the Academy of Sciences of the Azerbaijan SSR. Then the collections here were studied by the Institute's researchers G. I. Ione and M. M. Huseynov.[1]

During our trip, an additional collection was collected. The place of finds—the Shorsu region—is located directly below the mouth of a small river of the same name, which flows into the left side of the Kura River. The stones knapped by man were collected on the surface of a high terrace of the Kura, which was composed of alternating layers of clay and pebbles lying horizontally. The height of the find spot is approximately 35–40 m above the Kura. A large amount of pebbles lies on the surface, covering it either with a continuous layer or in the form of scattered inclusions in clay. The pebbles are very diverse in their petrographic composition and have, for the most part, an oval-rounded, flattened shape. Rare knapped stones are rather easily recognizable by the eye among many pebbles of rounded outlines.

At the same time, it should be noted that individual flakes and fragments of basalt and obsidian were collected during the investigation, although they were not found equally everywhere, rather they were concentrated on a certain limited area, forming a kind of accumulation. In terms of shape and knapping technique, the collected stone products look very primitive. All morphologically identifiable pieces can be divided into two groups: flakes and cores.

The flakes are small (2.5–6 cm). They have a subtriangular shape and wide and smooth striking platforms without secondary trimming. In spite of their small size, the flakes are massive and detached from the core by quite strong blows, due to which there are usually marked cones at the point of impact on the striking platforms. On the ventral surface, the bulb is large and occupies a large area. The striking platform and the ventral surface form an obtuse angle (Fig. A1.2: 1–4). There is no faceting on the platforms, and the occasional jaggedness of the edge

[1] Huseynov, M. M. (1955) On the first finds of Stone Age materials in the Shorsu valley. *DAN Azerbaijan. SSR* 1: 55–69.

looks like deliberate retouching, but is the result of later damage. In a word, the general appearance of the flakes indicates Lower Paleolithic, pre-Mousterian. Correspondingly, the core, discoidal in form, with bipolar reduction traces, also have a primitive appearance (Fig. A1.2: 5–8).

However, it is impossible to accept the definition of chipped stones from Shorsu as Lower Paleolithic both in terms of some features of their appearance and in terms of the conditions of their location. First of all, attention is drawn to the relatively fresh texture of these artifacts, the slight alteration of their surface in comparison with the finds of the Lower Paleolithic age known from Transcaucasia. Both by the nature of the patina and by the degree of wearing, the stone products from Shorsu do not seem to be particularly ancient. The material for their manufacture was mainly basalt of a dark gray greenish color, obsidian is less common, and siliceous dolomite is even less abundant. It is the latter, the least durable, that has the most weathered surface, which sharply differs from the surface in a fresh fracture.

Less pronounced, and often absent, surface patination on basalt fragments and, finally, there is absolutely no significant patina on obsidian. There are absolutely no fragments in Shorsu with a characteristic velvety patina that masks the completely natural glassy luster of obsidian. This kind of patina is characteristic of the obsidian tools of the Lower Paleolithic localities of Armenia, and during my in-

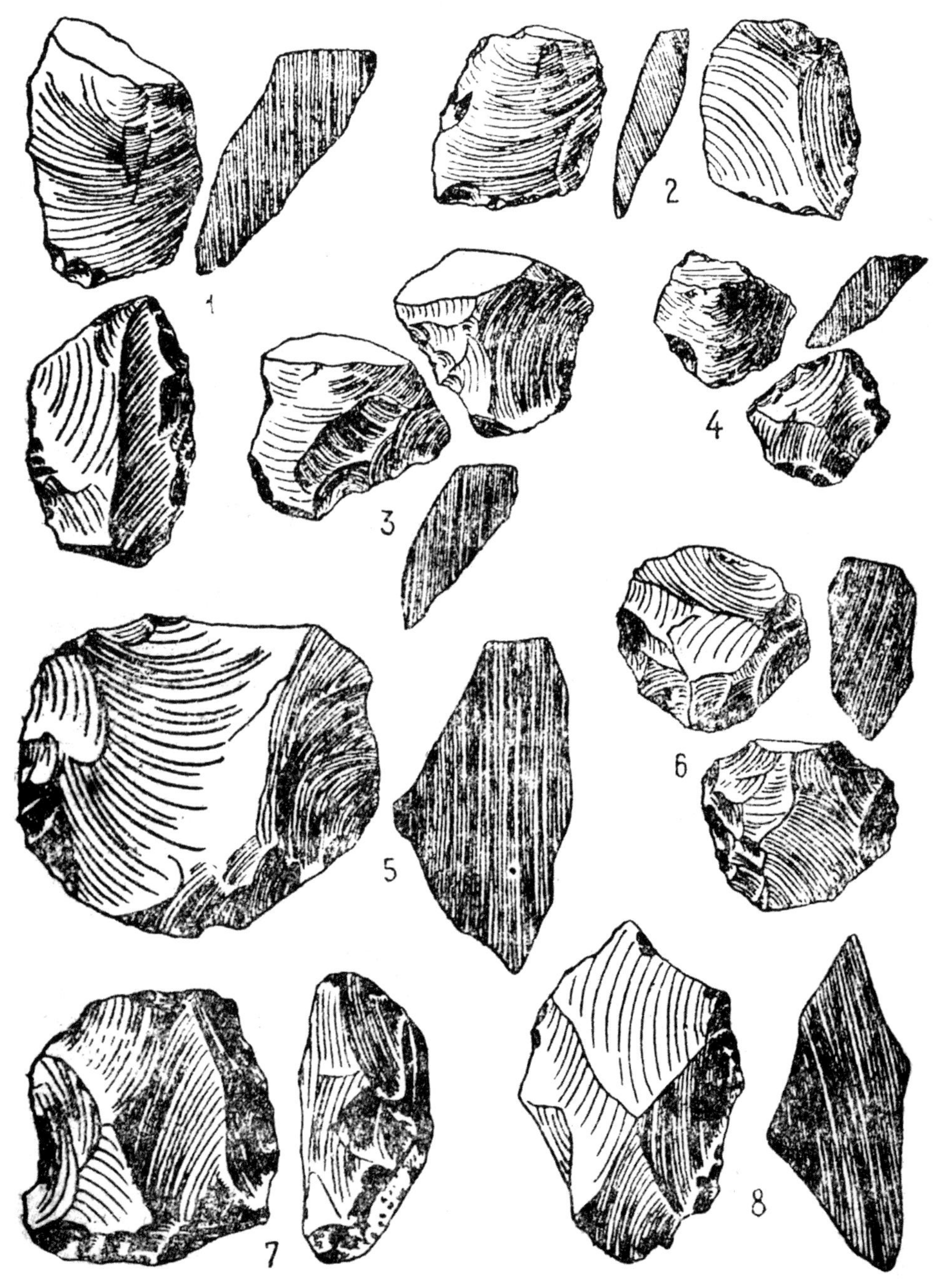

Fig. A1.2 Lithic artifacts from the Shorsu district. 1–4: Blanks; 5–8: Discoid cores.

vestigations in Shorsu, throughout the day, I paid a special attention to everything that even remotely resembled it.

There is also no noticeable traces of water rolling on the products from Shorsu, and it is not necessary to talk about their possible transportation by water to a more or less significant distance. Comparing the latter observation with the above-mentioned location of the finds in a limited area in separate spots, as well as with the insignificant size of the flakes collected in Shorsu, we can apparently assume that, firstly, these finds are confined precisely to the surface of the terrace, and not to the deeper layer and, secondly, that the knapping of the stone that took place here could be accompanied by selection, with larger flakes being carried away, while smaller, less suitable ones were left in place.

M. D. Gavrilov, a geologist of the Azerbaijan Geological Administration, whom I contacted after the trip with a question about the possible geological dating of the find site, familiar to him from personal research, kindly shared his thoughts on this matter. According to M. D. Gavrilov, the site in Shorsu, where the collection of stone products was made, represents the surface of the IVth terrace of the Kura, the accumulation of deposits which took place in the post-Khazar (early Khvalynian) time.

If we take into account that the Mousterian site in the Dry Mechetka gully, on the northern outskirts of Stalingrad, lies in the Moi fossil, developed on the surface of the Khazar terrace, and is covered by a twenty-meter thickness of deposits of the Acheulean and Khvalynian age, it becomes obvious that the Lower Paleolithic age of the finds in Shorsu should be considered unlikely. One can also recall here the old find of fossil remains of Quaternary slopes, described by Simonovich and Sorokin.[2] This find, made at the village of Krakhkesaman, on the left bank of the Kura, in sandy-clay soil, at a depth of 8 m, in all likelihood, in the outcrop above which the flakes considered here are collected, once again shows that the remains of the early Quaternary age must lie in other conditions than the latter.

Nevertheless, one has to regret that the collections in Shorsu are small in quantity, and observations on the conditions of the finds are fragmentary, because, regardless of dating, these finds, of course, are of interest. Refraining for the time being from attempts to detail the age of the collections from Krakhkesaman, which remains problematic, it is necessary to emphasize the need to continue work on this locality both in order to obtain more complete archaeological materials for its characterization, and to check on the spot by a specialist geologist the conditions of their occurrence.

In conclusion, it should also be noted that although the finds in Krakhkesaman did not bring an expected solution to the question of the presence of traces of the Lower Paleolithic in the territory of Azerbaijan, nevertheless, the likelihood of such finds, noted at the beginning of this report, is undoubted, and further information is brought to the accumulating information. Confirmation of what has been said is necessary. I will confine myself here to pointing out two more finds made after our trip.

Khojaly

In October 1954, in Tbilisi, N. Z. Kiladze showed me what she had collected in the vicinity of the village Kachreti, in eastern Georgia (about 50 km northeast from Krakhkesaman). They are three extremely archaic flint flakes, both in form and in texture, which are not in doubt at their Lower Paleolithic age (most likely, Acheulean). This find comes from collections on the surface, where dozens of later flints and obsidian without patina were also collected.

Another find was also obtained in the autumn of 1954 from the territory of Azerbaijan. We are talking about a cordiform flake, very primitive in technique, identified by A. A. Jessen near the village of Khojaly, in the pebbles of the river. With the kind permission of A. A. Jessen, I provide images (Fig. A1.3: 1) and a brief description of this interesting find. The dimensions of the flake are 58 × 40 mm, thickness 16 mm, oval outline. The material is an ordinary, red translucent carnelian, in some parts containing opaque inclusions of a granular structure of a light brown color. On the back of the flake and on the striking platform, a partially rough surface of the pebble from which it was made has been preserved. A massive, convex percussion bulb, occupying almost the entire ventral surface, bears a characteristic large flaw. The striking platform forms an obtuse angle with the ventral surface. Although the flake is heavily water-rolled, several rings are clearly visible on its ventral surface, dividing the faces obtained at the core, as well as a small damage on the edges. The latter is not in the nature of retouching, but rather traces of use or later damage. Taking the available signs of a single find as characteristic features of the technique of this period, it is necessary to assume that the flake is of pre-Mousterian age.

A. A. Jessen introduced me to another chance find one from the vicinity of Khojaly. This is an obsidian flake, not as archaic as the first one, but still quite primitive (Fig. A1.3: 2). It was discovered

[2] Simonovich, S. and A. Sorokin (1875) *Izv Kavkazsk, Department of the Russian Geographical Society* 1.

on the territory of the Cyclopean fortress located near the village. A flake of homogeneous in color, transparent smoky obsidian was made, the surface of which is covered with a light patina, which does not hide the original color and transparency of the material.

A calcareous crust adhered tightly to the lower plane of the tool, which formed during its initial occurrence in the layer. The dimensions of this specimen are 68 × 28 mm, the thickness is 8 mm. Although, according to the elongated outlines, this specimen represents a knife-like plate, but the latter is apparently not separated from the prismatic disc-shaped core, since its long axis is not perpendicular to the striking platform. On the longitudinal edges of the tool there is a rather rough sharpening retouch, applied by the separation of large scales.

The dating of the last find is somewhat difficult. Elongated proportions and retouching with large facets without scars are signs of a younger age, not the character of a workpiece chipped from an early-type core. I deliberately lingered in some detail on these occasional single finds on the surface, which are not sufficient to provide firm conclusions, yet give quite definite indications for setting further investigations.

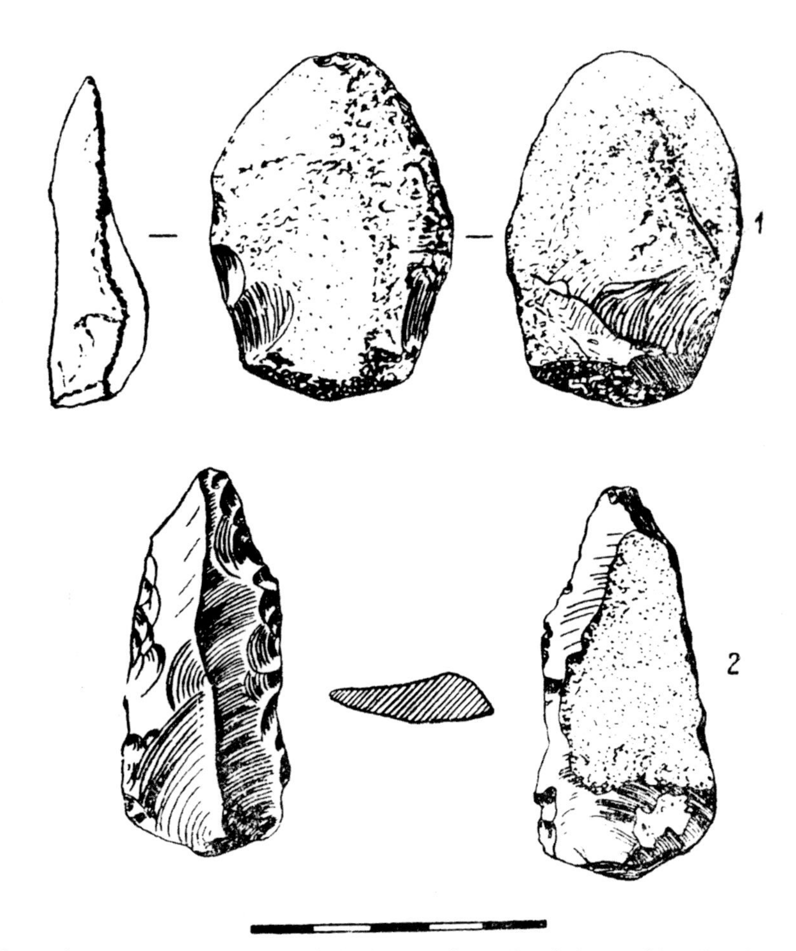

Fig. A1.3 Lithic artifacts found near the village of Khojaly. 1: Carnelian flake; 2: Obsidian flake.

Chapter A2

Paleolithic site in Azerbaijan: Report of 1957*

* Huseynov, M. (1957) *Paleolithic Site in Azerbaijan: Report of 1957.* Manuscript on file of the Institute of Archaeology, Ethnography, and Anthropology, Baku (in Russian).

A2.1 Introduction

Related to the study of the Paleolithic period, the caves on Avey Mountain in the territory of Dash Salahli village (Fig. A2.1), Gazakh region, have started to be studied since 1956. The first research here was launched in 1953 under the direction of Sergei N. Zamyatnin and with the participation of the author of this report. A small area was excavated in the Damjili cave for the purpose of sounding. As a result of several days of excavation, the remains of flint and obsidian artifacts belonging to the end of the Upper Paleolithic were discovered.[1]

Since 1956, excavations have been continued by the author under the auspices of the Azerbaijan History Museum.[2] In 1957, substantial archaeological excavations were carried out here. As a result, it was found that the Paleolithic site of Damjili was multi-layered but the stratified cultural layers were completely destroyed by nature and by medieval people. Cultural layers were mixed. Accordingly, it is impossible to provide a single stratigraphic sequence at this cave.

Fig. A2.1 Distant view of the Avey Mountains looking west.

1 Zamyatnin, S. N. (1956) *Reconnaissance on the Stone Age in Azerbaijan in the autumn of 1953.* Unpublished manuscript in the Archives of the Institute of History of the Academy of Sciences of Azerbaijan. SSR.

2 Huseynov, M. M. (1956) *The Report of 1956 on the Archaeological Excavation conducted in Damjili near the Village of Dash Salahli Gazakh Region.* Unpublished manuscript in the Archives of the Institute of History of the Academy of Sciences of Azerbaijan. SSR.

A2.2 Cave and stratigraphy

The cultural layers here were mostly destroyed by natural forces and later by medieval people. Damjili cave is more or less semi-circular in shape and faces directly to the east (Figs. A2.2; A2.3). The cave is 4 m high from the present ground to the ceiling at the entrance. The inside of the cave is 8 m wide and 18 m long. In general, the total size of the present area of the cave, including the destructed area, is 17 × 27 m (Figs. A2.4; A2.5). The ceiling of the cave has been cracked in three places (Fig. A2.6). The beautiful spring water drips continuously from these cracks. Drops poured from above are collected in a 50–60 cm-high pool built with stone and cement. The length of the front wall of the pool is 6 m, the side walls are 2.5 m on one side, and 5.5 m on the other. The water collected in the pool flows through a specially made limestone trough and pours into the valley in front of it.

The cave itself is made of flint limestone, and there is a thick layer of conglomerate at the bottom. Underneath the conglomerate is tuff-originated clay. As a result of the research, it was found that there are thick layers of tuff under those clays. A small ravine starts from the cave. The length of the valley is about 400 m and its depth reaches 15–25 m. From the top of Avey Mountain, another ravine hangs over the cave. That ravine cuts through the cave and joins with the opposite one. During heavy rains, by this ravine the strong floods that emerge in the mountain pour into the area right in front of the cave. A Paleolithic occupation was located in this area.

The torrents of rain and spring water flowing continuously inside the cave have washed and mixed the remains of different cultural layers of the Paleolithic occupations. As a result of the excavation, pits of different depths made by medieval people were also discovered on the ground surface of the cave. It can be said that the most valuable multi-layered Paleolithic site was completely destroyed and mixed up by the heavy destructive forces of nature for a long time, and later by people. Here, the layers belonging to different epochs have been erased and disappeared. Here it is possible to make a certain opinion only based on the obtained lithic artifacts.

The excavation revealed that the upper layer consists of organic soils of different colors. This layer contained various glazed glass fragments from the 9th to 13th centuries. The thickness of the layer was from 0.3 to 1.5 m inside the cave, and more than 3 m in the uncovered terrace area of the cave. Below this organic soil layer was a very thin layer of yellow clay up to 10 cm thick was found in the lowest part. Paleolithic remains were found in this layer. The yellow clay layer was mixed with sediment called breccia (Figs. A2.7–A2.10).

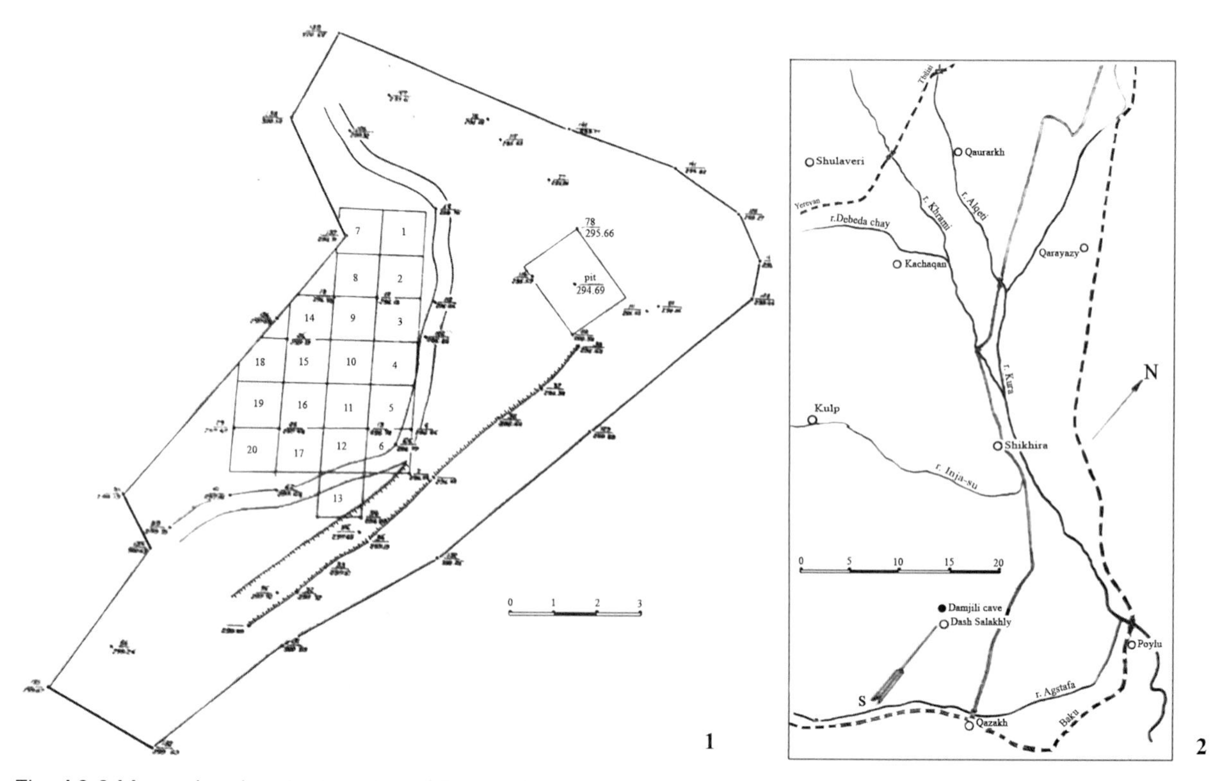

Fig. A2.2 Maps showing the location of Damjili Cave and its excavation area. 1: Distribution map of the excavated squares; 2: Map of the region where Damjili Cave is situated.

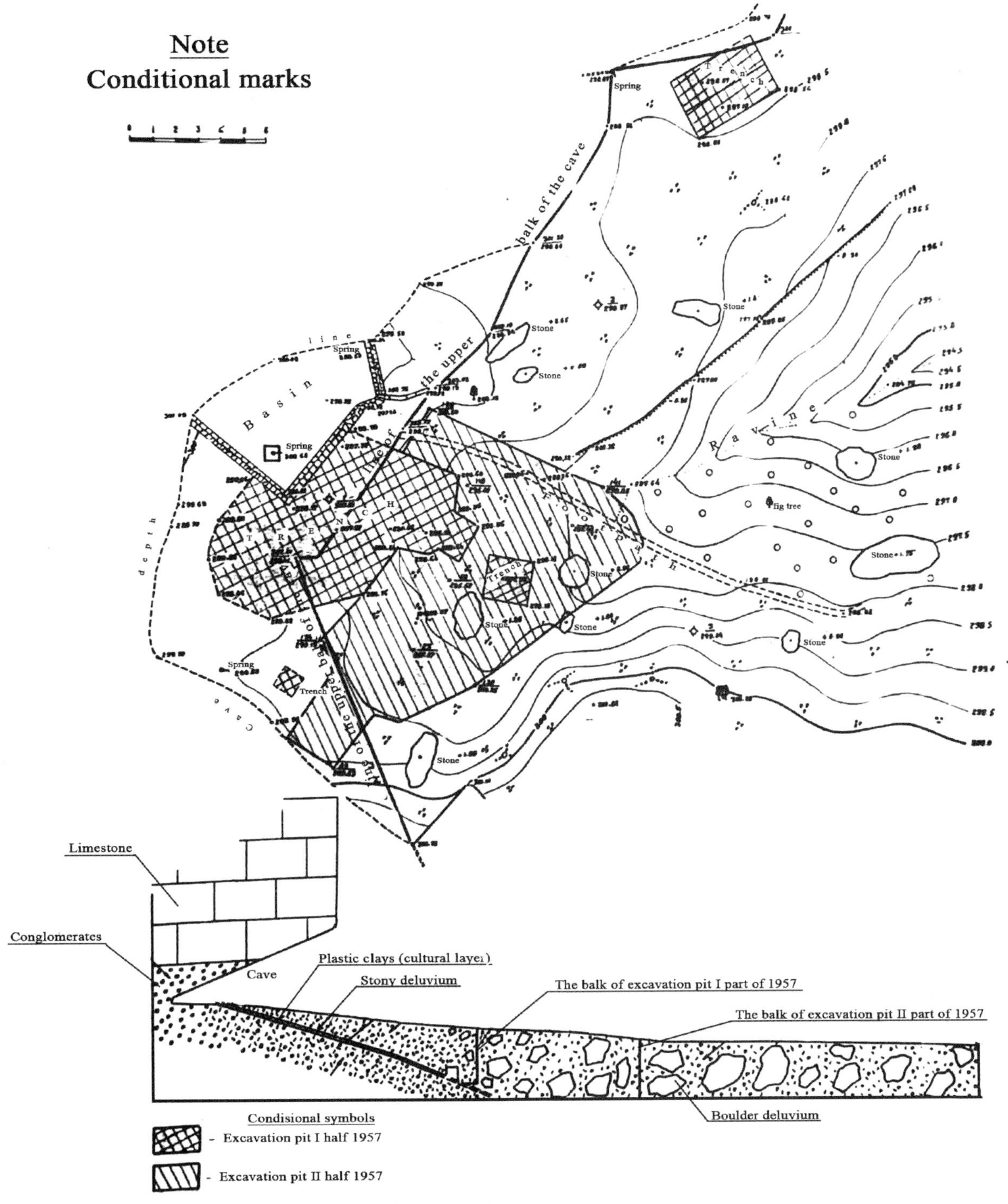

Fig. A2.3 Details of the excavation area and stratigraphy.

A2.3 Excavation results

As a result of the archeological excavation, various tools and debitage belonging to the end of the Acheulean, the beginning of the Mousterian, the end of the Mousterian period, the Upper Paleolithic, and even to the Neolithic period were obtained from that yellow clay layer. Various animal bones were also found together with stone artifacts. It is interesting that these cultural remains belonging to different periods of the Stone Age were obtained from the bottom layer altogether. This observation suggests that the water flowing, as mentioned above, gradually washed the cultural layers lying on top of each other, and consequently, the remains of various assemblages belonging to the different periods of the Stone Age collapsed. It should be noted that the destruction of these layers took place over several thousand years. Due to its multi-layered nature, Damjili Cave is considered one of the first archaeological sites in Transcaucasia. It is a pity that such a valuable ancient site was destroyed and disappeared without being preserved as it was.

Fig. A2.4 Damjili Cave, looking the entrance.

Fig. A2.5 Overview of Damjili Cave. Cave 1 is hidden in vegetation to the left. The white dirt soils to the right are from Cave 2.

Lower-Middle Paleolithic

In spite of all the above, the recovered remains can prove the existence of chronologically different prehistoric occupations. Among them, a roughly shaped basalt stone flake that does not have a standardized shape is noteworthy (Fig. A2.11: 1).

Fig. A2.6 Excavation of Damjili Cave.

Fig. A2.7 Excavation of Damjili Cave.

This specimen was detached from a rather large river stone with a round shape. It has a large and pronounced percussion bulb. There are several flake scars on the blank running in different directions and sizes. These scars are derived from knapping before being detached from the core. The dorsal surface retains a remnant of the natural cortex. This artifact differs from all other finds based on its shape and manufacturing technique. It should be noted that this flake was produced with the splitting tech-

Fig. A2.8 Overview of the excavation trench at Damjili Cave.

Fig. A2.9 Closer view of the excavation trench at Damjili Cave.

nique. This specimen is similar to the ones found in the Transcaucasia Paleolithic sites of the end of the Acheulean period.

One of the best prepared discoidal cores was also found together with that flake. This quite large core was reduced for blank production from both

Fig. A2.10 General view of the excavation trench at Damjili Cave.

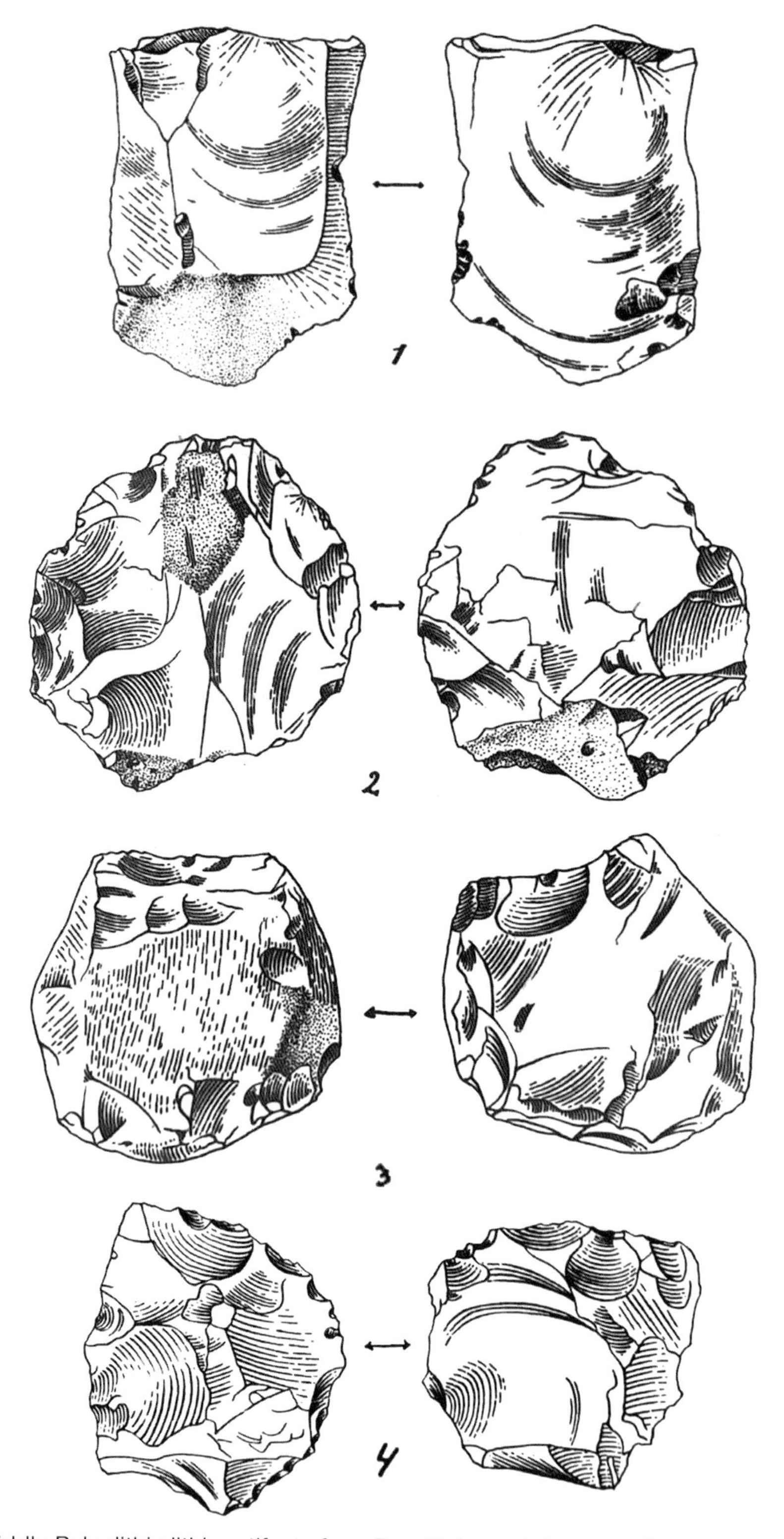

Fig. A2.11 Lower-Middle Paleolithic lithic artifacts from Damjili Cave. 1: Massive flake; 2–4: Discoidal cores.

surfaces along the edges towards the center. Flaking scars on both surfaces of the core are in a triangular shape. They demonstrate that prehistoric people chipped away flakes from the core for producing sharp-edged tools. Due to the fact that the flakes were removed from both surfaces of the core centripetally, the core edges got an indented-protruding shape. Both surfaces have almost completely lost their natural cortex, showing a thin layer of the same inside yellowish-tobacco color. This instrument is even naturally lustrous. According to the above-mentioned features, this specimen can be attributed to the beginning of the Mousterian period (Fig. A2.11: 2). Furthermore, two discoidal core fragments were also obtained (Fig. A2.11: 3, 4).

Among the interesting findings, we can refer

to Mousterian scrapers and sharp-pointed tools (Figs. A2.12; A2.13). Their blanks are mainly from the discoidal core. Sharp-pointed tools were also obtained. The resulting tools and debitage exhibit broadly and obliquely flaked surfaces and large impact bulbs (Fig. A2.14). The bulb extends up to half of the blank in relation to their size. This finding shows that those flakes were detached from a discoidal core by a heavy percussion. The dorsal scars on these blanks are triangular in shape. All these features indicate that these artifacts belong to the end of the Mousterian period. Some of the recovered specimens were utilized. In most cases, the Mousterian remains are so heavily reduced that they look like knapped from all sides.

Upper Paleolithic

Along with the above-mentioned findings, many pointed round and semi-circular scrapers, notched scrapers, cutting tools, awls, and prism-shaped cores were found. These tools differ from the Mousterian tools that we have shown above not only in terms of their preparing technique, but also in terms of their shape and the usage purposes. All of these tools consist of those made from a prism-shaped core using the consistent technical method. The edges of those tools are finely retouched. Among the tools found belonging to the Upper Paleolithic, the following can be shown.

(1) Pointed scrapers. These are made of variously colored flints and obsidian, on blanks made from prismatic cores. Their retouches are seen along single- or double-edges. The blanks are mostly oblong and sometimes shortened. Their lateral edges are blunted and the other edge is nibbled. Such tools are mainly attributed to the Upper Paleolithic in Transcaucasia (Figs. A2.15–A2.17).
(2) Scrapers in various shapes. Such tools were made of flint of different qualities and colors. They are slightly hamped on the dorsal surface, while the other surface is quite flat. In general, semi-circular edges are completely exhausted due to the heavy use (Figs. A2.18: 1–3, 15; A2.20: 5–13).
(3) High scrapers. These tools were made of thick blanks that are not very large. Their working edges are not straight, but have a very uneven shape. Their heights go down towards the used edges (Fig. A2.20: 1–4).
(4) Notched scrapers. They are made on oblong flat blanks, mainly of flint and obsidian. Such tools are of two types: elongated and short ones. Both edges of these tools are blunted (Figs. A2.19: 1–11; A2.21; A2:27: 32–35).

Such kinds of tools were found in the Upper Paleolithic sites in the regions of Arzni and Nurnus on the left bank of the Zangi River in Armenia. The above-mentioned different types of retouched tools were even found by Stefan Krukovskiy in

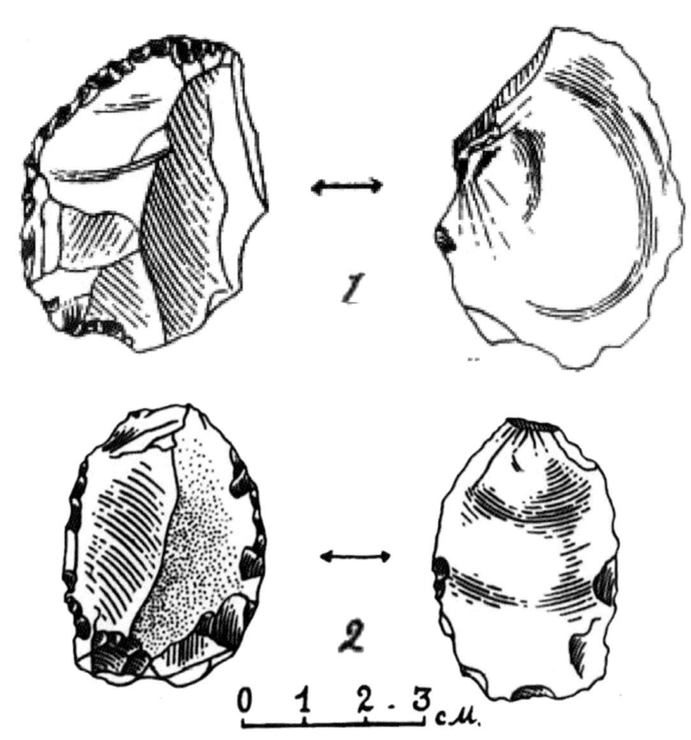

Fig. A2.12 Lower-Middle Paleolithic lithic artifacts from Damjili Cave (Scrapers).

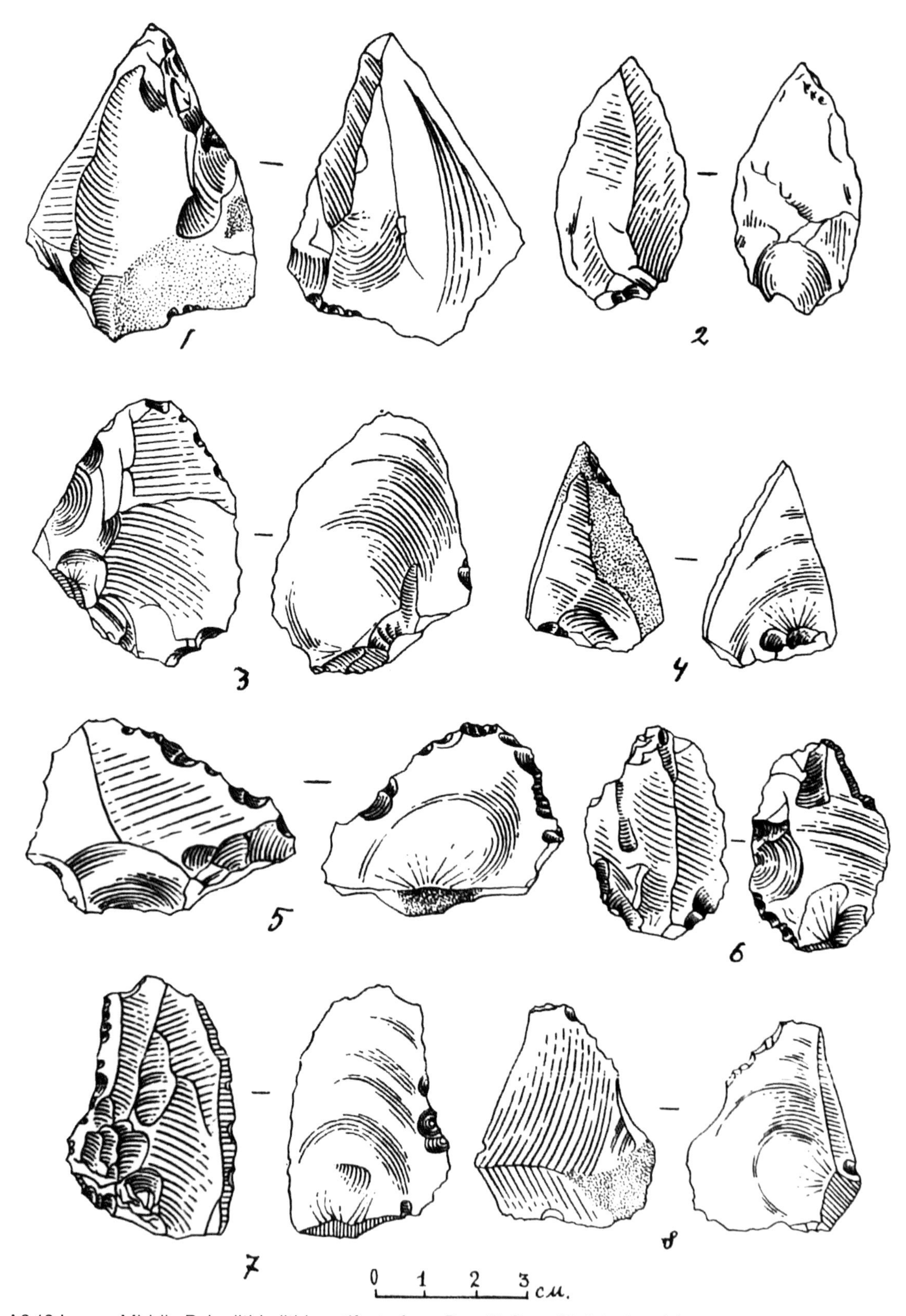

Fig. A2.13 Lower-Middle Paleolithic lithic artifacts from Damjili Cave (Pointed tools).

1916 in the Upper Paleolithic site in the cave called Gvarchilas-klde in Rgani.[3] Some of the tools found in Damjili are similar to Barmaksiz tools, which belong to the end of the Upper Paleolithic, found mainly in Georgia.[4] In general, the pointed tools found in Barmaksiz are comparable to the Upper Paleolithic

[3] Panichkina, M. Z. (1950) *Paleolithic Armenia* 1950 səh. 9.

[4] Krukovsky, S. (1916) Cave Gvardzhilas-kade in Pgani. *Izv. Cav. Museum* 10, v.Z., p. 253.

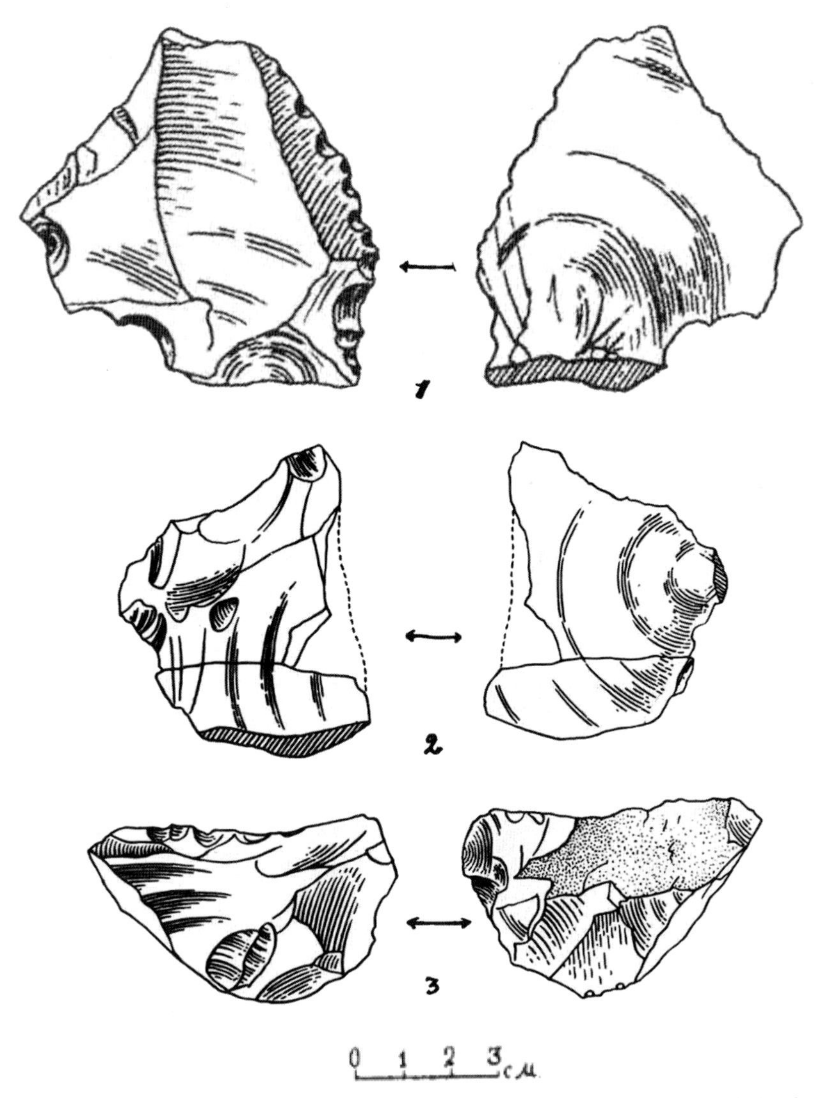

Fig. A2.14 Lower-Middle Paleolithic lithic artifacts from Damjili Cave (Mousterian type flakes).

tools from Damjili not only in terms of their shape, but also in terms of the fauna found there.[5]

The Damjili cave is one of the most valuable sites among the multi-layered Paleolithic sites in Transcaucasia, Crimea, Central Asia and Russia. Based on the lithic artifacts found at this cave, we can boldly compare this site to other Paleolithic sites. Professor E. A. Huseyinova, Director of the Institute of History, Azerbaijan SSR, writes in her letter about Damjili to A. H. Kalandadze: "In addition to the Dash Salahli Cave, there is a multi-layered Damjili site, where there are not only Middle and Upper Paleolithic but also Mesolithic and Neolithic inventory remains."[6]

Since the Paleolithic site at Damjili does not have a reliable stratigraphy, it is impossible to assign the recovered lithic tools into the Upper Paleolithic. As mentioned above, only Mousterian tools are certainly represented. Despite all this, the tools are comparable to the artifacts from different periods of the Paleolithic, according to the typological variability.

The shortened end-scrapers found in Damjili were discovered also in the Sakajia cave. Such tools are included by S. N. Zamyatnin in the groups belonging to the end of the Upper Paleolithic.[7] Along with this type of tools, the one-side burins were obtained. The burins were made of elongated obsidian blades. The retouch is made from the ventral surface of the blank (Fig. A2.22).

Besides the one-side burins, other types of burins were not encountered. Unlike other Transcaucasia Paleolithic sites, burins have been found from

5 Archaeologist. Foundation Museum of Georgia inv... 158, 159, 161, 22, 81.77.

6 The letter is currently kept in the special files of the scientific secretary of the Institute of History. The letter was received on May 16, 1958.

7 Zamyatnin, S. N. (1957) Paleolithic of Western Transcaucasia. *Collection Museum of Anthropology and Ethnography* XVIII, p. 472.

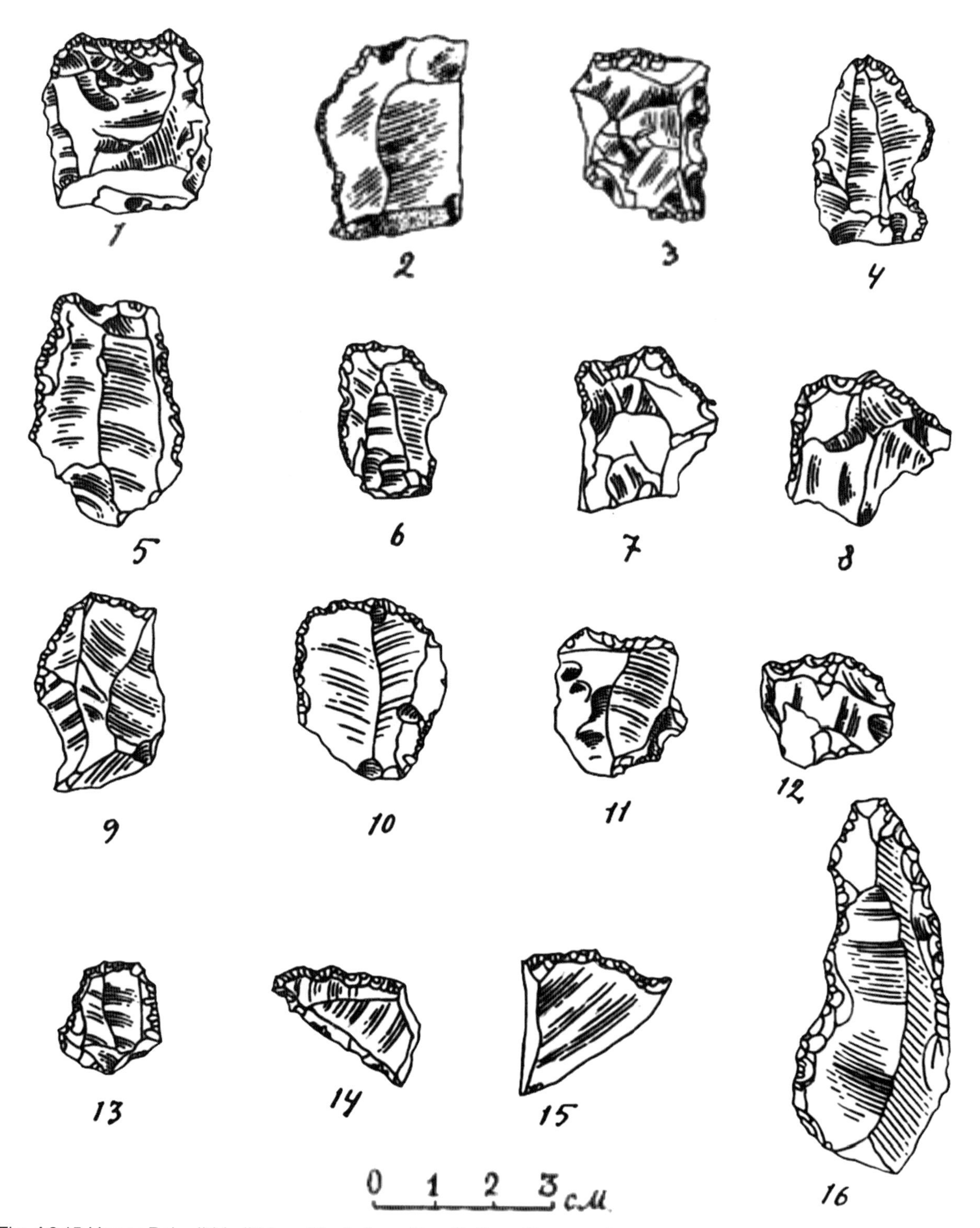

Fig. A2.15 Upper Paleolithic lithic artifacts from Damjili Cave (Scrapers).

Damjili in a small quantity. Nevertheless, one of the characteristic tools of the Upper Paleolithic is burins. Moreover, those tools are found in different shapes.

Thus, when the tools found in Damjili are compared with the Transcaucasian materials, it becomes notable that among the tools, short, semicircular, notched tools were uncovered in the Gvarchilas-klde cave.[8] Such tools were found in the Devs Khvreli site, which belongs to the end of Paleolithic.[9]

The above mentioned carinated end-scrapers and similar end-scrapers obtained from Damjili were also found in the site of Elin-Bor, belonging to the end of the Upper Paleolithic, in the central

[8] Zamyatnin, S. N. (1953) New data on the Paleolithic of Transcaucasia Sov. *Ethnography*, p. 118.

[9] Miradze, S. (1933) Paleolithic man Devs-Khvreli. *Proceedings of the Museum of Georgia*, U1, Tblisi, p. 66 (in Georgean language).

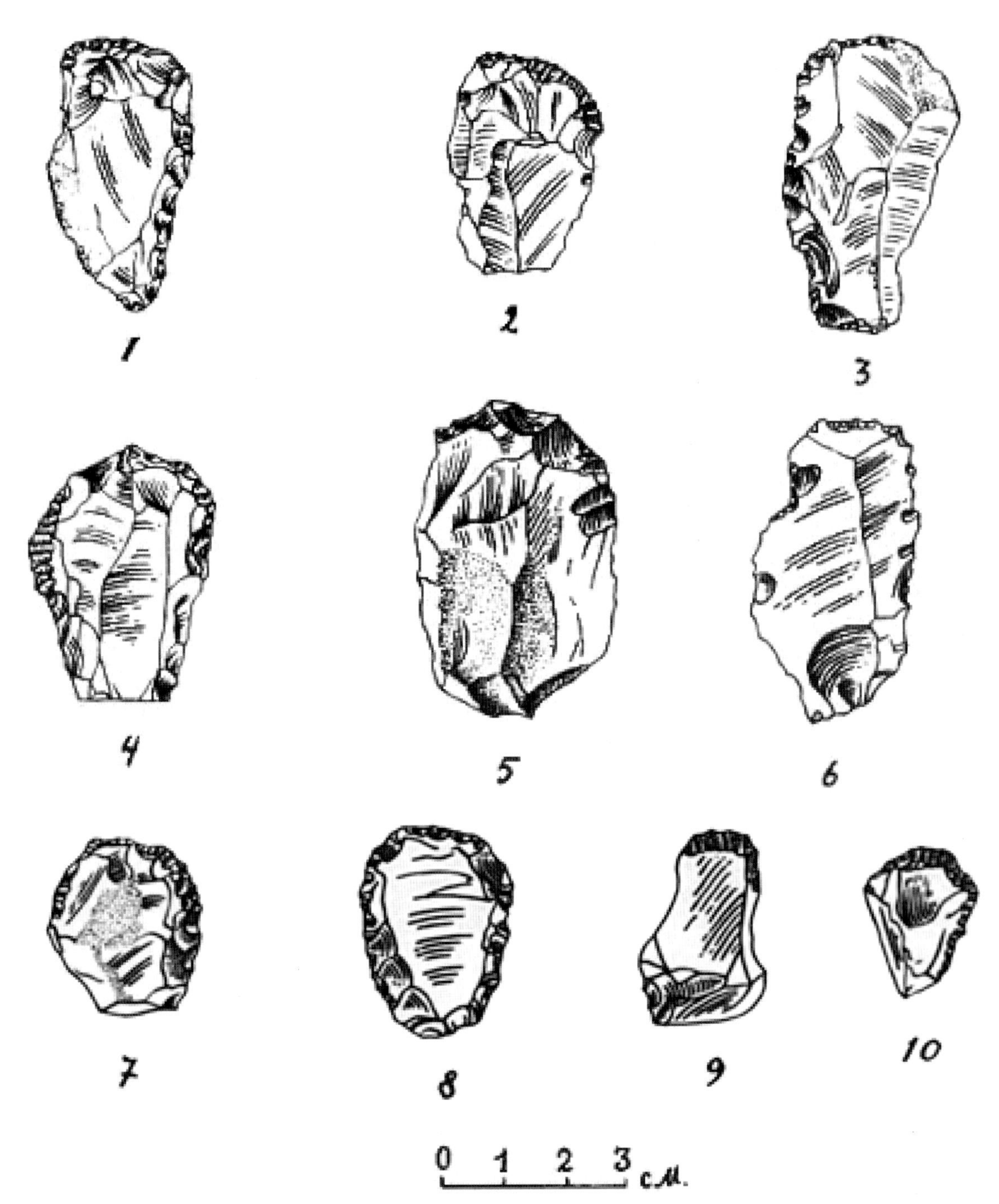

Fig. A2.16 Upper Paleolithic lithic artifacts from Damjili Cave (Scrapers).

region of the European part of the RSFSR.[10] Among the tools found in Damjili, there are two tools made of obsidian, similar to bird skulls, which are completely different from all other findings in shape. Interestingly, bottom parts of both of these tools are retouched from both surfaces. It is probable that they were made for inserting into a wooden handle. Due to their shape, these tools seem to have been used for piercing (Fig. A2.23: 3, 6). Such types of tools were found from the Upper Paleolithic sites of Armenia.[11]

Other findings include various types of notched end-scrapers were found (Fig. A2.23: 1, 2, 4, 5, 7–12). Similar tools were frequently obtained from Devs-Khverli Cave.[12] Also discovered from Damjili are some borers made of flint and obsidian. There are only 20 such tools. All borers were made on short and broad blanks. They are basically divided

[10] Vosvodsky, M. V. and P. I. Borikovsky (1937) Site Elin-Bor Owls. *Archaeology* 3, 1937, p. 84.

[11] Sardayan, S. A. (1954) *Paleolithic in Armenia*. Yerevan, p. 160.

[12] Nioradze, G. K. (1934) Paleolithic of Georgia. *Proceedings of the International Conference on the Study of the Quaternary Period of Europe*, vol. U, p. 226.

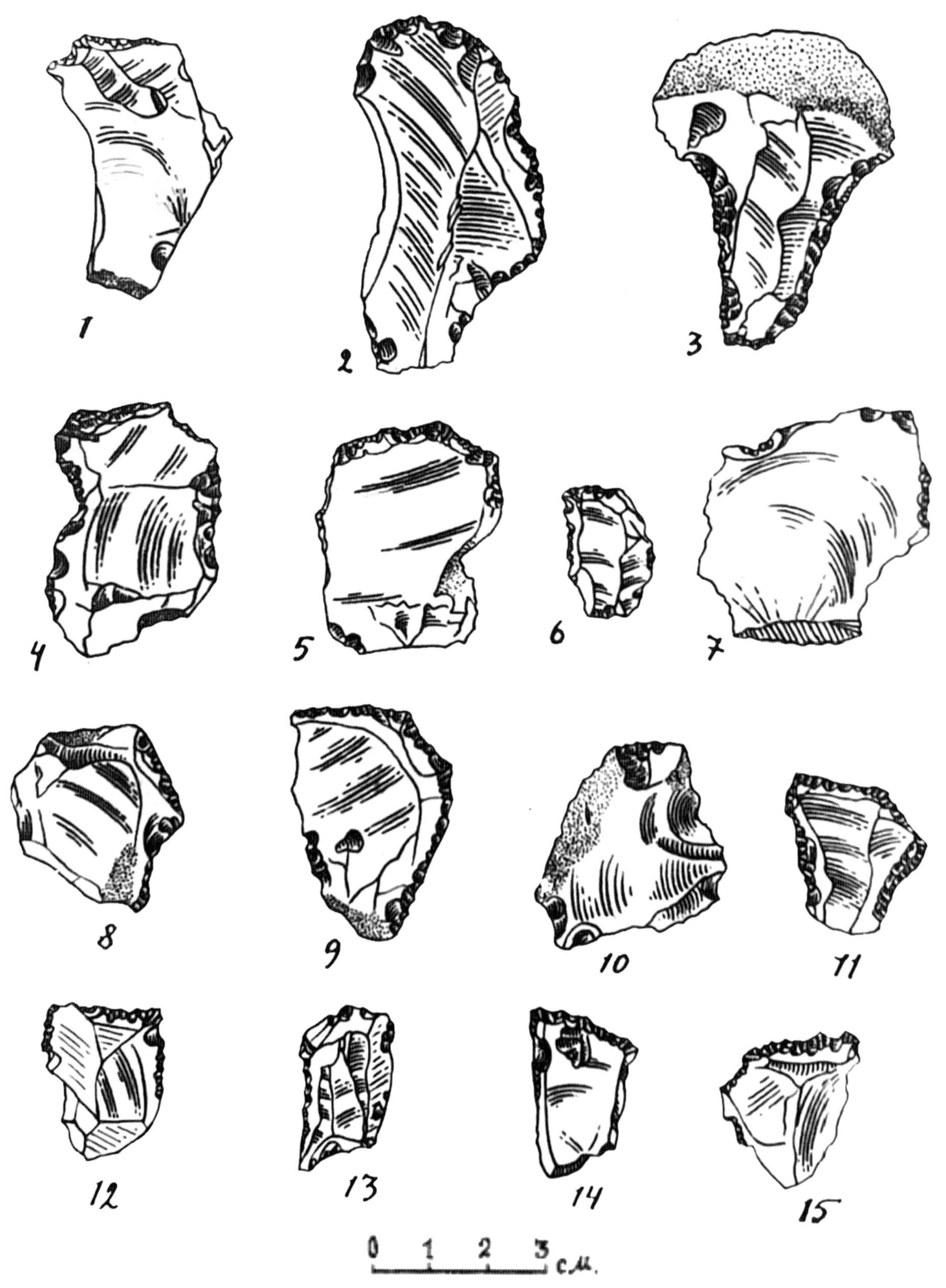

Fig. A2.17 Upper Paleolithic lithic artifacts from Damjili Cave (Scrapers).

into two types according to their shape. One is those with a round base and notched sides. Their tips are formed by a little retouching. The second type of borer is made of two triangular shaped blanks. Such borers have a wide surface and a large percussion bulb. These types of tools have changed their natural color and are covered with the same patina color. Their surface suggests washing and water rolling for a long time.[13] These types of tools are similar in shape to the Mousterian pointed tools. It is possible

[13] Patina is a color formed by oxidation on the stone. This color is different and is created as a result of several thousand years.

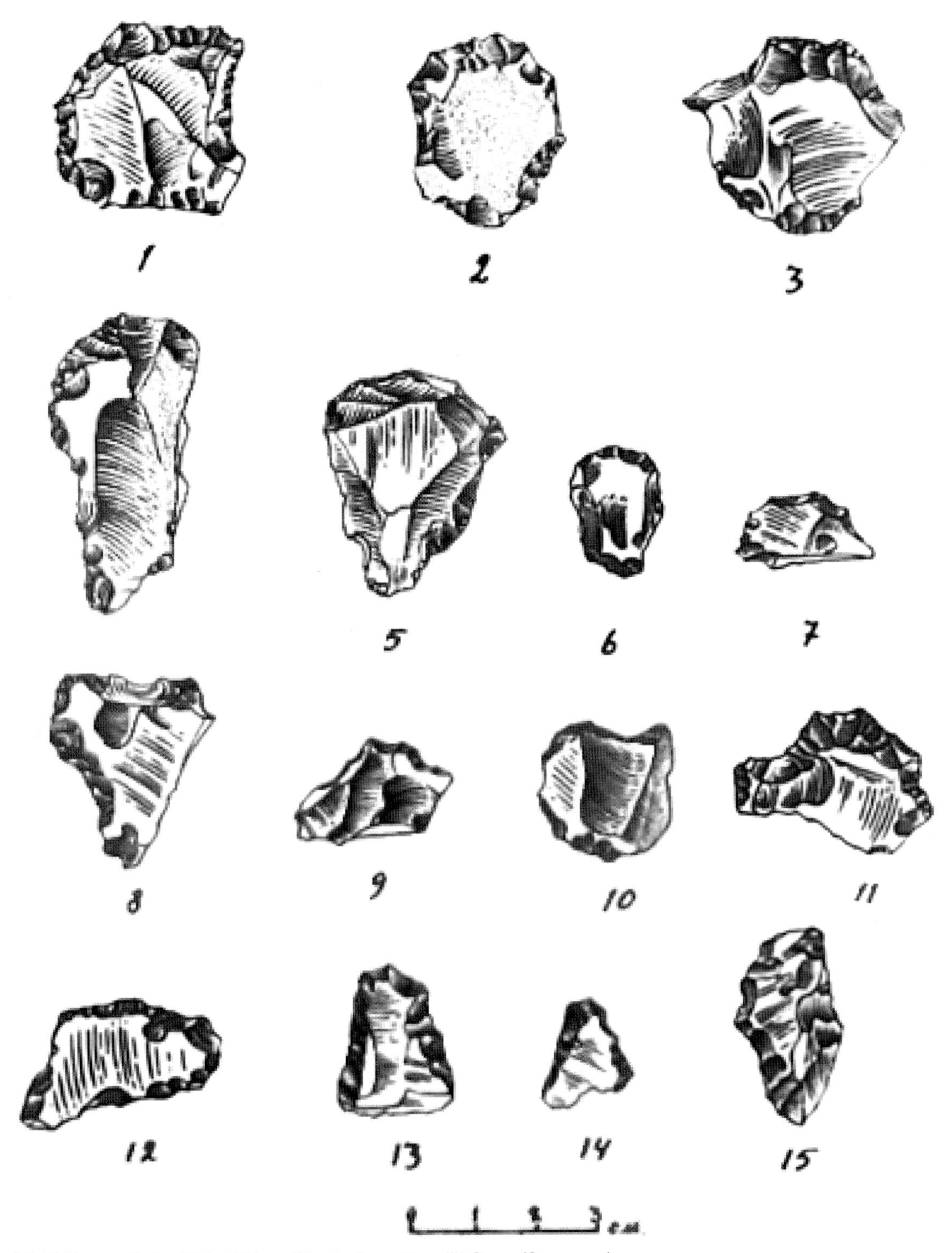

Fig. A2.18 Upper Paleolithic lithic artifacts from Damjili Cave (Scrapers).

that the Upper Paleolithic people found Mousterian blanks and later worked them into the shape of borer. Their length is from 3.5 to 4.5 cm. The width is 2.5 cm, and the thickness is 1 cm (Fig. A2.24).

The Barmagsiz assemblages include no tool in the type of Damjili borers. All of the Barmagsiz borers are from very thin and long blanks. It can only be noted that small-sized borers with a wide base were found from Sakajia, which are assigned by S. N. Zamyatnin to the middle phase of the Upper Paleolithic. However, the tools similar to the larger ones found in Damjili were discovered in the Taro-Klde site. These are attributed by S. N. Zamyatnin to the first phase of Upper Paleolithic.[14]

Damjili borers completely differ from these sites of the Transcaucasia Paleolithic (Mgvimevi,

[14] Zamyatnin, S. N. (1957) Paleolithic of Western Transcaucasia. *Collection Museum of Anthropology and Ethnography* XVIII, p. 472.

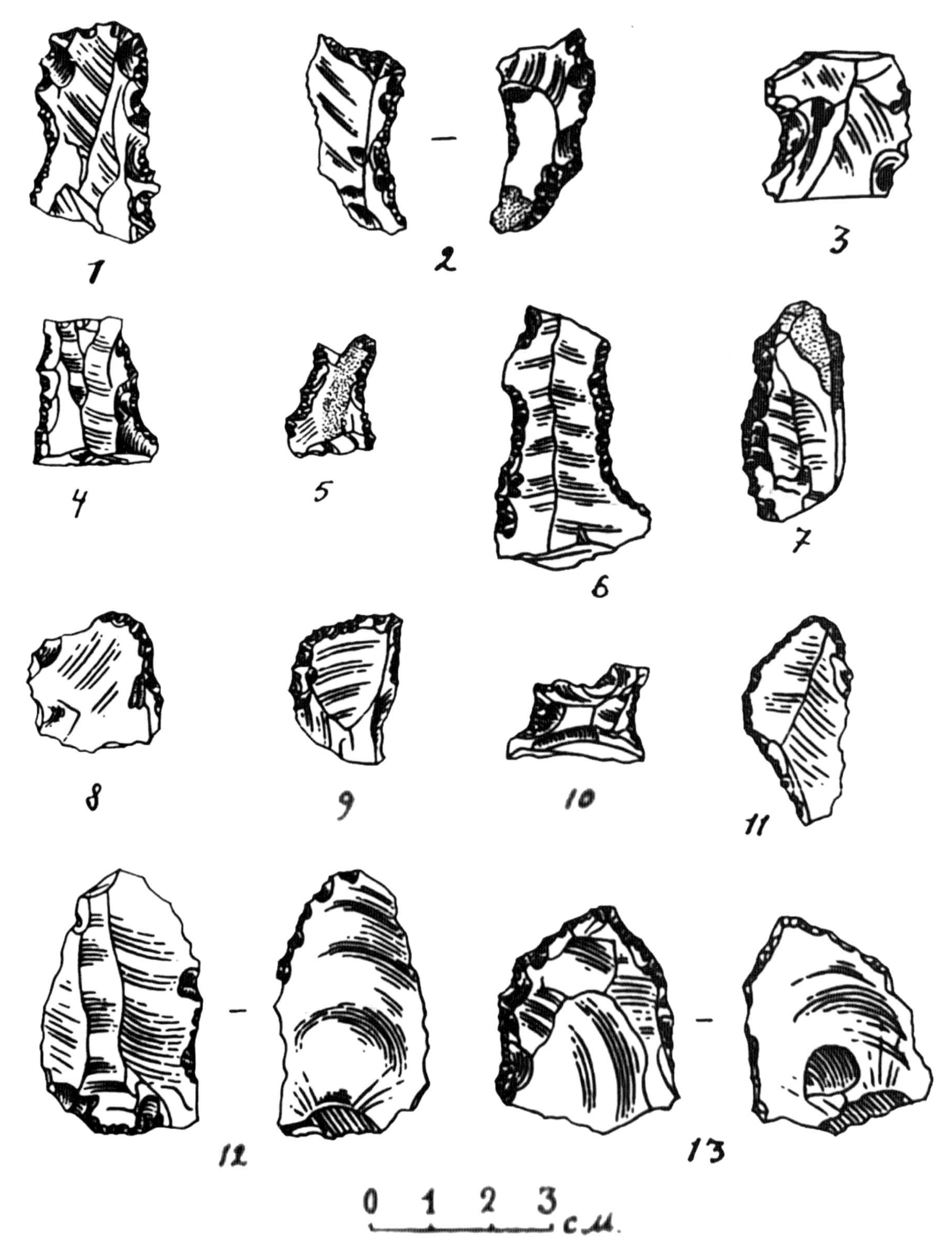

Fig. A2.19 Upper Paleolithic lithic artifacts from Damjili Cave (Scrapers).

Sakajia, Barmaksiz, etc.) in their form and in the fact that they were washed and smoothed a lot.

In addition to these tools, about four hundred prismatic core remains were found in Damjili. All cores consist of different colored flints. ninety percent of the cores remained in a used-up condition. They are quite small. Some of them are so used that they have even turned into a round shape. Prehistoric people did not leave the cores unused until the end.

Both ends of the cores have very well-preserved striking platform remains. Cores also exhibit several longitudinal ridges of negative flake scars. The cores indicate that knife-like blades were produced from them. The traces of those blank removals, showing wide and sometimes thin scars, prove this interpretation. Some of the cores are elongated and have an elaborate surface on both sides. Prehistoric people started using from one side of the core while

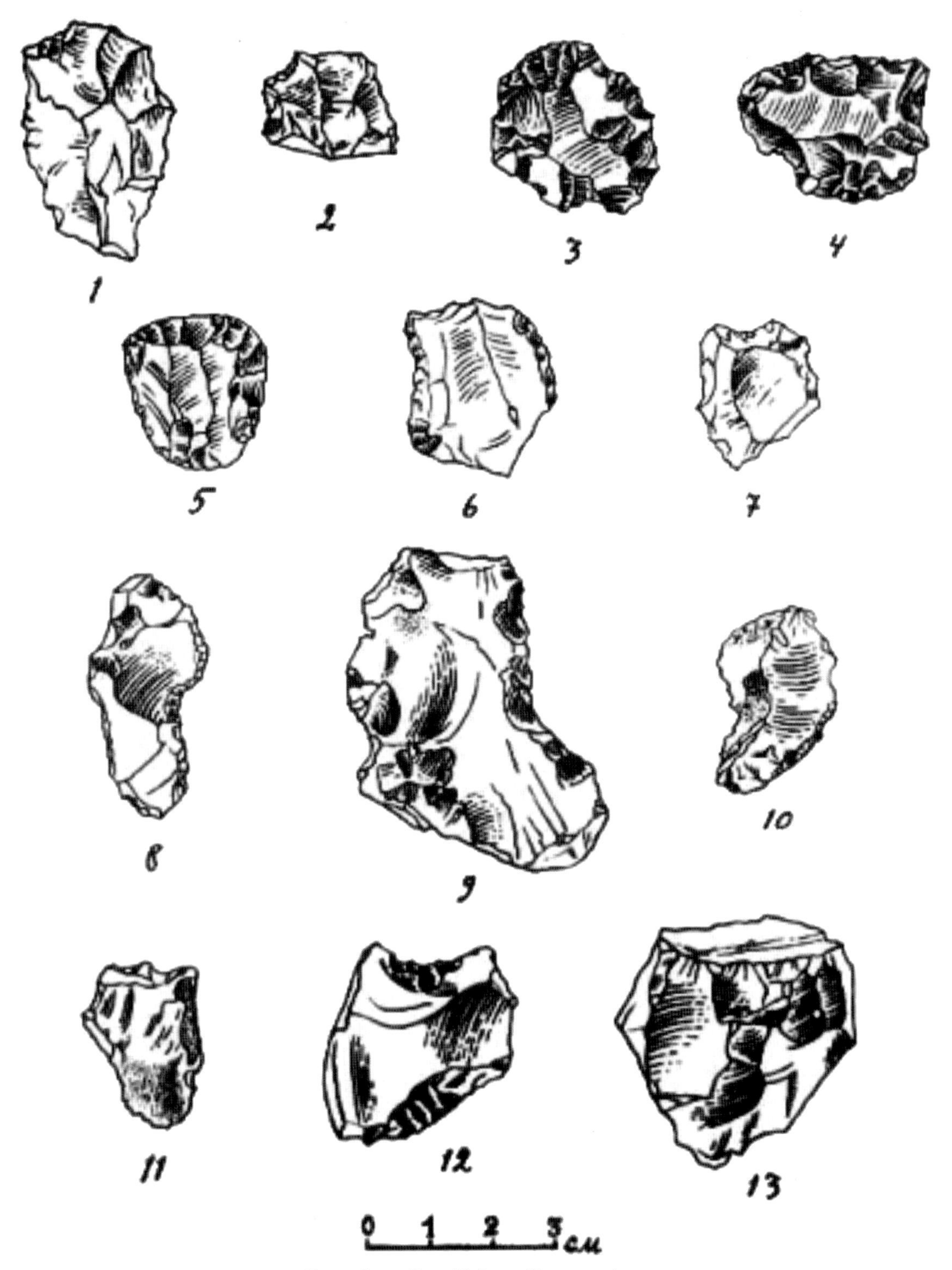

Fig. A2.20 Upper Paleolithic lithic artifacts from Damjili Cave (Scrapers).

making tools. The other side was kept for necessity (Figs. A2.25; A2.26).

Most of the cores consist of small round flints. Apparently, since there was no local material, prehistoric people collected and used whatever they could get their hands on. From all the obtained cores, it is evident that since there was no reliable flint necessary for prehistoric people living in Damjili, they collected all kinds of stones that they could see and used them to make all kinds of tools. Therefore, most of the available tools are designed for the right purpose. Due to the small size and poor quality of the collected stones, they were reduced into all kinds of blanks by the people of the Damjili Stone Age. Its parts were retouched and processed in the field of work. Almost 90% of the specimens obtained in Damjili have been utilized. These prove once again that since there were not as many appropriate rocks as desired, all kinds of stone artifacts were considered valuable for the prehistoric people

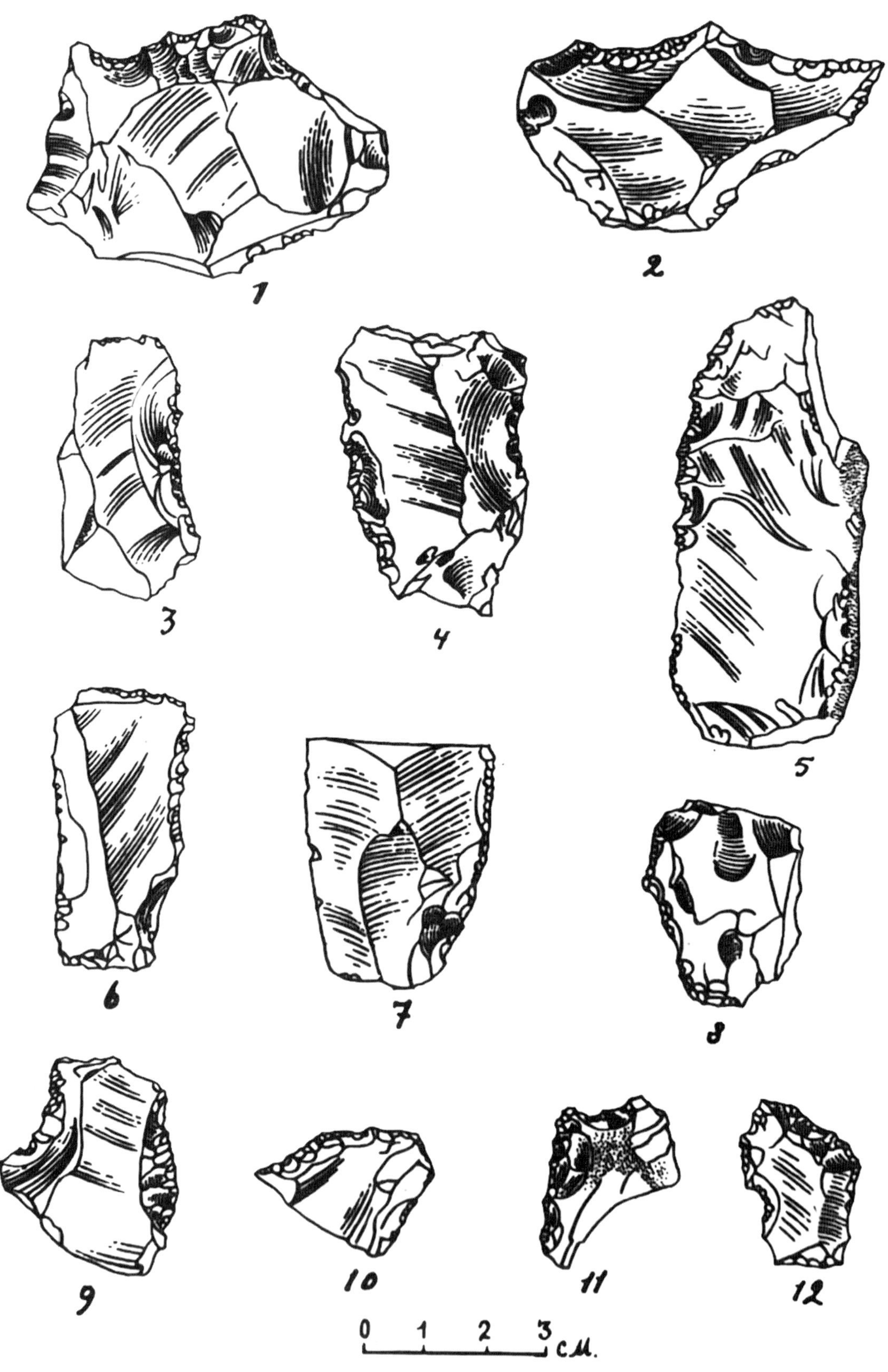

Fig. A2.21 Upper Paleolithic lithic artifacts from Damjili Cave (Scrapers).

living in Avey Mountain and they did not throw them out. It is interesting that they made tools out of tuff and limestone among the bad quality flints.

All these show the needs of the Paleolithic people for stone material in that period. Maybe because of this deficiency, they spent the whole day looking for raw materials to make tools with great difficulty. Despite the entire above-mentioned difficult situation, people were able to live in Damjili cave for a long period of time.

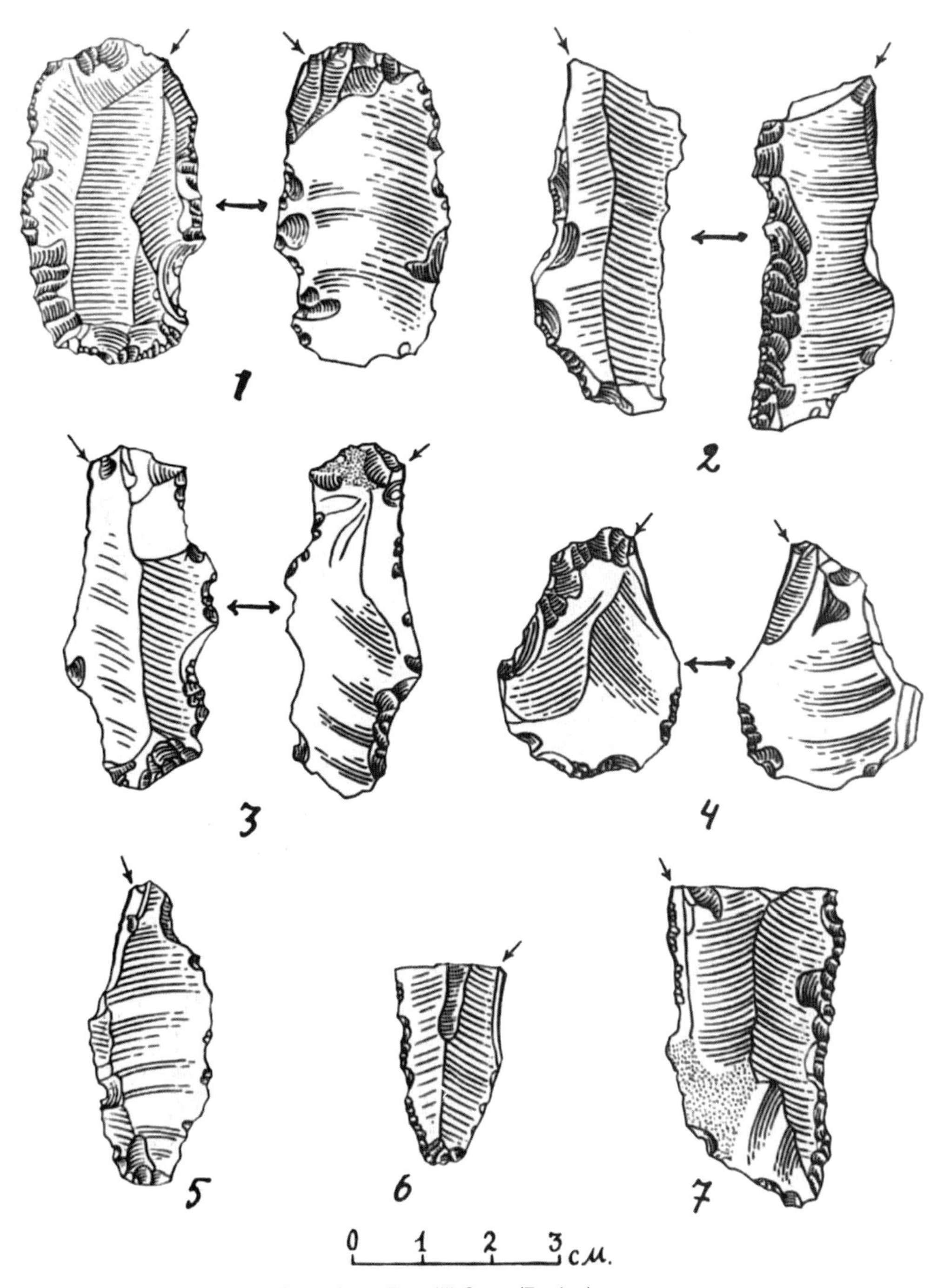

Fig. A2.22 Upper Paleolithic lithic artifacts from Damjili Cave (Burins).

The tools found at Damjili provide a clear piece of evidence of the Upper Paleolithic human occupations and even later periods. During the excavation, knife-like blades were found, one or both edges of which were completely blunted. Such blades were struck from the cores. The length of the blanks is not so long. Its size is 2 × 4 cm and the width is 1 cm (Fig. A2.27: 5–9). Some of them have a small notch on the lateral edge, which seems to have been done for a certain purpose. The blanks are mostly with one or two dorsal ridges. One surface is flat and the dorsal surface is pronounced. Of these, only one end of one of them has been retouched. One edge is blunted, and the opposite edge is serrated as a result of using. There may be such type of knife-like blades with one side blunted, which may have been used for cutting or splitting fish (Fig. A2.27: 19). Knives with notches on the sides or ends may have been manufactured for the purpose of production and used in pasture farming as they were needed to make bone tools. The notches of such tools are circular and made on one or both edges (Fig. A2.27: 15, 16, 20, 21).

In addition to the above-mentioned knife blades, there are also tools similar to arrowheads with sharp triangular ends. They have two retouched edges. It

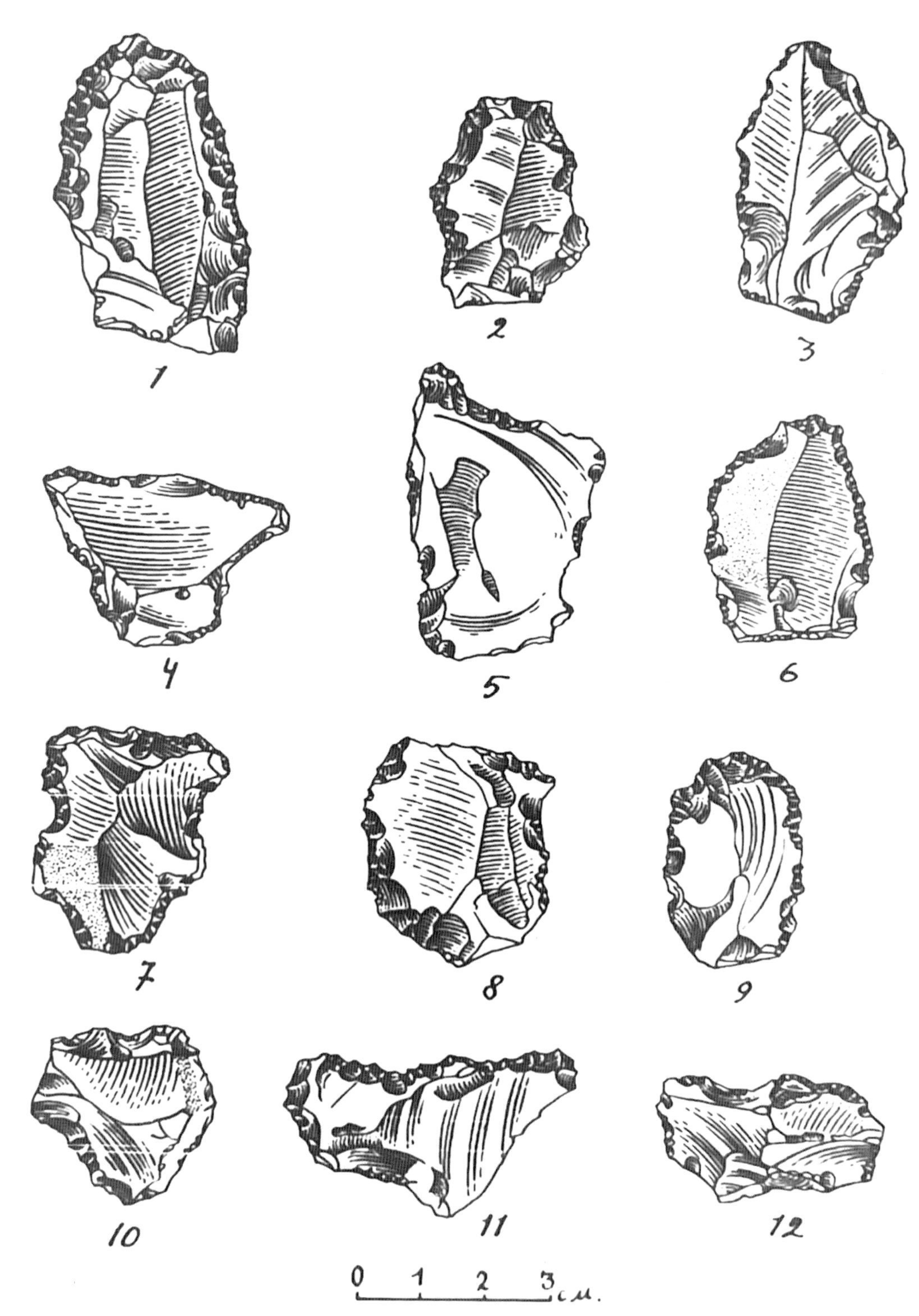

Fig. A2.23 Upper Paleolithic lithic artifacts from Damjili Cave (Notches).

also seems to have been done to stick into the shaft. On both sides of the other triangular tools, the retouches are made with a tiny flaking. It is 4 cm long and 1.5 cm wide. The length of its inserting part is 2 cm. Such tools seem to have been the weapons of the Stone Age people (Fig. A2.27: 13, 14, 17). Tools of this type have been found in Transcaucasia at Upper Paleolithic sites including the Taro-klde and Hergules-klde caves in Georgia. A blade of this type was found in the Migmimavi site with its sides retouched.[15] It was even found in sites like Deves-Xvreli and Sakajia.

Knife-like tools with retouched sides have been made from slabs with blunted sides, and are from an Upper Paleolithic site of Suyuren 1.[16] Similar tools were also found in the Lostyonka 5 and the Anasovka

[15] Zamyatnin, S. N. (1937) Cave campsite of Mgvimevi, near Chiatura, *Soviet Archaeology* 3, p. 65.

[16] Vekilova Suren, E. A. (1957) ... *Materials and Researches on the Archeology of the USSR*, 59, p. 281.

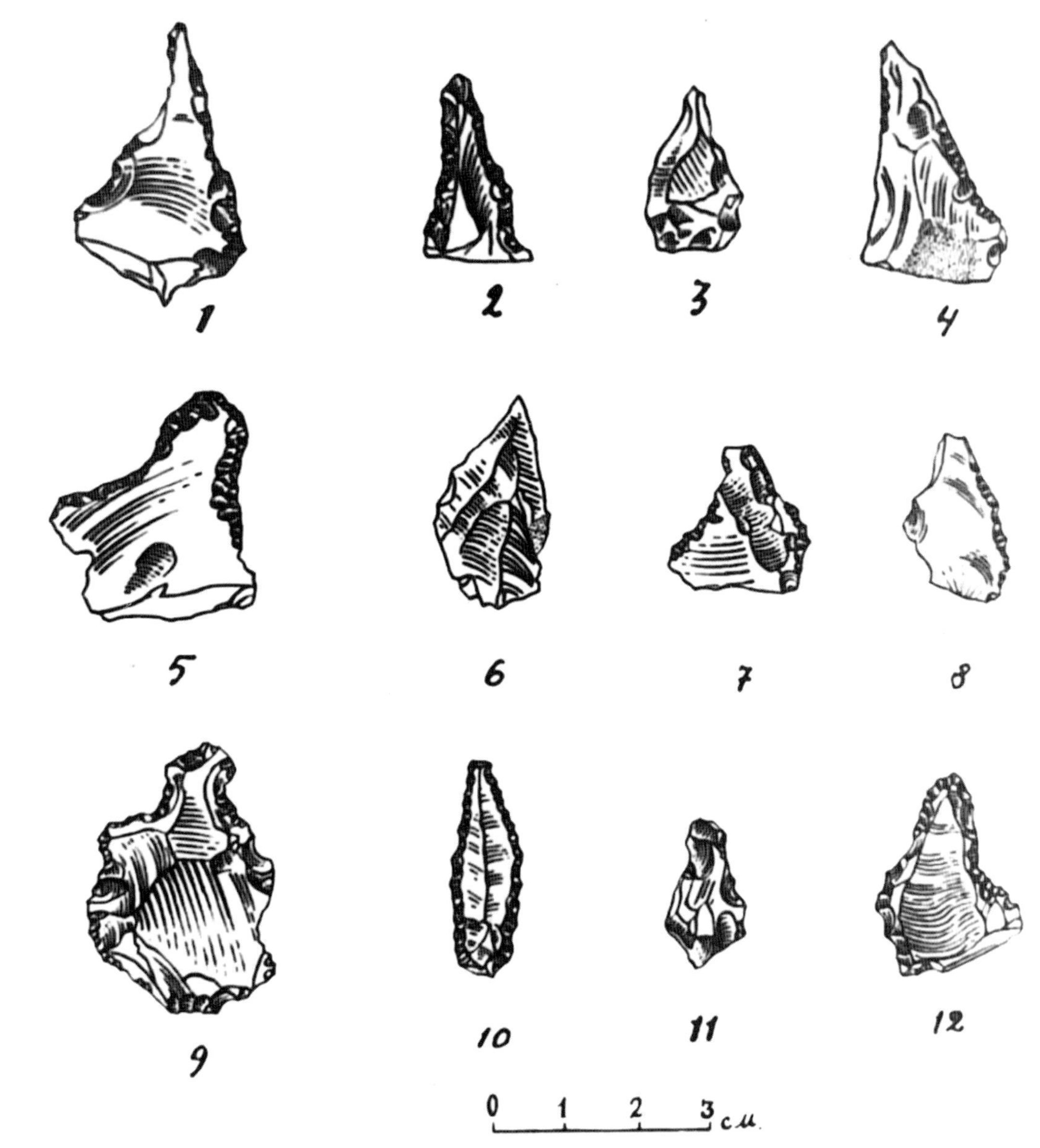

Fig. A2.24 Upper Paleolithic lithic artifacts from Damjili Cave (Borers).

2 sites in the Kostuyonka-Borsheva regions.[17] Such tools are considered special for Damjili, and their forms are also different from other tools.

The end of the Upper Paleolithic

In addition to the artifacts mentioned above, quite small thin tools were also found in the Damjili cave site. These artifacts have been uncovered all over the world from all Upper Paleolithic sites known to us so far. Such artifacts are called microliths or microlithic tools. Their lateral edges are parallel to each other and are serrated on both sides or on one side with a fine retouching method (Fig. A2.27: 13, 23). One of the microliths found here was blunted by semi-circular notches at three sides. The edge of one side of the blank has a sharp thin nose. The reverse side has been kept unretouched, but has been used, so some nibbled damages are noticeable. One of the examples that can prove this has preserved tiny edge damages.

Small nosed tools of this type are believed to have been used to chip and pierce some materials. (Fig. A2.27: 25). This assemblage includes only one very small tool called "crescent." Its one side is semi-circular and made of reddish flint. While reminding the shape of a sickle, one lateral edge of this specimen is directly blunted from one side to another, and the opposite thin edge is in the form of a sharp blade. This specimen is wider than the two ends from the middle. One of its ends has a thin sharp edge and the other end is smoothed and thinned. It is probable that it was made to be hafted to a wooden handle (Fig. A2.27: 29). Such type of

[17] Rogachev, A. N. (1957) Multi-layered site of the Kostenkovsko-Borshchevsky region. *Materials and Researches on the Archeology of the USSR*, 59, p. 95-97.

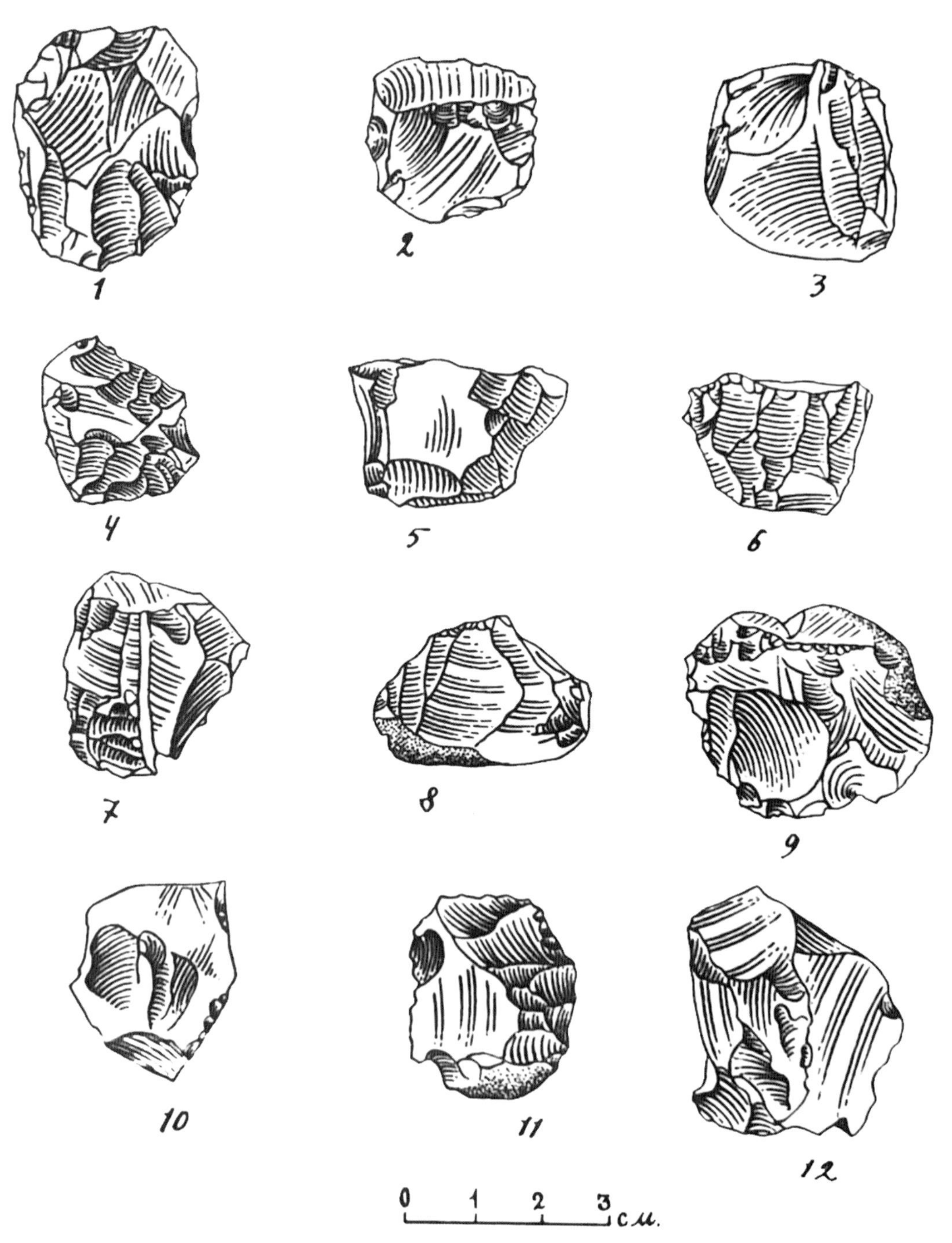

Fig. A2.25 Upper Paleolithic lithic artifacts from Damjili Cave (Cores).

tools may have been used for decoration on a certain object, to make scratches or to open the belly of a bird or fish. If it had a handle, it could have been used to skin any killed animal. One end of the segment is so sharp that it even resembles a surgical knife. Its length is 2 cm, its width is 0.5 cm, and its thickness is 2.5 mm.

Along with this, there is another crescent-shaped microlithic tool. It is made on a board, one side of which was retouched and blunted. The other side is left unretouched. This tool differs from the above piece only in that it is made of a thin blade. In addition, due to its smallness, it is not suitable for holding. It is still very difficult to speculate about what purpose it is used for. Its length is 12 mm, its width is 0.5 mm, and its thickness is immeasurable (Fig. A2.27: 30).

The most interesting ones among the microlithic tools are two obsidian tools. They were made on very small and thin blades. Both sides are retouched and the ends are blunted. There are two parallel ridges on the dorsal surface. The corners were cut off in their end. These retouches may have been made for piercing. The edges of both tools are so abrasive that they are completely smoothed. This demonstrates that they have been used for a long time in the work process. Their length is 2–2.5 cm, and their width is 5–6 mm (Fig. A2.27: 24–28).

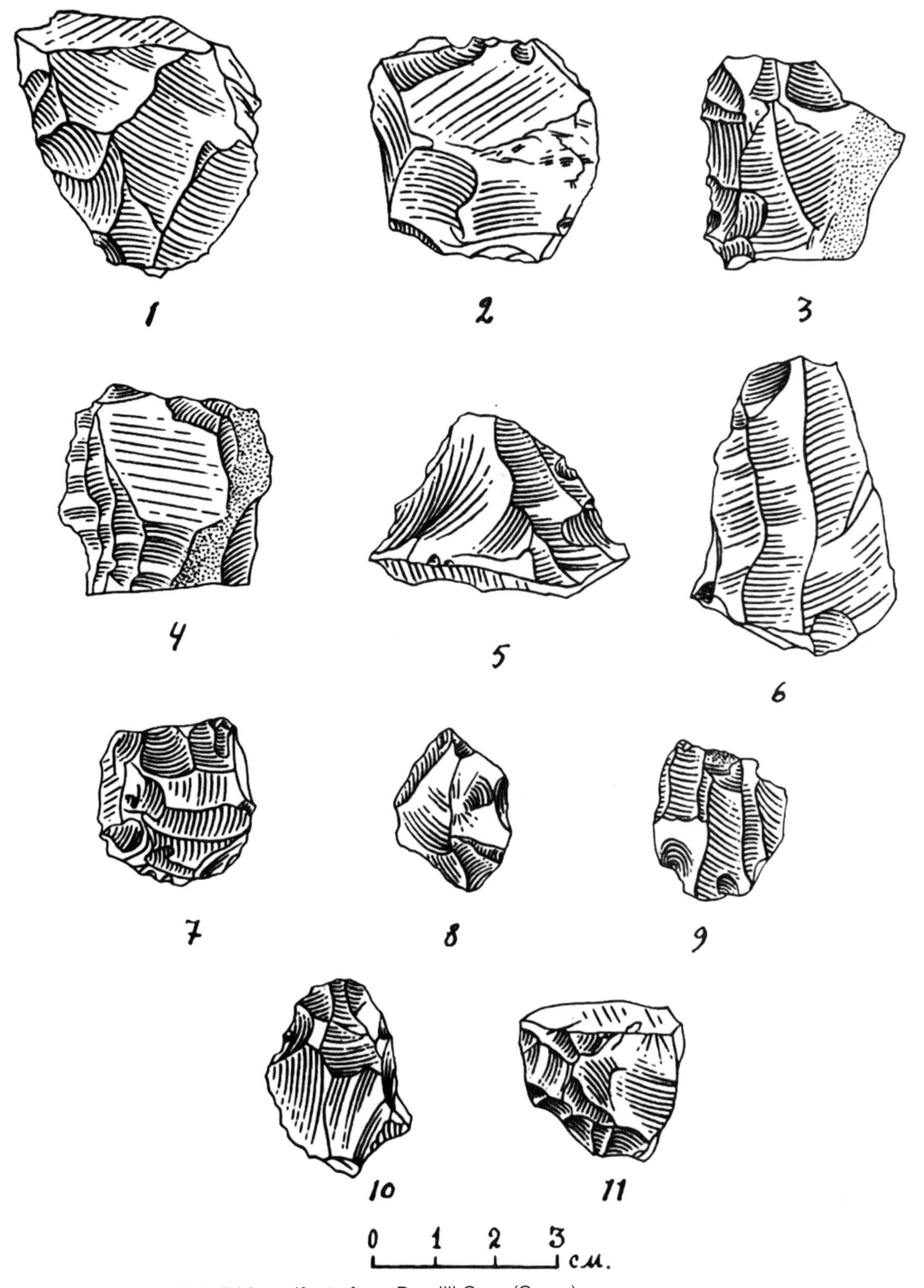

Fig. A2.26 Upper Paleolithic lithic artifacts from Damjili Cave (Cores).

All microlithic tools were made of obsidian. However, two segments were made of flint. These once again prove that the Stone Age people who were the owners of Damjili did not throw away all the debitage and fragments, even the smallest pieces, during the production of tools, but used them for various purposes. This can be the best proof that to what extent the people of the Stone Age could stockpile the materials.

Such microlithic tools are also known to us from Transcaucasia Paleolithic sites. Geometric shapes microlithic tools mostly crescents although in small quinteties have been found from the Gvarcilasklde site. They are well made and some are triangular in

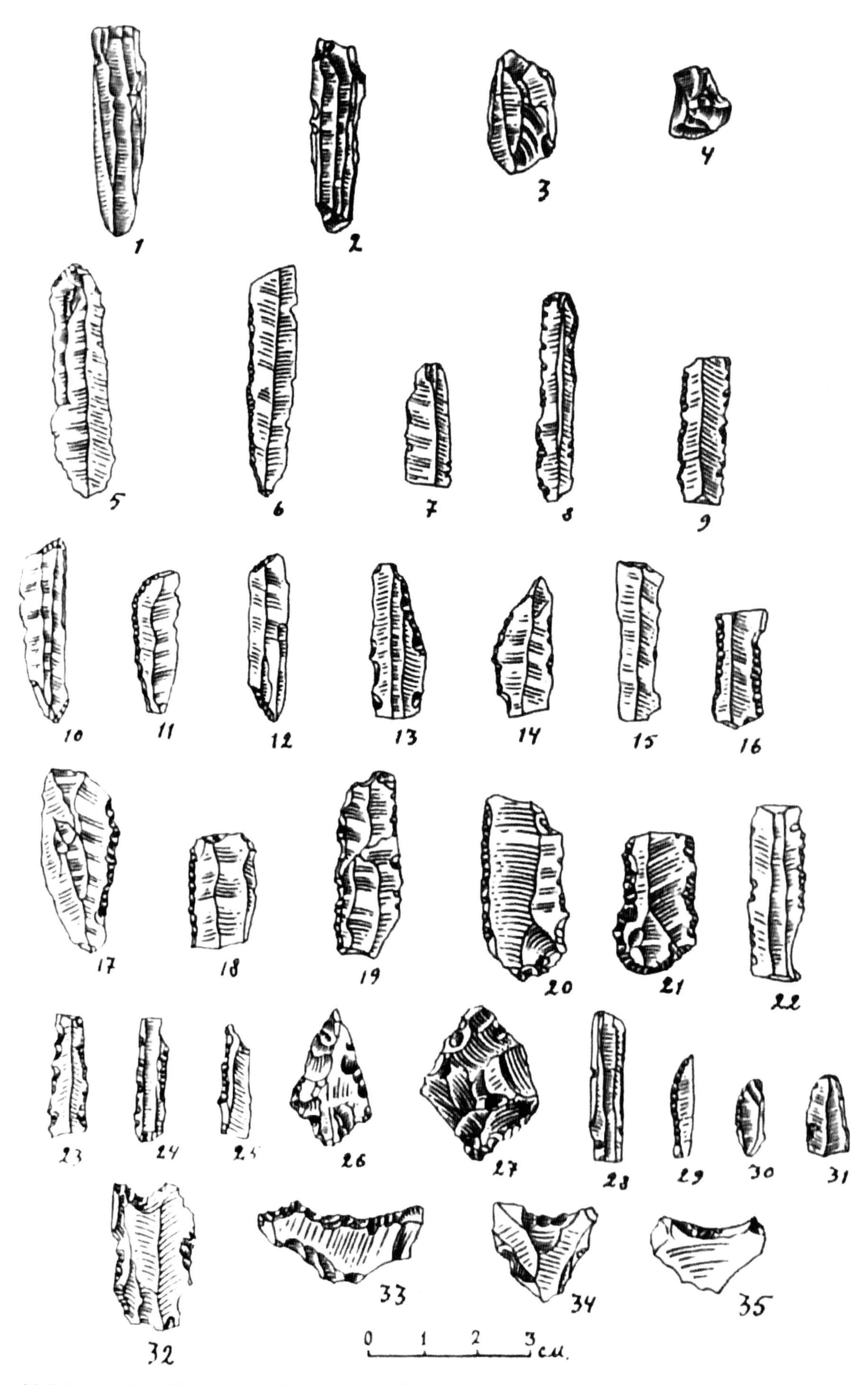

Fig. A2.27 Upper Paleolithic lithic artifacts from Damjili Cave. 1–4: Cores; 5–17, 23–25, 28–31: Microlithic tools; 18–22: Knife-shaped tools; 26, 27: Arrowheads; 32–35: Shavers. (Editors note: also see Fig. 3A.14 of this volume).

shape.[18] Various cutting tools, pointed tools made of large oblong blades, round scrapers, knife-like tools, blunted blades, oblong segments, etc. were discovered from this site.[19] Retouched microlithic blade tools were also found in Kostyonky Upper Paleolithic sites.[20] They were mostly found in Kostyonky II, III, and IV.[21]

It should be noted that not a single trapezoidal tool was found from the Damjili Paleolithic site. Trapezoids were mostly found in Transcaucasia and also in the Caucasus from the Neolithic period.

In addition to the microliths and crescent-shaped tools, the truncated bladelets were also found in the Damjili site. Such small tools were mainly made of flint and obsidian. They are geometrically shaped and have parallel angles. Their one end is truncated by retouch, that is why we call them truncated bladelets. The opposite end is straight truncated and thinned for piercing. The truncated bladelets have jagged fragments in their working edges, as they were used a lot in the work process. These are mostly thin and elongated (Fig. A2.27: 10–12).

Tools of this type were discovered in Upper Paleolithic sites also in Georgia. Among the stone tools found in the Sakajia cave, the main tool groups consisted the truncated bladelets.[22] Although the blades found in Sakajia are slightly wider than those in Damjili, the working technique and shape taking are the same. It is the same in size as the implements from Damjili. Their size is 35–60 mm, the width of all of them is 6–7 mm, and the thickness is up to 2–3 mm. Tools with a working edge at one end were also uncovered in the Upper Paleolithic sites of Devs-Khvreli.

Tools of this type are also found from Europe, North Africa, Senegal and the Gambia to the Red Sea, and thus not only in Central and South Asia, but also in India. Sites of this type can belong to the beginnings of the Neolithic or the Mesolithic period in Europe.[23]

In addition to the above-mentioned microlithic tools, crescents, truncated bladelets, a small number of special pencil-shaped cores have been found at Damjili. There are only five examples. Its length is 3.5 to 4 cm, and the thickness is 1 cm. Finding these cores here once again proves that all microlithic tools were produced in the Damjili site itself. These cores have a classic shape, keeping their original shape beautifully (Fig. A2.27: 1–4).

The pencil-shaped cores have a well-preserved striking platform with one end in the form of a pyramid, and the other end in the form of a prism. On the sides, parallel ridges extend from one end to another along the entire circle. All of them prove the production of microliths using these cores. Additionally, it should be noted that there are percussion marks on the circular edges of the striking surface to remove blades. It is very interesting that some of the obtained microliths were detached from the cores themselves. The length and narrowness of the cores fit the microliths in terms of their size.

Such microlithic tools are believed to have been produced in the Damjili site at the end of the Upper Paleolithic. In general, such types of tiny tools are pertained to the end of Paleolithic period across the world. Tools of this type with used edges found in a large quantity in the Magdalenian at Bruniquel and Raymond belong to the late Upper Paleolithic Magdalenian epoch.[24]

In addition to these tools, arrowheads were also found from the Damjili site. These arrowheads were made of flint and obsidian and are quite small in size. Only five arrowheads were found. However, one of them was lost during the excavation. These arrowheads have a very sharp point and a small tip for sticking into a certain shaft. They are wide in the middle and have sharp blade edges on the sides. (Fig. A2.27: 26, 27). Finding arrowheads shows and proves that Damjili primitive people had a very high level of hunting economy. Arrowheads of this type were also found in Georgia belonging to the same period.[25] "Now we know many Mesolithic sites from different countries, they all have got arrowheads which were used as a weapon by people who still

[18] Tushavramishvili, D. M. (1955) *Paleolithic Remains in the Kvardzhilas-klde Cave, Western Georgia*. Author's abstract, Tbilisi, p. 10.

[19] Kalendadze, A. N. and D. M. Tushavramishvili (1955) *New Excavations in the Area Kvardzhilas-klde Kr. Soovsh*. Institute of Archeology. Ukr. USSR in 4. Kiev, p. 155.

[20] Rogachev, A. N. (1953) New data on the sratigraphy of the Upper Paleolithic of the East European Plain. *Materials and Research on the Archeology of the USSR* 39, p. 47.

[21] Rogachev, A. N. (1957) Multilayer sites of the Kostenkovsko-Borshchiv region. *Materials and Research on the Archeology of the USSR* 59. issue 3, p. 30.

[22] Zamyatnin, S. N. (1957) Paleolithic of Western Transcaucasia. *Collection Museum of Anthropology and Ethnography* XVIII, p. 472.

[23] Efimenko, P. P. (1953) *Primitive Society*. Kyiv. p. 632.

[24] de Mortile, A. (1903) *Before Historical Life*. The third edition translated from French. St. Petersburg, p. 15.

[25] In 1956, at the session of Caucasian archeology convened in Yerevan, H. Mladez demonstrated the same type of arrowheads.

had no idea about pottery for thousands of years. And these (clay vessels) are not in the Mesolithic."[26]

A2.4 Conclusions

All the above facts show that the Damjili site has been a camp of prehistoric people for an unimaginable long time. The findings obtained from here indicate that people who started making tools from the discoidal cores a hundred thousand years ago lived throughout the three epochs of the Upper Paleolithic (Aurignacian, Solutrean, Magdalenian) and even in the New Stone Age, and life continued until the period of pencil-shaped cores.

At the end of the Upper Paleolithic, there is a change in the social and economic life of the primitive human society, which is called the new period of the history of human society—the Neolithic period (New Stone Age).[27] At the end of the Upper Paleolithic, i.e. at the transition to a new era—in the Mesolithic period, a new turn occurs in the territory of Azerbaijan as well as in the neighboring territories. Here, the hunting economy has gradually entered its new development. One of the evidences that can prove this is the Mesolithic camp found in Dagestan and very well presented.[28]

Thus, based on the above materials, it can be said that in Azerbaijan, as in other countries, the Paleolithic gives way to the Mesolithic, and during this long period, bows and arrows began to develop and spread.

[26] Artsikhovsky, A. V. (1954) *Fundamentals of Archeology*. Moscow, p. 42.

[27] *Primitive Culture* (The first edition). State Hermitage Exhibition Guide, 1955, p. 12.

[28] Kotovich, V. G. (1957) Chokhskaya site: the first monument of the Stone Age in mountainous Dagestan. *Scientific Notes* 1.3, Makhachkala.

Chapter A3

Paleolithic station at Damjili Cave: Report of 1957*

*Huseynov, M. (1957) *Paleolithic Station at Damjili Cave: Report of 1957.* Manuscript on file of the Institute of Archaeology, Ethnography, and Anthropology, Baku (in Russian). The figures and photographs referred to below have not been accessible to the editors. However, judging from the descriptions, *fig. 14* should be the same as Fig. A2.26 of this volume.

A3.1 Introduction

Related to the study of the Stone Age in Azerbaijan, the caves on Avey Mountain in the territory of Dash Salahli village, Gazakh region, have started to be studied since 1956. The first research here was launched in 1953 under the leadership of S. N. Zamyatnin and with the participation of the author of the report.[1] A small area was excavated in the Damjili Cave for the purpose of exploration. As a result of the excavation, the remains of tools made of flint and obsidian belonging to the end of the Upper Paleolithic were found here. Those tools include round scrapers, microlithic blades, retouched tools, and cores. A special type of knife made of obsidian that draws attention among these tools should be mentioned. This knife can be compared well with tools belonging to the end of the Upper Paleolithic in terms of its appearance and manufacturing technique. Comparing this knife with Rgani Paleolithic knives in Georgia, S. N. Zamyatnin calls it "Rgani knife." Such "Rgani knives" were found from the Barmagsiz site in Georgia and Seidi on the Mediterranean coast, etc.[2]

A3.2 The 1956 season

In 1956, the author conducted the second excavation in the Damjili Cave.[3] Since the surroundings of the cave were studied principally that year, a little excavation was carried out. Therefore, the work done in 1956 was exploratory.

Site setting

Damjili Cave is located at the foot of Avey Mountain, 2 km in the west from Dash Salahli village. The cave is situated at the mouth of a narrow gorge under a high flint-limestone cliff. The height of the rock is approximately more than 100 m. Above the cave (20–30 m) there is a second artificial cave. There is no way out there. But the manually carved door and small window of the cave are clearly visible. There is also a deep valley is located on top of the rock where Damjili cave. The flood emerged during the rainy season in Avey Mountain flows into that valley and pours directly in front of the Damjili cave. The valley in front of the cave was formed as a result of the floods from above.

The ceiling of the cave has been cracked in three places. The beautiful spring water drips continuously from one of those cracks. Drops poured from above are collected in a half-meter-high pool recently built with stone and cement. The length of the front wall of the pool is 6 m, the side walls are 2.5 m on one side, and 5.5 m on the other. The water collected in the pool flows through a specially made limestone trough and pours into the valley in front of it.

Since there is always water, fruit-bearing fig trees grow all around the cave and even along the valley ahead. Due to the abundance of these fig trees, it is very difficult to see the cave. There are two big springs at a distance from the cave. One of them, called "Qarabulaq," is on the left of the cave, at about 150 m, and the second one called "Korbulaq" is on the right, about 300 m. Both springs are at the same level as the cave and are located at the foot of the mountain.

During the exploration around Korbulaq, many pieces of reddish flint were discovered. Such pieces

[1] Zamyatnin, S. N. (1958) Investigating the Stone Age of Azerbaijan, autumn 1953. *Institute of History Bulletin* 13: 5–18.

[2] Zamyatnin, S. N. (1957) Paleolithic of Western Transcaucasia. *Collection Museum of Anthropology and Ethnography* XVIII, p. 472.

[3] Huseynov, M. M. (1956) *The Report of 1956 on the Archaeological Excavation conducted in Damjili near the Village of Dash Salahli, Gazakh Region.* Manuscript on file at the Archives of the Museum of History.

of flint were noticed on the sides of the cave. During the exploration, obsidian flakes were also found around the cave. Near Korbulaq, two stone tools were found from the discarded soil of the ditch dug by local people to carry water. The depth of the water channel is about 1 m. One of the tools was made in a triangular shape. It was produced from a disc-shaped core using a preparing technique. It has a lateral percussion surface and a large percussion bulb. Both edges of this tool are very nicely retouched. This retouch corresponding to the Mousterian period, gives the tool a good shape. That tool was made of basalt stone. It is covered with gray-green colored patina. This tool is the best Mousterian pointed tool due to its shape and knapping technique. That point tool is the first Mousterian tool found on the territory of Azerbaijan.

According to S. N. Zamyatnin, such pointed tools made of basalt stone are also found on the shores of the Black Sea in Abkhazia and Armenia. Along with that pointed one, a second tool fragment was found. It was made of white flint and its one edge is little retouched. The tool is triangular in shape. But one end is broken. It is covered with solid patina. This implement appears very old because of its patination and shape. However, since the retouch was made by pressure technique, the tool can be attributed to the post-Mousterian period but before the Upper Paleolithic (*fig. 1: 6*).

All these observations provided the best basis for starting the excavations at Damjili Cave.

Results of the 1956 excavations

The presence of springs and the abundance of flint there could not fail to attract the attention of Stone Age people. That is why we had to expand the area excavated by S. N. Zamyatnin in 1953 at Damjili Cave. A 4 × 5 m area was taken for the excavation.

The cave is semicircular, its area is 17 × 27 m, and its height is 4 m. During the excavation, it was determined that the top layer of the cave is black soil. Fragments of glazed pottery from the Middle Ages were discovered here. Below that, a layer of yellow clay continued to a depth of ... m. From this layer, 600 pieces of flint and obsidian tools and fragments were obtained.

Among these tools, it is worthy to mention the basalt rough scraper in a round shape. This scraper is knapped from a disc-shaped core. Its dorsal surface is quite convex, while the lower face is flat. The working edge is retouched all the way around. This specimen exhibits a very thick patination. Considering its shape and preparation technique, it can be considered typical of the Mousterian period (*fig. 1: 3*). It was made of the same material as the scrapers and pointed tool mentioned earlier. They were also flaked from a disk-shaped core and the retouch technique belongs to the method of Mousterian period.

In addition to the scrapers, one point tool was discovered from the same layer of the cave. This tool is triangular in shape and its blank was flaked from a disk-shaped core. The tool has wide and beveled butt with an expanding bulb of force. The bulb covers up to half of the ventral surface, as often seen on the artifacts from other Mousterian sites (*fig. 1: 2*). Along with these tools, up to thirty lithic artifacts derived from the Mousterian period were obtained (*fig. 1: 4, 5*).

Scrapers, pointed tools, and flakes similar to the artifacts found in the Damjili cave have been uncovered in the territory of South Ossetia. These tools belong to the Mousterian period. It is interesting that other types of tools have also been found from that layer. According to their shape, those tools were made on elongated, thin, blade-like blanks with one or more ridges on the dorsal surface. They include pointed, round, scrapers, borers, and knife-shaped blades. Among them, we can mention the following groups that are most characteristic of the Upper Paleolithic.

(1) End-scrapers. These tools were made from different colors of flint and one end of them is retouched. The dorsal surface is convex and the other surface is flat. Some of these have blunted lateral edges. (*fig. 2: 1, 2, 5*).
(2) Round scrapers. Those are mainly made of obsidian and flint. These tools were made of small flakes. Some of them show blunting on all edges (*fig. 2: 3, 4*) while others on two or three edges (*fig. 2: 6, 7*).
(3) Notched scrapers. These tools were made from large flakes. One edge of a flake has a retouched notch. Such tools could have been used in the household as a scraper-knife (*fig. 2: 8–12*).
(4) Borers. Such tools were made of rough flint. One end of the oblong blank was heavily retouched on both sides to produce a narrow tip (*fig. 2: 14, 15*).
(5) Knife-shaped blades. These obsidian blades were made on long and rough blanks. Both of their lateral edges were very strongly retouched (*fig. 3: 1–4*).
(6) Microlithic blades. These were made of very small, thin knife-like bladelets (*fig. 4: 1–7*).
(7) Core. A prisma-shaped core was found along with tools. Many flake scars were preserved on its all surfaces (*fig. 2: 13*).
(8) Burin. This tool was made of a small thin obsidian blade. Its one lateral edge was completely

retouched. The other edge was nibbled from its use as a burin (*fig. 2: 16*).

(9) Dual-purpose tool. There is only one specimen of this type. It is called dual-purpose because one edge is used as scraper and the other serves as a burin. This specimen was made on a thick flint blank (*fig. 2: 17*).

In addition to these lithic artifacts, burnt animal bones were also found in Damjili Cave in 1953 and 1956. Those bones were identified by Nasiba Alakbarova, senior researcher of the Museum of Natural History of Academy of Sciences of the Azerbaijan SSR. They mainly belong to the following animals: 1) Boar; 2) Bull; 3) Fish; 4) Mollusc; 5) Tortoise; 6) Sheep.

A3.3 The 1957 season

Based on the findings mentioned above, we continued the excavation in Damjili Cave in 1957 (*photos 1, 2*).

This time we started digging the entire area inside the cave. Despite the largeness of the excavation area, Paleolithic remains were discovered in the part of 20 m^2 only. As shown above, that is, we noted in the previous excavations that the layer concerning the Paleolithic remains consisted of a 10 cm thick layer of yellow clay. Compared to the previous excavations, this year, the cultural layer turned out to be deeper, approximately at a depth of 2 m. The cultural layer were mixed. Since the inside of the cave is very inclined, the layer is also sloping. On top of that thin layer there is black soil, large pieces of rock, and other stone fragments.

Paleolithic materials were discovered only in a thin layer of yellow clay. As a result of the excavation, approximately 2000 tools and 5000 debitage and cores were found from this small area of 20 m^2. There were also many burnt fragments of animal bones and hearth ash.

The tools found here were mainly made of flint, basalt, and obsidian. While reviewing all the materials obtained from Damjili Cave, these remains contain tools belonging to the Lower (Mousterian), Upper Paleolithic, and Mesolithic (end of Upper Paleolithic and beginning of Neolithic) periods according to their shape and flaked stone technology. This is due to the fact that the layers have been mixed here.

In this report, we have tried to present these findings in chronological order. This division was made on the basis of a comparative analysis of the materials uncovered in the Stone Age sites of Transcaucasia.

Lower Paleolithic (Mousterian)

The objects belonging to the Mousterian period mainly consist of scrapers, pointed flakes, and cores. In addition, the raw materials concerning this period differ from all the findings in terms of their shape and flaked stone technology. As in the last year's excavation, the Mousterian tools were mainly made of basalt, and a small amount made of flint. There are a few tools belonging to the Mousterian period, and relatively many debitage. The tools include the following:

(1) Scrapers. Total: two pieces. These were made of black basalt stone, knapped from discoidal core. Dorsal part is quite convexed while the other side is wide and flat. Both of tools exhibit well-preserved bulbs of percussion and butts. They retain a certain amount of patination. Their edges were very well retouched using a technique characteristic of the Mousterian period (*photos 2, 4*).

(2) Points are of the Mousterian type made on triangular flakes. They consist of flint and basalt specimens. Some of the points are retouched, and others are not. It should be noted that these retouches are made by pressure. This method allows us to attribute the tools to the very end of the Mousterian period. This interpretation is applicable only to the flint tools. One of the flint points was blunted on three sharp edges and made into a scraper shape. Of course, this blunting was done later (*photos 5, 6*). Those of basalt stone are probably more ancient, judging from their heavy patination and water-rolled surfaces (*fig 5: 1–4*). Points made of flint belong to a later period than those made of basalt (*fig 5: 5–8*).

(3) Flakes. They were made of basalt and were knapped from discoidal cores. They are triangular in shape and of different sizes. Dorsal surface is very convex and the other one is quite flat. They were so rolled over with water that they were completely smoothed out on all sides. The flakes are covered with thick patination. Excessive smoothness and patinated surface indicate that they belong to a more ancient period (*fig 6: 1–3*).

Among them, one flake is completely different from the others due to its shape, flaking and size. It was flaked from a basalt stone. Its butt is more than 1 cm thick and bulb of percussion is a 2–3 cm wide. The irregularly shaped quadrangular dorsal surface has preserved its natural cortex near the tip. On the dorsal surface, there are several flake scars of different shapes and sizes. Due to the large size of the butt and the bulb of percussion, the flake was likely flaked from a large round stone by very heavy percussion. This is completely covered with patination. The width of the flake is ..., length ..., and its thickness is ... cm (*photo 7*).

(4) The core is disc-shaped, with traces of flaking by the method of striking the edges towards the center; some of the scars look like triangles. The core edges are uneven. It was made of white flint and covered with yellowish milky colored thick patination. The surface is even shining like it has been varnished. This shows that it is more ancient. The diameter of the core is ..... cm. Photo...

We have shown above that a group of lithic artifacts from the Damjili Cave belongs to the Mousterian period. Both in terms of typology and raw material, they resemble the Mousterian artifacts found in the middle and high mountainous zone of the Greater Caucasus in the territory of South Ossetia.[4]

Mousterian specimens similar to Damjili Cave were obtained near Derbent in an ancient terrace on the shores of the Caspian Sea. Comparable artifacts were also discovered from the area called Yashtukh in the North Caucasus. A discoidal core made of basalt and other artifacts were found in the caves of the North Caucasus by S. N. Zamyatnin in 1936.[5] These types of artifacts and cores are also known to us in the territory of Armenia from the Artini Mountain and from one of ancient terraces of the area called Areni.

All above-mentioned findings prove once again that people started living in the Damjili Cave in the Mousterian period. It is interesting that other types of artifacts were discovered from the same layer. These tools are presented in the following groups.

Upper Paleolithic

The artifacts belonging to this period were made of different types of flint, obsidian, and other stones. They were mainly made of small and slightly convex flakes. They are sometimes in oblong shapes similar to blades, with serrated and/or blunted edges. Among these, the following groups can be mentioned.

(1) End-scrapers. These tools were mainly made of red and yellowish flint. They are elongated in shape and one lateral edge or sometimes both edges were retouched. The dorsal surface of these tools is convexed and has one or two ridges (*fig. 7: 10*).

(2) Scrapers in various shapes. These tools are round (*fig. 8: 1–3*), semi-circular (*fig. 8: 7, 9, 11, 12*), and oblong (*fig. 8: 4–6, 8, 10, 13–15*) in shape. Their blanks are flat flakes. The edges of the tools were quite well retouched. In contrast to these, there are also scrapers made of larger blanks. They were mainly made of obsidian and were of different (*fig. 9: 4, 5, 8*) and triangular shapes (*fig. 9: 1–3*). The edges of these tools were quite well retouched and blunted as a result of use.

(3) Notched scrapers. Such tools were made from different shaped blanks. A notch has been made from their lateral edges (*fig. 9: 7, 10, 11, 12*). Some of them were made of thick oblong blades with a large notch at one or both edges. Those notches were blunted by retouch. This type of scrapers played a necessary role in producing bone tools (*fig. 10: 1–11*). Some of the notched scrapers were made of extremely large and convex flakes, with wide notches made at two or three edges (*fig. 10: 9, 12*).

(4) Burins. These tools, mostly long and very thick, were made of obsidian blades. The lateral edges of the blanks were quite firmly retouched. These burins were made by burin blow applied only one edge of the blade. Therefore, such tools are called one-side burins. Very few examples have been found from the Damjili Cave. However, (in Transcaucasia and Crimea) they are among the most characteristic tools at Upper Paleolithic sites (*fig. 11: 1–7*).

(5) Cores. Many of the lithic artifacts found at Damjili Cave are cores. All of them were made of small cobble stones. They were mainly made of reddish flint. The obtained cores were exhausted. That is, it is no longer possible to obtain a flake usable as a tool. They are prismatic and sometimes spherical in shape. Flint of this type is more common at the foot of Avey Mountain, around Korbulaq (*fig. 12: 1*).

(6) Borers. These tools were mainly made of large thick flint and obsidian flakes. The lateral edges of the blanks were heavily retouched, and their end was pointed. There are flint (*fig. 13: 1–8, 10*) and obsidian borers (*fig. 13: 3, 11, 12*). Such tools were mostly used in household work, especially for piercing the leather, bone, etc.

These tools found at the Damjili Cave are most likely compared with the tools of the sites belonging to the last periods of the Upper Paleolithic of Transcaucasia due to their form. Manufacturing the

[4] Lyubin, V. P. (1956) Proceedings of *the Paleolithic in Ossetia*. Reports read by an archaeologistat the Sessions in Mountains. Yerevan, 1956.

[5] Zamyatnin, S. N. (1957) *Study of the Paleolithic Period in Kafkaz for 1936–1948*. Baku: Publishing Institute of Ethnography of Azerbaijan.

notches on the lateral edges of most of the scrapers found here is typical for this period. A majority of similar tools have been found in Armenia from the Upper Paleolithic sites called Arzni and Nurnus.[6] In 1916, the Polish scientist Stefan Krukovsky discovered tools, especially scrapers, cutting tools, and knives similar to the ones found at Damjili, from the Upper Paleolithic Gvarchilas-klde cave in Rgani.[7]

The materials obtained from the Damjili Cave by S. N. Zamyatnin in 1953 seem to be very close to the Upper Paleolithic site found at Barmagsiz Cave.[8] Indeed, the materials found at Damjili and Barmagsiz confirm this idea. Barmagsiz materials such as scrapers, burins, borers, and others stored in the State Museum of Georgia are the same as the tools of Damjili.[9]

The tools found from Damjili are compared not only with the sites of Gvarchilas-klde and Barmagsiz but also with other Upper Paleolithic sites of Georgia (H. Kiladze demonstrated arrowheads at the scientific meeting of Caucasian archaeologists convened in Yerevan in 1956). The stone artifacts found at the Upper Paleolithic sites of Devis-khvreli in Georgia are of varied types. They are more similar to Damjili remains. There are points, round and notched tools among the Devis-khvreli specimens.[10]

Mesolithic (the end of the Upper Paleolithic and the beginning of the Neolithic)

A group of lithic artifacts discovered at Damjili Cave pertains to the Mesolithic period. These were made finely and small in size due to their form and manufacturing technique. Those from Damjili mainly consist of microlithic tools, blades in the form of knives, arrowheads, scrapers, and cores:

(1) Microlithic tools. These tools were made of very small and thin blades. Dorsal sides of microliths are mostly single-ridged. Sometimes there are found two-ridged pieces. Microliths were retouched on one and sometimes both sides. There are also such microliths that have a small notch on their sides. One or both edges have been made laterally and serrated. A few of the microliths also look like tiny segments. It can be assumed that such tools were used for dissecting fish belly (*fig. 14: 5–17, 23–25, 29, 30, 31*).

(2) Knife-shaped tools. They are not very long, but are wide enough. Their one and sometimes two edges were completely blunted. The blanks were flaked from prism-shaped cores. Their length is 2 to 4 cm, and their width is up to 1 cm (*fig. 14: 18–22*).

(3) Arrowheads. They are mostly triangular in shape and have a very sharp and thin end. Both faces are convex in the central part. The wide end has a short tip for hafting to a wooden shaft (*fig. 14: 26, 27*).

(4) Scrapers (scobel). These were made of thin microlithic blades. A not too deep notch was made at one end and sometimes on one lateral edge. These notches seem to have been made and retouched for a special purpose. It can be assumed that such tools were used to make awls from bone (*fig. 14: 32–35*).

(5) Cores. They are basically pencil-like in shape. Therefore, archaeologically, it is called a pencil-shaped core. They are elongated and multi-faceted. The shapes and sizes of the microlithic tools found here indicate that they were knapped from these cores. Thus, all microliths were made of pencil-shaped cores. Their length is up to 5 cm, and their thickness is 1 cm (*fig. 14: 1–4*).

The occurrence of these types of artifacts shows that people of the Stone Age might have used them as weapons. The discovery of arrowheads demonstrates its development already during that period. It shows that people invented the bow and arrow. The appearance of microlithic tools, knives, and arrowheads proves that hunting and fishing came to the fore in the economy of the prehistoric human society precisely during this period, i.e. at the end of the Paleolithic. It is true that no fishing tools were found here, but fish bones were found.

These microlithic tools were mostly found at Late Upper Paleolithic sites. Similar tools were discovered from Devis-khvreli, Sakatla and Mgvimevi Paleolithic sites in Georgia.[11] It is no coincidence that geometrically shaped microliths and crescents similar to Damjili microliths were also obtained from the Gavarchilas-klde site.[12] The arrowheads discovered in the mentioned site were found in the

[6] Panichkina, M. Z. (1950) Paleolithic Armenia. səh. 9.

[7] Krukovsky (1916) Cave Gvardzhilas-klde in Rgani. *Izv. Cav. Museum* 10, v.Z, p. 253.

[8] Zamyatnin, S. N. (1957) Paleolithic of Western Transcaucasia. *Collection Museum of Anthropology and Ethnography* XVIII, p. 472.

[9] Archaeologist. Foundation Museum of Georgia inv. 158, 159, 161, 22, 81.77.

[10] Nioradze, G. K. (1934) Paleolithic of Georgia. *Proceedings of the International Conference on the Study of the Quaternary Period of Europe*. U, p. 226

[11] Zamyatnin, S. N. (1937) Cave canopies of Mgvimevi, near Chiatura, *Soviet Archaeology* 3, p. 65.

[12] Tushavramishvili, D. M. (1955) *Paleolithic Remains in the Kvardzhilas-klde Cave, Western Georgia*. Author's abstract, Tbilisi, p. 10.

site belonging to the end of the Upper Paleolithic in the territory of Georgia.[13]

In general, as far as we know, microlithic tools have been found everywhere in the sites of the late periods of the Paleolithic. In Bruniquel and Raymond (France) tools of this type belong to the late Madeleine period of the Paleolithic. Microlithic tools distribute from Svekal in North Africa to Gambia and from there to the Red Sea and then to Central Asia and India. In Crimea, Ukraine, Turkmenistan, Transcaucasia and Europe, they are mainly attributed to the beginning of the Neolithic or the Mesolithic period.

Other remains

During the excavation of 1957, along with the above-mentioned lithic artifacts, the fragments of animal bones were found. These remains are burnt. The bones are generally so fragmentary and damaged that it is very difficult to identify them. Nevertheless, it was possible to determine which animals were present based on the remains of teeth, astragals, and long bones found in the bone assemblages.

The bones were identified by N. Alakbarova. The result is as follows: 1) Horse; 2) Gazelle; 3) Deer; 4) Bull; 5) Onager (a breed of donkey); 6) Saiga antelope; 7) Goat; 8) Red deer. These animal bones are undoubtedly those of the animals hunted by the Stone Age people.

It should be noted that there are also fragments of glazed pottery and pits dating back to the Middle Ages in the cave. Even at the bottom of the cave, a two-meter-deep pit with a narrow mouth and a wide middle part was discovered. This pit contained ashes and animal bones.

As a result of the excavation, it was found that the potsherds pertaining to the Middle Ages were discovered in the upper soil layer. The pits destroyed the Paleolithic layer. On the whole, the current layers of the cave were frequently destroyed and mixed up by medieval people.

A3.4 Conclusions

As we mentioned above, all lithic artifacts and bones belonging to the Stone Age were discovered from a thin layer of clay with a thickness of 10 cm. The lithic artifacts belong to various stages of the Stone age. Then, a question arises, how did it happen that the materials of different periods were located in one thin layer and not in different layers.

1) It can be assumed that the reason for this is that the strong flooding from the above cliff and the spring water from the inside of the cave washed away cultural layers, and floated tools and tool fragments. A small amount of the material of the upper layer has settled down and got mixed up with the material of the lower layer. As a result of these water activities, the cultural layer has become thinner.
2) Pits dug by the Middle Ages communities inside the cave destroyed all the layers.
3) One of the reasons for the mixing of cultural layers was the excavation of a 40 m^2 area in the cave and the construction of a cemented wall for the pool due to the use of spring water in the recent past. In that area, the cultural layer was not only mixed, but also dug up and completely thrown away.

As a result of all the above-mentioned reasons, this precious and multi-layered monument in the Damjili Cave was destroyed. Therefore, it was not possible to determine the stratigraphy through our excavation. In spite of these, the Damjili site can be considered one of the valuable sites of the Stone Age in Transcaucasia due to its findings. This site is the first one of the Stone Age studied in Azerbaijan, and it is of great importance.

[13] In 1956, at the session of Caucasian archaeology convened in Yerevan, H. Mladez demonstrated the same type of arrowheads.